The Rudolph's second house just for books, purchased in 1987, was located adjacent to their Arcadia Avenue home, at 2650 Indianola Avenue, Columbus
(Taken August 1992 by Ronald L. Stuckey; in the collection of RLS)

Hollyhocks, brought to the United States by European settlers, "remind us of man's attempt to preserve the familiarity of his surroundings," a statement Rudolph made in one of his Wellesley College Chapel talks (Paper No. 32)
(Taken 11 July 1993, at an old house (since removed) along the side of road, Chatfield, Ohio; by Ronald L. Stuckey; in the collection of RLS)

EMANUEL D. RUDOLPH'S STUDIES IN THE HISTORY OF NORTH AMERICAN BOTANY

Emanuel D. Rudolph in living room of his home at 384 Arcadia Avenue, Columbus, Ohio (Taken 1991, by Kevin Fitzsimmons; in the collection of RLS)

Emanuel D. Rudolph surrounded by Botany textbooks (upper l. and lower r.) and biology textbooks (lower l.), all on shelves in the Rudolph Memorial Reading Room at the Herbarium of The Ohio State University (Taken 5 November 1995; by Ronald L. Stuckey; in the collection of RLS)

EMANUEL D. RUDOLPH'S STUDIES IN THE HISTORY OF NORTH AMERICAN BOTANY

With an Appendix:
Relationships between Science and Religion

Edited by Ronald L Stuckey

Ronald L. Stuckey 16 October 2000
The Ohio State University

and

William R. Burk
University of North Carolina

Introductions by Ronald L. Stuckey

2000

ISSN 0833-1475
ISBN 1-889878-05-7

Sida, Botanical Miscellany, No. 19

Sida, Botanical Miscellany
Founded by Wm. F. Mahler and Barney L. Lipscomb, 1987

Editor
Barney L. Lipscomb

Botanical Research Institute of Texas
509 Pecan Street
Fort Worth, Texas 76102-4060, USA
Telephone: 817/332/4441
Fax: 817/332/4112
E-mail: Sida@brit.org

Sida, Botanical Miscellany, No. 19
Arrangement and sequence by Ronald L. Stuckey
Published in collaboration with
RLS Creations
P.O. Box 3010
Columbus, Ohio 43210-3010, USA

Computer page layout by Ricki C. Herdendorf
EcoSphere Associates
1507 Cleveland Road East
Huron, Ohio 44839-9503, USA

Cover photograph:
Alcea rosea by Dr. Alice Brown Russell, © 1993.
Photographed at Tryon Palace Historic Sites and
Gardens, New Bern, North Carolina.

BRIT

Foreword

How often do you meet a man who widened horizons, enriched possessions, all the while radiating the joy of living? Emanuel D. Rudolph studied lichens, zealously collected books, explored Antarctica, and animated the history of biology in the classroom and at scientific meetings. Some suggest his greatest legacy was showing in paper after paper, the ways men and especially women, created biology in this country. Following Frans Verdoorn's recommendation, Rudy taught history of biology horizontally. Its methodology is concerned not only with what, but also where, when, why, and with whom. By answering these questions, for example, one can understand what Asa Gray's influence was then and now!

Rudy was always a warm friend. Professionally he was a lichenologist, a well-versed bookman, and a biohistorian who advocated documentation. It was his ready pleasure to help others. As Peter Collinson helped John Bartram or William Darlington assisted Asa Gray, so Rudy served all of us as a clearing-house of information. What a gift!

Joseph Ewan
Missouri Botanical Garden

In Memory of Rudy[1]

You have left us, but we still remain
To walk life's road alone
Our thoughts of you will live again
Of all the happy days we've known
In spring we'll wait for roses red
And watch the lilacs fade to blue
In early autumn when brown leaves fall
We'll catch a glimpse of you.

You have left us, but we still remain
To recall the many books you've sought
Each one you've touched along the way
Is now a hallowed thought
We'll hear your voice, we'll see your smile
Though blindly we may grow
The memory of your helping hand
Will carry us onward, we know.

You have left us, but we still remain
To continue with your plan
Although shadows may appear, and pain,
Your name will live as a great man
We've known so much happiness
We've had our cup of joy
And memory is one gift we have
That death cannot destroy.

You have left us, but we still remain
To ask you one more thing to do
As you hear us call your name
Walk slowly down that long, long path
That we may walk the same
As you go down that long lonely road
We'll want to know the steps you've taken
For some day, we too, will follow you.

Ronald L. Stuckey,[2] Professor Emeritus of Botany

---

[1] Adapted from a recitation, *Should You Go First and I Remain*, which is used in country music with the song, *Beyond the Sunset*; taken from the Hank Williams, Sr. version. The original poem written by Albert K. Rowswell. *Rosey Reflections*. Published by the author, Pittsburgh, Pennsylvania. 1945, pp. 78–79.

[2] Ronald L. Stuckey [ed.]. 1992. Memorial for Emanuel D. Rudolph, 9 September 1927 – 22 June 1992, Tuesday, 7 July 1992, The Ohio State University Fawcett Center for Tomorrow, 2400 Olentangy River Road, Columbus, Ohio. *The OAS Newsletter In The Ohio Journal of Science* 92(5, supplement): 164F–164L. (Reproduced with permission)

Emanuel David Rudolph
9 September 1927 – 22 June 1992

(Photo Archives, The Ohio State University)

Known affectionately as "Rudy," he was a most kind, generous and modest gentleman, who displayed good humor and good will. He was a truly valued friend, respected colleague, and eminent scholar, who gave much thoughtful counsel to his students and friends and showed genuine enthusiasm for life, science, and books.

Ronald L. Stuckey
Columbus, Ohio

Dedication

To Joseph and Nesta (Dunn) Ewan[1]

Joseph[2] and Nesta (Dunn) Ewan are very special people because for more than fifty years they have been writing about the history of biology and biologists. Their contributions to scholarship are represented by numerous lectures and seminars, many insightful book reviews and notes, and important monographs and books. The primary focus of the Ewans' scholarship is biographical, rather than institutional or philosophical. Their fascination and interactions with people and settings, make the Ewans' writings come alive. People make history, and that information is exciting to read.

Joseph Ewan, trained as a naturalist and taxonomic botanist, is one, who in the tradition of George Sarton, the founder of the American history of science field, and Ernst Mayr, the distinguished Harvard zoologist, has become a major interpreter of the history of science, especially of American natural history. Born in Philadelphia, Ewan received his undergraduate education at the University of California, Los Angeles and Berkeley (B.A. 1934). At Berkeley, he was a research assistant (1933–1937) to Willis Linn Jepson (1867–1946). Joe then taught at the University of Colorado (1937–1944). Following Ewan's service as a botanist in Colombia (1944–1945), he worked at the Smithsonian Institution (1945–1946) and the United States Department of Agriculture (1946–1947). Then he taught at Tulane University, New Orleans, where he was designated the Ida A. Richardson Chair of Botany (1972). When he retired (1977), he was accorded Emeritus status.

Ada Nesta Dunn Ewan, born in Canada, met Joe at UCLA and they married in Reno, Nevada (1935). She was a practicing librarian, and over the years, she assisted Joe in assembling an outstanding personal library centered on their interests in the history of natural history, biographies of scientists, systematics, and exploration, particularly of the North American West and Latin America. This unique library of over 4500 volumes, with its interwoven manuscript finders' guides and related correspondence, came to Missouri Botanical Garden, St. Louis in 1986, where the Ewans curated and continued to use it in their writings. Their library was later purchased by the Missouri Botanical Garden.

Well-deserved honors have been bestowed on Joe Ewan. Among them are two honorary doctor of science degrees: one from the College of William and Mary in Virginia (1972); and a second from Tulane University (1980). He was awarded a Guggenheim Fellowship (1954), named a fellow of the Linnean Society of London (1976), and presented the Founders' Medal, Society for the History of Natural History, London (1986). Ewan has served as visiting professor at the University of Hawaii (1967, 1974), the University of Oregon (1978, 1981), and as distinguished visiting professor at The Ohio State University (1982).

---

[1] Emanuel D. Rudolph. 1989. Introduction, pp. [7–8]. *In* Keith Crotz (comp.). *Ewaniana: The Writings of Joe and Nesta Ewan. American Horticultural Series* No. 2. The American Botanist, Booksellers, Chillicothe, Illinois. [64] pp. Permission to reproduce Rudolph's Introduction courtesy of Keith Crotz. [Modified and edited for use as Dedication here by Ronald L. Stuckey].

[2] Deceased 5 December 1999.

Joseph and Nesta Ewan

(From a photograph that was displayed for many years
in the Rudolph living room; taken 1970 by Bob Coleman;
in Dinwiddie Hall on the campus of Tulane University, New Orleans;
in the Rudolph photograph collection)

I have had the privilege of the Ewans' friendship for over 35 years and have been greatly rewarded by it. Joe and Nesta Ewan are excited about their search for information and connections in the history of natural history. In Joe's lectures, writings, and correspondence, tantalizing footnotes suggest interesting avenues beckoning to be traveled, and provocative problems aching to be solved. Envelopes arrive with clippings or short notes suggesting something that needs to be searched or providing answers to questions long sought. Joe and Nesta Ewan are scholars in the true sense of that word, because they are constant seekers after knowledge that they generously share.

Emanuel D. Rudolph
Columbus, Ohio

Preface

Ample evidence exists in Emanuel D. Rudolph's research files and from evocative discussions with Stuckey that early in his studies of the history of botany, Rudolph intended to write a book on the popularization of botany. In later years that intention took lower priority while he researched his invited book on the scientific history of the Missouri Botanical Garden. Our mission has been to assemble, from his unpublished manuscripts and published papers, a memorial contribution that, at least in part, preserves Rudolph's ideas and the results of his accomplished research in the history of botany, a discipline to which he was deeply devoted.

The idea for this memorial volume of Rudolph's papers in the history of botany was conceived when, at Stuckey's invitation, William R. Burk came to Columbus, Ohio, for a few days in August 1992 to provide help with the Rudolph estate. We soon realized the wealth of unpublished research information that Rudolph had in his files and agreed that these items should be made available to the botanical community. After locating Rudolph's manuscripts that were reasonably complete, Stuckey made editorial changes and corrections; Burk verified all quotations and references and also provided editorial assistance.

To make the book more representative of Rudolph's total research effort, after this initial editing phase, we added summaries for most of the topics on which Rudolph had already published papers. The summaries were derived from his abstracts, partially completed manuscripts, notes used in delivering presentations at scientific meetings, and selected paragraphs from his publications. For certain topics on which Rudolph did considerable research, but left only partial manuscripts, notes, or outlines, Stuckey used that information, coupled with his own knowledge, additional research, and Burk's bibliographic assistance, to prepare the final papers. These papers are identified with appropriate credit statements. Stuckey wrote the general introduction, "Emanuel D. Rudolph's Studies in the History of North American Botany," and all of the subsequent shorter introductions to the papers. These introductions not only highlight Rudolph's endeavors in the history of botany, but they also incorporate Stuckey's collaboration with him on various botanical history projects. Collectively, these introductory essays provide a view of research in botanical history in the United States and at The Ohio State University from 1961 to 1992.

Burk's bibliographical and editorial contributions extended to all quotations and references in the footnotes of the general introduction, the shorter individual introductions, and the publication notes provided at the termination of each printed paper. Burk completed the word processing of many of the early drafts of the individual papers, as well as the final word processing for the entire manuscript. We have had assistance in word processing from Maureen Marais, J. Perry Edwards, and Kathy Royer at The Ohio State University. At the invitation of *Sida, Botanical Miscellany* Editor, Barney Lipscomb, Rudolph's presentations on the relationships between science and religion that he gave in the Houghton Memorial Chapel at Wellesley College between 1955 and 1960 are printed here from the original typescripts. Thirty-eight papers have emerged from Rudolph's research files to comprise this memorial volume.

Ronald L. Stuckey
William R. Burk
Chapel Hill, North Carolina

Acknowledgments

Throughout the preparation of this book as a memorial to Emanuel D. Rudolph, we have made regular use of the archival and library collections at The University of North Carolina—Chapel Hill, The Ohio State University, Duke University, and North Carolina State University. In particular, we acknowledge with thanks the special assistance from librarians, Judith Warnement, Botany Libraries, Harvard University; Anita Karg, Hunt Institute for Botanical Documentation; Connie Wolf, Missouri Botanical Garden Library; Bruce Leach, Biological Sciences/Pharmacy Library, The Ohio State University; Julie Walker, Burrage Library, Olivet College, Michigan; and Bea Beck, Rancho Santa Ana Botanical Garden Library. Susan Whitfield and David Miller of the Photo-Laboratory, The University of North Carolina—Chapel Hill, provided technical assistance with numerous illustrations.

We are also grateful for the time and assistance that individuals have devoted in critically reading either the entire manuscript or individual papers. Those reviewing the total manuscript were Parker Bauer, Ocala, Florida; Henry Brecher, Columbus, Ohio; John W. Frederick, Columbus, Ohio; and Vassiliki Betty Smocovitis, University of Florida, Gainesville. Those reviewing individual papers were John Burnham and the late June Z. Fullmer, both of The Ohio State University; the late William C. Dickison of The University of North Carolina—Chapel Hill; and Scott LaGreca and Richard Weinstein, both of the Harvard University Herbaria. Contributing information on specific items were William Culberson of Duke University; Charles E. and Ricki C. Herdendorf of Huron, Ohio; Dale E. Johnson of Timber Press; Norbert Kilian of the Botanic Garden and Botanical Museum, Berlin-Dahlem; David Romito of The University of North Carolina—Chapel Hill; and Edward G. Voss of the University of Michigan.

Financial support for the production and printing of this Rudolph memorial book was provided by the interest income from two endowments at The Ohio State University: the Ronald L. Stuckey Herbarium Fund and the Emanuel D. Rudolph Botanical Fund; and a contribution from John and Chara Haas of Villanova, Pennsylvania; and a contribution from the Botanical Research Institute of Texas, Fort Worth. We are especially grateful to Barney Lipscomb, editor, for accepting these papers of Emanuel D. Rudolph as a contribution to *Sida, Botanical Miscellany*.

Ronald L. Stuckey
William R. Burk

Emanuel D. Rudolph in his outfit for exploration in Antarctica, Hallett Station (Taken 1962; in the Rudolph photograph collection)

Emanuel D. Rudolph at his table of books for sale, Central Ohio Book Fair, Columbus (Taken by Ann Rudolph, 13 October 1990; in the Rudolph photograph collection)

TABLE OF CONTENTS

TITLES OF PHOTOGRAPHS IN THE FRONTISPIECES TO THE NUMBERED RUDOLPH PAPERS

TITLES OF TABLES IN THE GENERAL INTRODUCTION

TITLES OF APPENDICES IN THE PAPERS

TITLES OF COLOR PHOTOGRAPHS

TITLES OF PHOTOGRAPHS NOT IN FRONTISPIECES

GENERAL INTRODUCTION
TO THE RUDOLPH PAPERS

By Ronald L. Stuckey

General Introduction

Emanuel D. Rudolph's Studies in the History of North American Botany

Ronald L. Stuckey

When one's life unexpectedly comes to an end in the autumn of an active creative career, much unfinished work is usually left in various stages of completion. And so it was with my close friend and academic colleague, Emanuel David Rudolph, whose death at age 64 occurred on 22 June 1992, in Columbus at The Ohio State University Medical Center, following a traffic accident of 19 June near Wooster, Ohio. To his close friends and professional colleagues, Emanuel D. Rudolph was known affectionately as "Rudy." A most kind, generous and modest gentleman, Rudy displayed good humor and good will. He was a truly valued friend, respected colleague, and eminent scholar, who gave much thoughtful counsel to his students and friends and showed genuine enthusiasm for life, science, and books. Even though a multiplicity of scholarly interests and writing projects occupied his time, Rudolph accomplished much in life through his finely tuned organizational skills conducted in a most calm, patient, unobtrusive, and efficient manner.[1]

For 37 years, Emanuel D. Rudolph was a very dedicated educator in botany, first at Wellesley College in Massachusetts, and then at The Ohio State University in Columbus. As a research scientist, Rudolph was a highly motivated investigator, who became the first botanist to undertake extensive research on the vegetation of the Antarctic continent. He made five field expeditions to Antarctica in the 1960s, and also spent two field seasons on the Arctic tundra of Alaska in the 1970s. From these efforts and his publications, he became internationally recognized for his diverse experimental field studies conducted toward understanding the total biology of lichens in both polar regions. Rudolph also engaged in research on stress conditions in lichens and fungi of coal strip mine areas in Ohio from 1972 to 1981, and was particularly concerned with the effects of sulfur dioxide and other pollutants, heavy metal accumulation, and the role of mycorrhizal fungi in mineral cycling.

Later, Rudolph became nationally known for his broad interests in the history of nineteenth and early twentieth century biology and botany. Rudolph's unprecedented

[1] Ronald L. Stuckey. 1994. Emanuel David Rudolph (1927–1992): Polar lichenologist and historian of botany; including, Appendix, Publications and reports of Emanuel D. Rudolph (arranged chronologically) by William R. Burk and Ronald L. Stuckey, pp. 443–446. *The Bryologist* 97: 437–446.

studies of children's books were windows toward the understanding of the popularization of biology, while in other extensively documented publications, he defined the role of women in the development of both popular and professional botany. At his death, Rudolph was engaged in writing a book-length manuscript on the scientific history of the Missouri Botanical Garden, of which he had completed the first chapter. These historical interests of Emanuel D. Rudolph comprise the subject of this compilation.

Educational Endeavors (1939–1955)[1]

Born 9 September 1927 in Brooklyn, New York, a young Emanuel D. Rudolph, at about age 12, took an interest in reading books, and began to learn about birds, plants, and natural history. He commenced to buy and collect books while in high school. At that time, student Rudolph was influenced by Dr. Phyllis S. Busch, a nationally recognized authority in the teaching of elementary science. She emphasized observation of natural phenomena and methods of scientific discovery, and encouraged him to pursue studies in the natural sciences. To make school life even more interesting, he joined a group of elderly bird watchers and regularly participated in their Saturday morning excursions to the parks and cemeteries in New York City. When birds were scarce, the group studied wild flowers, trees, seashore animals, mushrooms, and lichens. Young Rudolph was often seen crawling over the ground looking for the smallest plants—the mosses, liverworts, and lichens.

Rudolph attended Erasmus Hall High School in Brooklyn, where he was a member of the Audubon Club, the Gardening on Campus group and its Program Committee, and the Biological Research Club, for which he served as president. He graduated in 1945, receiving a certificate for being the outstanding student of the biological sciences in the graduating class. After attending New York University during the summer, Rudolph joined the United States Army on 30 October 1945, shortly after the end of World War II. He was stationed at Santa Fe, New Mexico, where he served as a medical laboratory technician at Bruns General Hospital. After discharge on 13 December 1946, he returned to his collegiate pursuits the next year.

As an undergraduate, Rudolph matriculated at Brooklyn College (1947), Hunter College (1948–1950), and New York University (1945, 1947–1950). He was elected a member of Beta Lambda Sigma and Phi Beta Kappa and graduated *cum laude* with a B.A. in biology from New York University (1950). During summers while attending college, Rudolph worked as a volunteer in the lichen herbarium at the Brooklyn Botanic Garden. Dr. George S. Avery, Jr. (1903–1994) the director, gave him this opportunity, not only to encourage him to learn more about these kinds of plants, but also bring some organization to the collection. From June 1950 until August 1951, Rudolph was employed in the Education Department of the Brooklyn Children's Museum. He also worked in the Natural History Department as an assistant the two previous summers teaching classes in general science, natural history, and social studies and aiding in the preparation of various natural history exhibits that were loaned to schools and other educational institutions. Rudolph was an active member of the Torrey Botanical Club, and when in New York during summers, he led field trips as a volunteer. In 1951 he served as business manager for the Club.[2]

Rudolph left Brooklyn for graduate work in St. Louis, Missouri, where he earned the Ph.D. in Botany in 1955 from the Henry Shaw School of Botany at Washington University and the affiliated Missouri Botanical Garden. His dissertation in the discipline of mycology, entitled "Revisionary Studies in the Lichen Family Blasteniaceae in North America, North of Mexico," was completed under the direction of mycologist Carroll W. Dodge (1895–1988), but never published. His committee consisted of other well known botanists, Edgar Anderson (1897–1969), Henry N. Andrews, Jr., Barry Commoner, Robert E. Woodson, Jr. (1904–1963), and Hugh C. Cutler.

Academic and Research Career (1955–1992)[1]

Rudolph's teaching career in the biological sciences began in the Department of Botany at Wellesley College (1955–1961). He taught courses in general plant science, plant biology, basic horticulture, plant structure, physiology, and plant diseases, while holding the ranks of instructor (1955–1959) and assistant professor (1959–1961). He continued his research on fungi and lichens, and participated in research projects sponsored by the National Science Foundation, first, at the University of Wisconsin (1959), where he conducted research on physiology and developmental morphology in the mold genus *Aspergillus* and examined herbarium specimens of arctic lichens in the family Teloschistaceae; second, at the University of Iowa (1960), where he studied radiation biology; and third, at American University (1965), where he focused on the history and philosophy of science and mathematics. Among Rudolph's other scholarly endeavors was his writing 382 abstracts for *Biological Abstracts*[3] and 195 book reviews for 28 serials, involving 79 reviews for *Choice*,[4] a serial devoted to the publication of reviews.

At The Ohio State University, Rudolph held appointments as an assistant professor (1961–1964), associate professor (1964–1969), professor (1969–1989), and professor emeritus (1990–1992). He served as director of the Environmental Biology Program (1974–1978), and chairperson of the Department of Botany (1978–1987). Throughout these 31 years, he was concurrently a research associate with the OSU

[2] Emanuel D. Rudolph. 1955. Letter of 5 April to Delaphine G. R. Wyckoff, Chairman, Department of Botany, Wellesley College, Wellesley, Massachusetts. 2 pp.

[3] William R. Burk. 1996. Emanuel D. Rudolph's role for Biological Abstracts. *Sida* 17: 243–249.

[4] William R. Burk. 1997. Emanuel D. Rudolph as a book reviewer for Choice. *Sida* 17: 803–808.

Institute of Polar Studies (now the Byrd Polar Research Center), of which he was director (1969–1973).

As will become obvious from the papers in this book, Rudolph had a life-long teaching and research interest in the history of botany and biology. At The Ohio State University, he taught a course on the history of botany during 1965 and 1967, and a course in the history of biology every year from 1970 until 1992, completing his last class one week before his fatal traffic accident. It is not known when Rudolph became interested in the history of botany and biology, but this interest probably grew from his passion for reading and collecting books. During Rudolph's lifetime he and his wife Ann amassed a library of over 53,000 books (Fig. I.1). Ann Rudolph died 26 April 1991,[5] and following her death, he prepared a will for their possessions. According to this will, the entire library and manuscript collection was donated to the Libraries of The Ohio State University, and was fullfiled by his named executor, Ronald L. Stuckey.[6,7,8,9]

Building a Book Collection (1942–1992)[6,7,8,9]

Rudolph collected books since high school in the early 1940s.[10] Between classes while at New York University, he would walk to Greenwich Village to purchase books in the second-hand book district, one block from campus. Initially his acquisitions were books in botany and natural history, including both basic and classic botanical reference sources concerned with cryptogamic plants, and books about the polar regions. He obtained books from many sources—from new and used book stores while traveling throughout the country, book catalogues issued by publishers and second-hand bookdealers, and thrift, Goodwill, and Salvation Army stores. As an indication of the size of his library in 1955, when Rudolph

---

[5] Ronald L. Stuckey. 1995. Ann Waterman Rudolph (1934–1991): Her life with plants, books, and buttons. *The Michigan Botanist* 34: 24–45.

[6] Ronald L. Stuckey. 1995. Emanuel David Rudolph (1927–1992): Book collector and library friend. *The Michigan Botanist* 34: 4–23.

[7] Ronald L. Stuckey. 1995. The Rudolph botanical book collection in the herbarium library at The Ohio State University. *The Michigan Botanist* 34: 46–54.

[8] Tod F. Stuessy and Ronald L. Stuckey. 1997. Botanical libraries and herbaria in North America. 1. Introduction. *Taxon* 46: 639–641.

[9] Tod F. Stuessy, Ronald L. Stuckey, Wesley L. Boomgaarden, and William R. Burk. 1997. Botanical libraries and herbaria in North America. 2. The Rudolph natural history library and its acquisition by The Ohio State University. *Taxon* 46: 643–648.

[10] Nancy Rollins. 1984. Professor fills life, house with book collection. *The Ohio State Lantern*, Tuesday, 27 November 1984, p. 5.

Fig. I.1. Emanuel D. and Ann W. Rudolph in the main library room of their home, 384 Arcadia Avenue, Columbus, Ohio. The books on the shelves behind them are part of their collection on children's natural history.

(Taken October 1983, by Anne Lubrecht; in the Rudolph photograph collection)

moved from St. Louis to Wellesley, "54 cartons of books arrived by truck and were put in his office by the janitor."[11]

At Wellesley College, Rudolph's interests in libraries and book collecting continued, and he became involved in a cooperative project of the college library and the Education Committee of the Wellesley Conservation Council. As a member of the Committee, he participated in the establishment of a "Conservation Shelf" in the college library, where books and papers on conservation topics were displayed and readily made available to library patrons.

Rudolph (1979)[12] wrote about what collecting books and acquiring a library meant to him.

> Book collecting provides me with excitement, adventure, intellectual stimulation, and pleasure. The excitement and adventure come from searching

[11] Delaphine G. R. Wyckoff. 1955. Letter of 8 September to Emanuel D. Rudolph, St. Louis, Missouri. 2 pp.

[12] Emanuel D. Rudolph. 1979. Children's scientific books. *Friends Line. Newsletter of the Friends of the Libraries of The Ohio State University*. [Columbus]. No. 16: 4–5. February 1979. (Quote, p. 5).

Northwest corner of main library room showing biography and reference books (Taken August 1992, by A. E. Spreitzer; in the collection of RLS)

Anne and Harry Lubrecht, book friends in the Rudolphs' main library room (Taken October 1983, by Ann Rudolph; in the Rudolph photograph collection)

The entire length along the south wall of the main library room added to the house in 1972 (Taken July 1992, by Ronald L. Stuckey; in the collection of RLS)

Northeast corner of main library room showing the children's natural history books (Taken August 1992, by A. E. Spreitzer; in the collection of RLS)

Emanuel D. Rudolph in the main library room in front of his desk with the children's natural history book collection in the background (Taken January 1988; in the Rudolph photograph collection)

Main library reading room showing one of the double-sided book cases that projects into the room (Taken August 1992, by Ronald L. Stuckey; in the collection of RLS)

> out books that will be integral to the collection, and the intellectual stimulation from studying books that open windows to readers' and writers' minds. The more books the more windows and the greater the illumination. Pleasure comes from seeing the collection grow and mature and from examining books, some beautiful and made with care, and each, whether beautiful or not, showing individuality and personality. My books are a never-ending source of pleasure.

The largest portion of the Rudolph library consisted of books in botany, zoology, natural history, polar science, and exploration. Another major portion consisted of approximately 8,200 titles representing the largest and most diversified private collection of children's natural history books known in this country. In addition, he had a huge collection of general textbooks in natural science, biology, and botany. The Rudolph library served many purposes. Not only was it available as a resource for his teaching of botany and the history of biology and for his research on polar lichens and botanical history, but also it was a focal point in the Rudolphs' home for visits by student classes, entertainment of distinguished guests, and their own private enjoyment (Figs. I.2, I.3). The Rudolphs were truly exceptional ambassadors for the world of books (Rollins, 1984).[10]

Fig. I.2. Winter view of the Rudolph home; house, main library room, and garage at 384 Arcadia Avenue, Columbus, Ohio.

(Taken February 1977, by Ann Rudolph; in the Rudolph photograph collection)

Fig. I.3. View of the closed-in front porch of the Rudolph Arcadia Avenue home filled with mostly agricultural and economic botany books.

(Taken August 1992, by Ronald L. Stuckey; in the collection of RLS)

History of Science Club at The Ohio State University (1963–1968)[13]

When Rudolph joined the faculty of The Ohio State University in 1961, he began interacting with his colleagues who had strong and active interests in the history of science. In the Department of Geology, certain faculty members had participated for some years in a "club" interested in the history of that discipline with Professor Aurele LaRocque (1909–1990) serving as the club's leader. A much broader interest in the history of science existed on campus, as represented by faculty in history, chemistry, physics, astronomy, botany, geology, pharmacy, and medicine. LaRocque compiled and circulated to these individuals a list of 37 faculty members' names and their affiliated departments. These faculty and other known individuals not on the list, were invited to attend a meeting on 25 May 1963 for the purpose of forming a History of Science Club. Those who attended were asked to write LaRocque

[13] Compiled from various information sheets and program announcements on the History of Science Club at The Ohio State University for the years 1963–1968.

about their particular interest in the history of science. On 27 July 1963, Rudolph[14] replied, saying that he: (1) had forwarded to Provost Albert Kuhn a list of useful books in the history of biology that were lacking in the University Libraries, (2) had not "published any papers or books on history of science although [he had] a keen interest in Erasmus Darwin and hope[d] eventually to publish something about his writings;"[14] and (3) intended to offer a course in the history of botany in 1963–1964. Rudolph did not write any papers on Erasmus Darwin, and his course in the History of Botany was not offered until 1965.

The History of Science Club met informally during its first five years. The members worked to stimulate interest in the history of science on campus and within their own disciplines. They wanted to: (1) identify those journals and books relevant to the history of science and to be certain that the University Libraries acquired them, (2) hold special public meetings at which club members or invited scholars from off campus would deliver lectures on a special topic in the history of science, and (3) determine those courses within individual departments relevant to the history of science, in an effort to establish an identifiable program of study for students.

Over the next several years substantial literature in the history of science was added to the libraries. One of the Club's projects was to replace the missing volumes in various journals. Rudolph, history professor June Z. Fullmer (1920–2000),[15] and Director of Libraries Hugh Atkinson (1933–1986) met with Provost Kuhn to discuss this proposal. Kuhn, a member of the Club, recognized the situation and appropriated $10,000 per year to acquire relevant literature for the University Libraries. Fullmer and Rudolph administered the funds, and purchased books and missing journal volumes as suggested by various Club members. In time, as general library funds diminished, this project also suffered reductions, and eventually the appropriation disappeared. One of the major purchases was the "Landmarks of Science Readex Microprint" edition (1966–1975), composed of micro-reproductions of scientific monographs and journals from the fifteenth to the early twentieth centuries.

In the lecture programs sponsored by the Club, Rudolph spoke at least twice. On 25 February 1965, he was one of five faculty members who commented briefly on courses of instruction in the history of their disciplines. Rudolph described the course in the history of botany that he was then offering. On 14 November 1968, his lecture to the Club, "Is Biology More Than Physics and Chemistry? The Checkered History of Molecular Biology," was based on research he conducted from 1966 to 1968. He already had presented this topic at an Ohio Academy of Science Meeting (1966) and to his NSF History of Biology Institute (1968), but

[14] Emanuel D. Rudolph. 1963. Letter of 27 July to Aurele LaRocque, Department of Geology. The Ohio State University, Columbus. 1 p.

[15] June Z. Fullmer. 1996. Letter of 24 March to Ronald L. Stuckey, Department of Plant Biology and the Herbarium, The Ohio State University, Columbus. 3 pp.

neither notes nor manuscripts for these presentations are known to exist. Rudolph also participated by serving as an informal secretary, and by preparing and mailing the notices for these special lectures. Among the lectures presented during 1966 and 1967 were those by June Z. Fullmer, "Humphrey Davy's Biographies: Notes on Scientific Biography," (2 March 1966); John C. Greene, then of the University of Kansas, "American Science in the Time of Thomas Jefferson," (20 April 1966); Joseph Ewan, then of Tulane University, "Archivia, Trivia, and Bibliophilia, or the Three Ingredients of Biohistorical Investigation," (26 April 1967).

An identifiable program of study for students interested in the history of science emerged from the Club's efforts. In 1963–1964, the members developed a list of 16 courses in nearly as many departments to serve as a beginning for the program. In 1966 the Department of History announced a graduate program in the history of science. As advertised,[16] the core courses in the program were taught by John C. Burnham, professor of history; Earle R. Caley (1900–1984), professor of chemistry; June Z. Fullmer (1920–2000), associate professor of history; Philip C. Keenan, professor of astronomy; Aurele LaRocque (1909–1990), professor of geology; and Emanuel D. Rudolph (1927–1992), professor of botany. Twelve other consulting faculty, including Ronald L. Stuckey, were also listed. Although Rudolph provided counsel and assistance to several history of science graduate students in the Department of History, no students received a degree in this discipline under his direction. He did serve on the doctoral dissertation committee of Don R. Baesel. At the time of his fatal accident, Rudolph was working with graduate students Peter Rillero in the history of science education, and Denise Adams, a member of his last class, who continued in the history of horticulture. After Rudolph's death, Ethan Stock, an undergraduate, completed an honors thesis, "Themes of Order and Sources of Authority in Popularized Victorian Natural History 1844–1889."[17] His study was based on 300 of Rudolph's children's books, now housed in the Rare Books and Manuscripts Collection of The Ohio State University Libraries. Following the retirement of the above listed faculty members mostly during the 1970s and early 1980s, the informal History of Science Club and the identifiable graduate program in the history of science became inactive.

[16] Department of History. 1966. Graduate Work in the History of Science at The Ohio State University. Announcement, Department of History, 1 p.

[17] Ethan Stock. 1994. Themes of Order and Sources of Authority in Popularized Victorian Natural History 1844–1889. Senior Thesis. [Department of History, The Ohio State University, Columbus]. 62 pp.

First Three Papers Presented in the History of Botany (1965–1967)

After Rudolph had acquired a substantial number of general textbooks in natural science, biology, and botany, and of children's natural history books, he began in 1965 to organize his information from these sources, and to write papers on the general theme of the history of the popularization of botany in the United States. A review of the various topics on which he wrote, and the scientific meetings at which he made presentations constitute the remaining portion of this general introduction. His methods of obtaining and organizing information were simple. He hand copied information from books, journal articles, letters, and other sources onto yellow tablet paper, or he photocopied items from these sources. In the early years, when preparing for an oral presentation, Rudolph used these collected notations and handwrote his complete presentations double-spaced on yellow tablet paper. In later years he wrote two or three paragraphs, or prepared an outline of the talk, and then filled the remaining sheets with notes or quotations that comprised the body of the presentation. Many of his early handwritten manuscripts remained in that condition, or they were typed, either by himself, his wife Ann, or a departmental secretary. One or more handwritten and/or typewritten versions are available for most of his unpublished manuscripts, suggesting that none or only a very few of these writings were discarded. Manuscripts or outlines have been located for most of his presentations, and those items that appear to be missing may yet be found among his archives.

Rudolph intended to write a book on the popularization of botany. He communicated this idea not only to me at various times, but also to *OSU Lantern* reporter Nancy Rollins (1984)[10] who wrote, "Rudolph is planning to write a book about the popularization of Natural Science. His own collection of books on the subject will be his resources." How he would have organized and analyzed these materials in total content, and the conclusions that would have been presented can now only be imagined. Many of the components of his projected book, however, reside in several of his journal publications, his published and unpublished abstracts, and his manuscripts written for presentations at meetings of professional societies.

Rudolph's first effort in the history of botany pertained to an overlooked American edition of a book on the popularization of botany by the German botanist, Matthias J. Schleiden (1804–1881). Rudolph first presented this topic 23 April 1965 at the Annual Meeting of The Ohio Academy of Science, held at Ohio University in Athens, and later on 16 August 1965 at the Historical Section Meeting of the Botanical Society of America at the University of Illinois, Champaign-Urbana. His manuscript is printed here. **(Paper No. 1)**.

In the fall of 1965, following this first presentation, George H. M. Lawrence (1910–1978), Director of the Hunt Library in Pittsburgh, sent Rudolph a little book, *History of Botany* (1965),[18] containing a paper, Herbals: Their history and significance. After reading it with "great pleasure," Rudolph replied with a letter of thanks saying that the book would be useful to him and his students. As a beginner unfolding the pages of the history of botanical science, his enthusiasm showed, when in the next paragraph he[19] wrote:

> As seems to be the case too often, I find myself occupied with research and teaching outside of the area of history of botany. Someday I should like to take a whole year and do nothing else. I thank you for providing the excuse for taking a little time to read some history of botany now.

Rudolph's second effort in the history of science involved a project that seemed to have considerable fascination for him. The topic concerned the development of molecular biology. During the 1960s, the content of biology was changing rapidly, largely because of the great advances in molecular biology. More specifically, this change resulted in the knowledge of the structure of some of the large molecules occurring in organisms, the roles of DNA and RNA in protein synthesis, the kinetics of enzyme activity, and the details of cell structures as revealed by electron microscopy, among other topics.

Rudolph prepared a presentation, "Molecular Biology: Retrospect and Prospect." He made this presentation the next year on 22 April 1966 at the Annual Meeting of The Ohio Academy of Science, held at The Ohio State University in Columbus. The abstract, in mimeographed form, survives, but no manuscript has been located and no paper was published. Rudolph also made a presentation on this topic, under the title, "The Origins of and Reactions to Molecular Biology," on 19 July 1968, to his NSF-supported History of Biology Institute held at The Ohio State University. More will be written about this lecture in the introduction to Paper No. 12. Rudolph's abstract[20] follows:

> Molecular biology with its origins in biochemistry and biophysics has in recent years assumed a key position in biological thought. In the period of

---

[18] George H. M. Lawrence. 1965. Herbals: Their history and significance, pp. 1–35 in *History of Botany*. Two papers presented at a symposium held at the William Andrews Clark Memorial Library, December 7, 1963. George H. M. Lawrence: Herbals, their history and significance. Kenneth F. Baker: A plant pathogen views history. Clark Memorial Library, University of California, Los Angeles. 70 pp. [and] The Hunt Botanical Library, Carnegie Institute of Technology, Pittsburgh.

[19] Emanuel D. Rudolph. 1965. Letter of 1 November to George H. M. Lawrence, The Rachel McMasters Miller Hunt Botanical Library, Pittsburgh. 1 p.

[20] Emanuel D. Rudolph. 1966. Molecular biology: Retrospect and prospect, p. 1. *In* Abstracts of Papers Presented before Special Session on the History of Science, The Ohio Academy of Science Meeting at The Ohio State University, April 22, 1966. [Columbus]. Mimeographed. [Abstract].

> its history, the concepts of molecular biology have influenced the classical fields of biology in various ways. The claim of some that molecular biology will in time become the only biology is analyzed in terms of the interrelationships of the different levels of biological organization and in terms of the relationships among the various sciences. The thesis is presented that molecular biology will take its place as one of the divisions in the biology of the future, rather than replace or encompass all of the other divisions.

Rudolph's third presentation was based on a manuscript concerning Massachusetts-born botanical illustrator Isaac Sprague (1811–1895), who drew plants for Asa Gray's publications. The manuscript was read twice, first on 21 April 1967 at the Annual Meeting of The Ohio Academy of Science at the University of Dayton, and then on 28 August 1967 at the Historical Section Meeting of the Botanical Society of America at Texas A & M University in College Station. The manuscript was not completed and submitted for journal publication until 20 years later. The published paper appeared in 1990, and an abridged version of it is printed here. **(Paper No. 18)**.

Writing on Two Special Projects (1969)

History of Bryology and Lichenology in the United States[21]

As a young scholar, Rudolph was respected for his knowledge in the disciplines of bryology and lichenology, as well as the history of botany. In the tradition of previous International Botanical Congresses, members of the Historical Section of the Botanical Society of America—in particular Jerry Stannard, founder of the Section—proposed that a small book on the history of American botany be prepared for the Eleventh Congress. Invited by Stannard, then of the University of Kansas, and editor Joseph Ewan, then of Tulane University, Rudolph was among 12 botanists selected to write one of the 13 chapters for the book. Ewan[22] wrote to Rudolph, stating that "The nature of the 'Short History' demands a concise undocumented approach . . . there will be a select bibliography of review articles . . . The names and dates and subjects should send the interested reader to modern bibliographies and indexes."

Rudolph's 8-page chapter, "Bryology and Lichenology," according to a review by Delevoryas (1970)[23] was "well organized and trace[d] the development of . . . the

---

[21] Joseph Ewan, (ed.). 1969. *A Short History of Botany in the United States*. Hafner Publishing Co., New York and London. ix, 174 pp.

[22] Joseph Ewan. 1968. Letter of 24 September to Emanuel D. Rudolph, Department of Botany and Plant Pathology, The Ohio State University, Columbus. 1 p.

[23] T[heodore] Delevoryas. 1970. Plant science chronicle. [Review of] A Short History of Botany in the United States. *Science* 169: 1194.

subject by periods," spanning the time from before 1838 to the 1960s. Rudolph divided the history of these disciplines into four periods:

The Beginning Period: Before 1838
The Sullivant-Tuckerman Period: 1838–1888
The Transition Period: 1888–1947
The Recent Period: 1947–

Concerning the transition period, Rudolph (1969)[24] wrote:

> Popularization of moss studies begun by Mrs. [Elizabeth G.] Britton in her articles in *The Observer* (1894), and by [Abel J.] Grout in *Mosses with a Hand-lens* (1900), together with the activities of the Sullivant Moss Society, stimulated serious amateurs to collect and study bryophytes. Other popular works that appeared in the latter part of the period included Elizabeth Dunham's *How to Know the Mosses* (1916, rev. 1956), [Emmett T.] Bodenberg's [*Moss Flora*] *of Mt. Rainier National Park* (1939), and H. S. Conard's *How to Know the Mosses* (1944; rev. including liverworts, 1956). . . . Popularization of lichenology in this period was less sustained than in bryology, perhaps because of the lack of a simple usable guide like *Mosses with a Hand-lens*, and a leader like its author. . . . Yet amateur lichenologists have never been numerous.

When Rudolph[25] submitted his manuscript to editor Ewan, he stated in an accompanying letter: "I tried to emphasize our national contributions and thus may have slighted the more general aspects in which our contributions are only a small part of the whole." Resulting from this project was *A Short History of Botany in the United States* (1969),[21] which became available in the summer of 1969 for the members attending the Eleventh International Botanical Congress in Seattle, Washington.

History of Botany in The Ohio State University[26]

During 1969, Rudolph also was invited to compose a short *History of Botany in The Ohio State University* as part of the University's Centennial Celebration in

---

[24] Emanuel D. Rudolph. 1969. Bryology and lichenology, pp. 89–96. *In* J. Ewan, ed. *A Short History of Botany in the United States*. Hafner Publishing Co., New York and London. ix, 174 pp. (Quote, pp. 94–95).

[25] Emanuel D. Rudolph. 1968. Letter of 14 September to Joseph Ewan, Department of Botany, Tulane University, New Orleans, Louisiana. 1 p.

[26] Emanuel D. Rudolph and Ronald L. Stuckey. 1969. History of Botany in The Ohio State University. Prepared for the Centennial of the University, 1970. Columbus, Ohio. [Typewritten manuscript, Columbus, Ohio. 22 pp. + Table I, regular faculty members . . .1925–1970. Copies in the University Archives, the Hunt Botanical Institute for Botanical Documentation, and in the possession of Ronald L. Stuckey].

1970. This 22-page typewritten account was written by Rudolph, with the sections on the Herbarium abridged from a more detailed unpublished manuscript on the history of the Herbarium earlier prepared by then Curator Ronald L. Stuckey (1969).[27] This centennial contribution, written nearly 30 years ago, is not reproduced in this memorial volume, because much of the information in the paper was incorporated into the book, *Botany at The Ohio State University: The First 100 Years* (1983), by Bernard S. Meyer (1901–1987),[28] chairman of the Department from 1946 to 1966.

Presentations at Meetings of Scientific Organizations (1965–1991)

Although Rudolph did not make presentations on topics in the history of botany every year, he presented a remarkable number of them to the Plant Sciences Section (1965, 1983–1991) and to the special history of science meetings (1966–1979) of The Ohio Academy of Science, the Historical Section of the Botanical Society of America (1965–1991), the Ohio Academy of Medical History (1970–1988), and the Midwest Junto for the History of Science Society (1975, 1981). The titles of Rudolph's presentations at these meetings are listed with the description of each organization below.

Rudolph was invited to give presentations at symposia and special programs in the United States and England. These invited symposia or contributed presentation sessions were: The Joint Atlantic Seminar in the History of Biology (Boston, 1973); the History of Science Society (Madison, Wisconsin, 1978); the Bicentennial Symposium for the birth of the pioneer midwestern physician Daniel Drake, at The Ohio Academy of Science (Cincinnati, 1985); the Royal Institution Symposium on Laboratories, the Place of Experiment (London, 1986); the Thirty-Sixth Annual Systematics Symposium of the Missouri Botanical Garden (St. Louis, 1989); the Gideon Mantell Bicentenary Symposium of the Society for the History of Natural History (Brighton, England, 1990); and the Teaching Biology, Making Biology Symposium of the International Society for the History, Philosophy & Social Studies of Biology (Evanston, Illinois, 1991). The presentations delivered at these special meetings are discussed at their appropriate places in this volume.

---

[27] Ronald L. Stuckey. 1969. History of The Ohio State University Herbarium. Typewritten, Manuscript. 9 pp.

[28] Bernard S. Meyer. 1983. *Botany at The Ohio State University: The First 100 Years. Ohio Biological Survey Bulletin. New Series* 6(2): i–vi, 1–177.

The Ohio Academy of Science (1965–1991)

The Ohio Academy of Science held special meetings for presentations on the history of science during the 1960s and 1970s. This interest in promoting the history of science within the Academy was initiated by Emanuel D. Rudolph, Ralph W. Dexter (1912–1991), and Ronald L. Stuckey. These gentlemen constituted an informal committee that arranged for two sessions of contributions at the Academy's annual meetings in April of 1966 and 1967. A third special history of science session of presentations was organized primarily by Stuckey and held 10 February 1968, apart from the annual meeting, at the Battelle Memorial Institute, Columbus. Additional meetings, all held separately from the annual meetings, were convened continuously through 1979, except for 1978, covering a total of 13 years.

Rudolph was active in presenting papers at these history of science meetings sponsored by The Ohio Academy of Science. In addition to providing assistance in the arrangements, he presided at many of the sessions, and delivered seven presentations during the period from 1965 to 1979. Most of these presentations, which he read, were concerned with the popularization of botany. The title of each presentation, the year given, and the paper number assigned for this book follow:

1965. "M. J. Schleiden's Contribution to the Popularization of Botany: *Die Pflanze und ihr Leben*." (Paper No. 1)

1966. "Molecular Biology: Retrospect and Prospect." [Not counted in Table I.1].

1967. "Isaac Sprague's Role in the Ascendency of American Botany during the Nineteenth Century." (Paper No. 18)

1968. "Botany in American Children's Books: The Period before 1900." (Paper No. 6)

1971. "The Origin and Development of the Botanical Textbook and Its American Aspects." (Paper No. 2)

1972. "Natural History Chapbooks: Throwbacks to Pliny and the *Physiologus* as Well as Windows to 18th Century Biology."

1979. "The Contributions of Women to the Development of 19th Century American Botany." (Paper No. 20)

When the special history of science meetings were no longer held, Rudolph continued to make presentations in the Plant Sciences Section at the annual meetings of the Academy, as follows:

1983. "A Brief History of Plant Personification." (Paper No. 10)

1985. "Daniel Drake as an Educator." (Special Symposium). (Paper No. 15)

1986. "Bruce Fink, Midwestern Botanist and Educator." (Paper No. 17)

1991. "Henry Shaw's Path from Sheffield to Mill Hill School and the Founding of the Missouri Botanical Garden." (Paper No. 24)

Because of Rudolph's interest in history, he served as chairperson of the Necrology Committee from 1984 to 1992, and in that role wrote 12 obituaries that were published in *The Ohio Journal of Science*. As chairperson of the Academy's Centennial Celebration Committee from 1987 to 1991, Rudolph was instrumental in planning several special programs. Working in conjunction with this Committee, he was a contributor to the selection of 25 Centennial Honorees, chosen for their contributions to science, education, and the Academy. Among botanists, only two were named, Ronald L. Stuckey and Clara G. Weishaupt (1898–1991).[29] Rudolph's published centennial note, "One Hundred Years of Plant Science 1891–1991, with an Ohio Perspective" (1991),[30] provided a concise and thoughtful review of this discipline. The portion of the contribution pertaining to this history at the national level is edited and reprinted here. **(Paper No. 30)**.

Historical Section, Botanical Society of America (1965–1991)

The forum most utilized by Rudolph for giving his presentations was the Historical Section of the Botanical Society of America. Founded and organized in 1964 by Jerry Stannard,[31] this section held its first session of presentations in August 1965, when the Society met at the University of Illinois in Champaign-Urbana. At this meeting, Rudolph read again his manuscript about Schleiden's overlooked book on plant popularization that he had presented in April of that year to The Ohio Academy of Science. **(Paper No. 1)**. The 17 presentations that Rudolph made at the contributed papers sessions of the Society's Historical Section, from 1965 through 1991, totaled more than were given by any other presenter in the history of the Section. The titles of the presentations are listed here with the years given, and their assigned numbers for this book.

1965. "M. J. Schleiden's Contribution to the Popularization of Botany: *Die Pflanze und ihr Leben*." (Paper No. 1)

1967. "The Role of the Botanical Illustrator Isaac Sprague in the Ascendency of American Botany during the 19th Century." (Paper No. 18)

---

[29] The Ohio Academy of Science. Centennial Celebration Commission. 1991. Centennial Awards Dinner in Recognition of Centennial Honorees, Friday April 26, 1991, Holiday Inn on the Lane, Columbus. The Ohio Academy of Science, Columbus. 11 pp.

[30] Emanuel D. Rudolph. 1991. One hundred years of plant science 1891–1991, with an Ohio perspective. *The Ohio Journal of Science* 91: 104–107.

[31] Jerry Stannard. 1965. Historical Section of the Botanical Society of America. *Plant Science Bulletin* 11(1): 5–6.

1968. "Botany in American Children's Books before 1900." (Paper No. 6)

1970. "Nineteenth Century American Schoolbooks and Textbooks of Botany." (Paper No. 2)

1972. "Botany in English and American Chapbooks." (Paper No. 7)

1976. "The Americanization of Germanic Botany, 1865–1915." (Paper No. 2)

1977. "Botany by Rote: Early Nineteenth Century Catechisms for Teaching about Plants." (Paper No. 8)

1978. "The Role of Women in Nineteenth Century Botany." (Paper No. 20)

1979. "A 150 Year History of the Botanist as a Career Model for American Children." (Paper No. 11)

1980. "The Introduction of the Natural System of Classification to American Students in the Nineteenth Century." (Paper No. 26)

1982. "Personifying Plants from Linnaeus to the Present." (Paper No. 10)

1983. "The Role of Almira Hart Lincoln Phelps in the Spread of Botany in Nineteenth Century America." (Paper No. 20)

1984. "The First American Plant Anatomy Book and Its Author." (Paper No. 21)

1986. "Coverage of Lichens in American Introductory Botany Textbooks 1860–1985." (Paper No. 4)

1987. "The Development of Teaching Laboratories for Botany in Nineteenth Century America." (Paper No. 13)

1990. "The Missouri and New York Botanical Gardens, Contrast and Similarity of Origins and Early Development." (Paper No. 23)

1991. "Why Henry Shaw Established a Public Botanical Garden in St. Louis and How He Developed It." (Paper No. 24)

Ohio Academy of Medical History (1970–1988)[32]

The Ohio Academy of Medical History, established for the purpose of providing a statewide working program for the discovery and preservation of materials pertaining to the medical history of Ohio, was founded 5 April 1952 at the Ohio State Museum in Columbus. The organization was an outgrowth of a group called the Ohio Committee of Medical History and Archives, inaugurated in 1937. In the

[32] [Patsy A. Gerstner]. [1978]. Ohio Academy of Medical History. [Historical Sketch]. Typewritten, 1 p.

1960s, the Academy, comprised mostly of veteran physicians, had its headquarters in central Ohio, and held a meeting each spring to give presentations on various aspects of the history of medicine. Others from related disciplines were also members or participants, examples being biologist Ralph W. Dexter (1912–1991) of Kent State University; and scholars from The Ohio State University: botanists, Emanuel D. Rudolph, Ronald L. Stuckey, and Adolph E. Waller (1892–1975); historian John C. Burnham; and medical historians Richard L. Meiling (1908–1984) and Charles L. Wooley. An informal atmosphere prevailed among the small group of about 20 individuals that usually attended the presentation sessions. These meetings provided an outlet for the non-physicians to deliver presentations about physicians who had botanical or other scientific interests or that concerned medical topics from philosophical viewpoints.

When Rudolph first began attending these meetings is not known, but he was probably introduced to them by retired Ohio State University botanist Adolph E. Waller, who had taught the course in the history of botany prior to 1961, when Rudolph joined the University. At Rudolph's invitation, I first attended with him at the meeting of 15 April 1967 at the Granville Inn in Granville, Ohio. It was customary for the meetings to be held every spring at this location near the center of the state. That practice was continued until the early 1970s, when the meetings were moved to other locations in an attempt to broaden its geographical base and add new members. Since 1977 the organization has had its headquarters at the Howard Dittrick Museum of Medical History, Cleveland, with Patsy Gerstner, Ph.D., as secretary. At the 1967 meeting, Waller made a presentation on "Dr. George Engelmann and the St. Louis Road to Modern Botany," an example of a presentation by a non-physician. Over the next several years Rudolph and I attended regularly. Rudolph gave his first presentation on several aspects of the physician-botanist, Matthias Jacob Schleiden (1804–1881) at the 1970 meeting. That year, Rudolph became secretary-treasurer of the organization and retained that position until 1977, when he was elected the president for the next year. As president, he presided at the 1978 meeting, as was the practice within the organization.

Rudolph made eight presentations at meetings of the Ohio Academy of Medical History. Their titles with dates and places of presentation are as follows:

"Matthias Jacob Schleiden (1804–1881): Lawyer, Doctor, Botanist and Anthropologist." 18 April 1970, Granville.

"Natural History Chapbooks: Throwbacks to Pliny and *Physiologus* as well as Windows to 18th Century Biology." 8 April 1972, Columbus.

"The Drift Away from Teaching Botany Mainly for Doctors in America in the 19th Century." 29 March 1975, Granville.

"The Elusive Doctor John B. Newman, M.D., of New York." 27 March 1976, Granville.

"What 19th Century Children Were Taught about Physiology." 29 April 1978, Cleveland.

"Some Medical Doctors Who Taught Nineteenth Century American Students of Botany." 26 April 1980, Granville. (Paper No. 16)

"The Career of Carl Linnaeus [1707–1778] as a Medical Doctor." 9 May 1981, Zanesville.

"Dr. Alexander Anderson (1775–1870), First American Engraver on Wood." 12 March 1988, Zanesville.

According to the organization's practice, individuals who made presentations at these Ohio Academy of Medical History meetings did not prepare abstracts of their papers. Rudolph did prepare manuscripts that survive for his presentations on natural history chapbooks, medical doctors who taught students, what children were taught about physiology, and the medical career of Linnaeus. A notebook of information is available concerning the life of the elusive Dr. John B. Newman (fl. 1846–1875). A partial handwritten manuscript of the introductory paragraphs, in addition to numerous quotations and notes, exists for his presentation on Dr. Alexander Anderson (1775–1870). The only contribution that Rudolph read at these meetings that is printed here is "Some Medical Doctors Who Taught Nineteenth Century American Students of Botany" **(Paper No. 16)**.

Midwest Junto for the History of Science (1975, 1981)

The Midwest Junto for the History of Science, founded in 1957, represents a group of historians of science located in the midwestern United States who meet annually to give presentations on teaching and research in the history of science. When the meeting was to be held in April 1972 at the University of Wisconsin, Madison, Rudolph[33] wanted to attend the event. In a letter to Department Chairman John A. Schmitt (1925–1994) he requested a small travel grant of $135. It evidently was to be Rudolph's first time to experience one of these meetings, and he listed three reasons why it was important for him to participate:

1) I teach a course in the history of biology and thus need to be aware of the latest research in the field.

2) My research on the popularization of biology in the eighteenth and nineteenth centuries requires that I consult with others who have research interests that are related.

3) The CIC [Committee on Institutional Cooperation of the Big Ten universities and the University of Chicago] has a cooperative program in the history of science of which I am part, and it is valuable for me to meet others cooperating in this program.

[33] Emanuel D. Rudolph. 1972. Letter of 28 February to John A. Schmitt, Department of Botany, The Ohio State University, Columbus. 1 p.

Table I. 1.
Numbers of Presentations by Topic Delivered at Scientific Meetings by Emanuel D. Rudolph, summarized by Stuckey

Meetings	Totals
Ohio Academy of Science	9
Historical Section Botanical Society of America	17
Ohio Academy of Medical History	8
Midwest Junto for the History of Science	3
Symposia	5
Total Number of Meetings and Topics	42
Number of Meeting Presentations Published	11
Number of Abstracts Prepared for Meetings	30

Rudolph's annotated program is sufficient evidence that he indeed attended the meeting. Later, he presented two papers at the Midwest Junto meetings: "The Botanical Textbook in Nineteenth Century America as a Reflection of Botanical and Cultural Trends," at Kansas State University in Manhattan (11 April 1975), and "Women as a Constituency in Nineteenth-Century American Botany, a Quantitative Analysis," at the University of Akron, Ohio (11 April 1981). No abstracts are known to have been written for these presentations. The manuscript for the presentation on the "Botanical Textbook" is incomplete. The introductory portion was written in full, but most of the manuscript consists of an outline of information that was to be highlighted during the presentation. Portions of this manuscript were used to develop the paper printed in this memorial volume on the "Origin and Development of the American Textbooks of Botany" **(Paper No. 2)**. The manuscript on "Women as a Constituency in Botany" has not been located, but it certainly is the basis for Rudolph's paper on "Women in Nineteenth Century American Botany; A Generally Unrecognized Constituency," published 1982 in the *American Journal of Botany*.[34]

[34] Emanuel D. Rudolph. 1982. Women in nineteenth century American botany; A generally unrecognized constituency. *American Journal of Botany* 69: 1346–1355.

Textbooks and Popularization	Children's Natural History	Educators, Illustrators, Laboratories	Women in Botany	Missouri Botanical Garden
2	2	3	1	1
4	5	3	3	2
0	2	6	0	0
1	1	0	1	0
0	2	2	0	1
7	12	14	5	4
1	3	3	3	1
6	9	8	4	3

Summary of Meeting Presentations

Rudolph had a remarkable record of presentations on the history of botany at meetings of state and national organizations (Table I.1). He delivered a total of 42 presentations, subdivided into 17 at meetings of two state organizations in Ohio, 20 at meetings of two national organizations in America, and five at symposia. Of the total 42 presentations, 17 of them, the largest number, were made at the Historical Section of the Botanical Society of America. Among the subject categories, 14 concerned educators, illustrators, and laboratories, and 12 pertained to children's natural history. Abstracts were prepared for 30 of the 42 presentations, but only 11 of the presentations were developed into journal publications. If Rudolph had prepared manuscripts or extensive notes for the remaining number of meeting presentations, these would total 31, nearly equal to the number of botanical history papers printed in this book.

Projects Lacking Papers (1987–1992)

During the late 1980s, Rudolph participated in three history of botany projects. They are discussed below under the headings "Organizing a Botanical History Symposium" (1987), "Creating a Readers Guide to Rees's Cyclopaedia"[35] (1987), and "Exhibiting Botanical Illustrations" (1988–1992). No accompanying presentations, however, are available with these titles.

Organizing a Botanical History Symposium (1987)

Rudolph, who loved to attend meetings and to share information by making presentations, was also an organizer of proposed projects to create added benefits in teaching and research. When the American Institute of Biological Sciences held its 1987 annual meeting on the Columbus campus of The Ohio State University, Rudolph and Stuckey organized a Symposium on "Early Botany in the trans-Allegheny Region" for the Historical Section of the Botanical Society of America (11 August 1987). This event marked the first time in the 22-year history of the Section that a symposium was on the program; introduced by Rudolph (Fig. I.4).

The presentations considered various botanical and other investigations in the western portion of the new American nation. Contributions were given by Joseph Ewan on Benjamin Smith Barton's influence on the region; Henry D. Shapiro on Daniel Drake as a new western naturalist, which was read at the meeting by Ann Rudolph (1924–1991); Charles Boewe on C. S. Rafinesque as a field naturalist in the region; Ronald L. Stuckey on William S. Sullivant and his central Ohio botanical associates; and Relda E. Niederhofer on Edwin Lincoln Moseley as an internationally-known Ohio naturalist (Fig. I.5). The contributions were subsequently edited by Stuckey, although not stated, and published in *Bartonia*, the journal of the Philadelphia Botanical Club (1988).[36]

In the symposium's published introduction Rudolph (1988)[36] asked if a uniqueness or distinctiveness existed in early nineteenth century botany and science as it was practiced in the trans-Allegheny region. Among the answers to this question, Rudolph, paraphrasing from Craven,[37] pointed out that geographically the region

[35] Abraham Rees. 1806–1822. *The Cyclopaedia; or, Universal Dictionary of Arts, Sciences, and Literature*. Samuel F. Bradford, Philadelphia. 41 vols.

[36] Emanuel D. Rudolph. 1988. Introduction [to] early botany in the trans-Allegheny region, a symposium organized by Emanuel D. Rudolph and Ronald L. Stuckey. *Bartonia* 54: 26–27.

[37] Avery Craven. 1934. The advance of civilization into the Middle West in the period of settlement, pp. 39–71. *In* Dixon Ryan Fox (ed.), *Sources of Culture in the Middle West, Backgrounds Versus Frontier*. The Appleton-Century Historical Essays. D. Appleton-Century Co., New York, London. 110 pp.

Fig. I.4. Emanuel D. Rudolph at the lectern introducing the symposium on "Early Botany in the trans-Allegheny Region," at the 1987 Columbus meeting of the Historical Section, Botanical Society of America.

(Taken 19 August 1987, by Ronald L. Stuckey; in the collection of RLS)

coincided with the Ohio Valley. It was shaped like a long wedge extending from the Alleghenies to Missouri, and could be a unit thought as a first Middle West, in addition to the territory embracing the remainder of the Old Northwest. Slightly modified portions of Rudolph's introduction follow:

> Scientific endeavors in the Colonies and early United States, according to most historians, are pale reflections of European, particularly British,

Fig. I.5. Symposium speakers on "Early Botany in the trans-Allegheny Region." l. to r. Joseph Ewan, Charles Boewe, Ronald L. Stuckey, Relda E. Niederhofer, Emanuel D. Rudolph. OSU Campus, Columbus, Ohio.

(Taken 19 August 1987, by Robert Bartolotta for RLS; in the collection of RLS)

science.[38,39] Such activity was practiced by a population living mainly on the eastern seacoast in the larger cities, such as Philadelphia, Boston, New York, and Charlestown, where libraries, scientific societies, publication opportunities, and other resources were available to naturalists. No wonder that investigations by naturalists west of the Allegheny Mountains should be cast by historians as even dimmer reflections of Eastern United States and European type science, and thus warranting even less of their critical study. However, I believe a "Western" or "frontier" attitude was reflected because the naturalists did more questioning and were more assertive, therefore making them unique.

. . . two hundred years ago [1787] Congress enacted the Northwest Ordinance that established a pattern for governance of the lands north and west of the River Ohio and provided for the orderly development of new states, on an equal status with the existing ones. This liberal and foresighted attitude of the Federal Government provided a healthy optimism in settlers of these territories as well in those south of the Ohio River.[40] Such optimism is reflected in the itinerant and especially in the resident naturalists who collected and described plants, animals, fossils, Indian artifacts, and everything around them in these Western lands. Although they were few in number, and often worked in isolation because large cities with resources

[38] John C. Greene. 1984. *American Science in the Age of Jefferson.* The Iowa State University Press, Ames. xiv, 484 pp.

[39] Robert V. Bruce. 1987. *The Launching of Modern American Science 1846–1876.* Alfred A. Knopf, New York. x, 446 pp.

[40] Robert V. Van Trees. 1986. *Ordinance of Freedom.* The Author, Fairborn, Ohio. x, 102 pp.

> were lacking, though developing,[41] they believed nature had provided ample novelty for their enthusiastic investigations. Mostly they collected specimens and described species, but some of them challenged the interpretations of the Eastern establishment.

"Those were exciting times . . . that combined scientific curiosity, patriotic devotion, regional pride, with a religious fervor for illuminating and admiring God's work," according to Rudolph.[36] He quoted from Daniel Drake (1820):[42]

> It is not the gratification of curiosity alone that delights us in studying the productions of the earth. The beneficent Creator has mercifully given us capacity for deriving pleasure . . . of those portions of his mighty works which immediately surround us, and with which we are most familiar.

Rudolph noted three apparent themes in the papers: (1) the naturalists had complicated networks of cooperating individuals, particularly with those on the East Coast; (2) the collections of preserved natural history specimens by individuals and institutions in the region were perceived to be important manifestations of the scientific achievements and were to be useful in public education; and (3) the discovery of species new to science, the finding of useful organisms, and the understanding of the regional flora and fauna were major stimuli for those who studied nature. Rudolph further stated that an understanding existed in the importance of contributions from universities, schools, societies, and museums, as a part of the development of science in the new West. Because most of the early naturalists earned their livelihood through non-scientific professions, it must have been difficult to interest new recruits in becoming attracted to the scientific endeavor, although a group of unheralded educators or the naturalists themselves may have been involved in the recruitment of new investigators.

Creating a Readers Guide to Rees's **Cyclopaedia (1987)**

During 1987 Rudolph joined as a volunteer in a project of producing a "Readers Guide" to Abraham Rees's *Cyclopaedia*. The *Cyclopaedia: or, Universal Dictionary of Arts, Sciences and Literature*, begun in London in 1802 and completed in 1820, under the direction of Abraham Rees (1743–1825) (Fig. I.6), consisted of 39 volumes of letterpress containing more than thirty-nine million words, five volumes of plates, and a one-volume atlas. An American edition, begun in Philadelphia in 1806 and completed in 1822, added information of special interest for Americans. This publication was expanded to 41 volumes with five volumes of plates and a one-volume atlas (Fig. I.7). The pages are unnumbered, and many of the topics are

[41] Walter B. Hendrickson. 1976. Science and culture in the American Middle West, pp. 33–47. *In* Nathan Reingold (ed.), *Science in America since 1820*. Science History Publications, New York. 334 pp. [Reprinted from *Isis*, Sept. 1961].

[42] Daniel Drake. 1820. On the utility and pleasures of the study of Mineralogy and Geology. *The Port Folio* (Philadelphia), 4th ser. [ser. 5], 9: 86–93. (Quote, p. 92).

contained under obsolete or unusual headings making it difficult to retrieve information. Consequently, the information often has been ignored.

The idea of preparing a "Readers Guide," an ambitious project, to Rees's *Cyclopaedia* was conceived and developed earlier in the 1980s by June Z. Fullmer (1920–2000),[43,44] in the Department of History at The Ohio State University. Through correspondence[45] and published notices[46] she brought together a group[47] of 22 American and British volunteers agreeing to prepare a two-volume publication. The plan was to have an analytical subject index, a "key" to the biographical literature about the contributors, a discussion of the major topics, and a history of publication of both the English and American editions. The "Readers Guide" was intended to be a useful and efficient tool for opening windows to early nineteenth century scholarly achievements as recorded in Rees's *Cyclopaedia*.

Rudolph was interested in the *Cyclopaedia* because, as he stated in his research report[48] of 7 July 1987 following a visit to the American Philosophical Society in Philadelphia, the treatise "was very important in popularizing science and technology at the beginning of the nineteenth century in England and America. Many problems are present relating to the authorship of the unsigned articles, to the changes made in the American edition and their editors, as well as the general information presented in these volumes." Rudolph's[49] projected responsibilities toward preparation of the "Readers Guide" were to (1) determine the publication dates of each volume, (2) discuss the history of publishing the American edition, including profiles of the publisher, Samuel F. Bradford (1776–1837) and associate

---

[43] June Z. Fullmer. [1987]. The Rees Project: A Readers Guide to Rees's *Cyclopaedia*. 1 p. [Manuscript].

[44] Robert Boyce. 1976. Fullmer's work a benchmark Encyclopaedia evolution explored. *On Campus, The Ohio State University*, 28 February 1976, p. 2.

[45] June Z. Fullmer. [1987]. Collaborators Invited. Typewritten, 1 p. [Manuscript].

[46] [June Z. Fullmer]. 1987. Rees's *Cyclopaedia* (1802–1820). *Newsletter, Archives of Natural History* 14(1): 112.

[47] June Z. Fullmer. [1987]. Appendix N. Working group on Rees's *Cyclopaedia* (September 1987), 2 pp. *In* June Z. Fullmer and Emanuel D. Rudolph. "Creation of the "Readers' Guide to Rees's *Cyclopaedia*." [Grant submitted 30 October 1987; not funded].

[48] Emanuel D. Rudolph. 1987. Organisms in 19th and 20th Century Children's Books: An Analysis [+ Note of Information about Rees's *Cyclopaedia* at the American Philosophical Society, Philadelphia, 8 May 1987]. Report of 7 July 1987, Office of Research and Graduate Studies, The Ohio State University, Columbus, Ohio. Typewritten, 2 pp.

[49] Harry Coles, J. Fullmer, E. Rudolph, and Nicolas Velos. [1987]. Tentative Outline for Paper about American Edition of Rees. Typewritten, 4 pp. [Manuscript].

Fig. I.6. Abraham Rees (1743–1825).

(From the American edition of Rees's *Cyclopaedia*,
Samuel F. Bradford, Philadelphia)

THE

CYCLOPÆDIA;

OR,

UNIVERSAL DICTIONARY

OF

ARTS, SCIENCES, AND LITERATURE.

BY

ABRAHAM REES, D.D. F.R.S. F.L.S. *S. Amer. Soc.*

WITH THE ASSISTANCE OF

EMINENT PROFESSIONAL GENTLEMEN.

ILLUSTRATED WITH NUMEROUS ENGRAVINGS,

BY THE MOST DISTINGUISHED ARTISTS.

FIRST AMERICAN EDITION,

REVISED, CORRECTED, ENLARGED, AND ADAPTED TO THIS COUNTRY,

BY SEVERAL LITERARY AND SCIENTIFIC CHARACTERS.

IN FORTY-ONE VOLUMES.

VOL. I.

PHILADELPHIA:

PUBLISHED BY SAMUEL F. BRADFORD,

AND

MURRAY, FAIRMAN AND CO.

PETER A. MESIER, NEW-YORK; BLAKE & CUNNINGHAM, BOSTON; CUSHING & APPLETON, SALEM; HEZEKIAH HOWE, NEW-HAVEN; CHARLES WHIPPLE, NEWBURYPORT; JOHN W. ADAMS, PORTSMOUTH; STEPHEN PATTEN, PORTLAND; FIELDING LUCAS, BALTIMORE; DANIEL RAPINE, WASHINGTON; JOSEPH MILLIGAN, GEORGETOWN; JOHN A. STEWART, ALEXANDRIA; D. HENDERSON, FREDERICKSBURG; WM. H. FITZWHYLSONN, RICHMOND; RICHARD COTTOM, PETERSBURG; JOHN HOFF, CHARLESTON; WILLIAM T. WILLIAMS, SAVANNAH; HOBBY & BUNCE, AUGUSTA; CRAMER, SPEAR, & EICHBAUM, PITTSBURG; AND JAMES W. PALMER, LEXINGTON, KENTUCKY.

Fig. I.7. Title page of the American Rees's *Cyclopaedia*, published 1805–1822.

(From Rudolph's copy of Rees's *Cyclopaedia* in the Rare Book Room, Herbarium, The Ohio State University, Columbus)

editor, Alexander A. Wilson (1766–1813), (3) reveal modifications in the biographies of significant Americans, and (4) record information on all botanical subjects, including specifically names of plants, names of species new to science, contributors of botanical articles, and geographical information pertaining to the North American distribution of plants. Rudolph left handwritten notes on some of these topics, in addition to a file of photocopied articles on these subjects. In October 1987 Fullmer and Rudolph[50] submitted a grant proposal requesting over $280,000 to the United States National Endowment for the Humanities for preparing the "Readers Guide," but unfortunately the grant was not funded, and the project was laid aside.

Rudolph's American set of Abraham Rees's *Cyclopaedia*, complete with the exception of volumes 11 and 23 and the volumes containing the illustrations and maps is in the rare book room at the Herbarium of The Ohio State University (Fig. I.8). He purchased this set in December 1988 for $120.00 from the Ohio Book Store in Cincinnati. In addition, he acquired some of the volumes containing plates from other sources at different times. In an effort to determine the dates of publication of the various volumes of the American edition, Rudolph searched systematically through two of the then contemporary published journals, The *Port Folio* and the *North American Review*. He recorded the information from these two sources in a table, dated 17 June 1987, as noted on the page. Because publication dates are extremely important in botanical papers containing names and descriptions of taxa new to science, Rudolph's research determining publication dates of the American edition of Rees's *Cyclopaedia* is of considerable use to taxonomic botanists. Printed here from Rudolph's research files is his list of determined publication dates (Table I.2). The titles of his unpublished notes and indices, except for the two given in footnotes 49 and 50, are as follows.

Rudolph, Emanuel D. [1987]. ["Index of Genera of plants, *Abroma* to *Amarantheides*, in the British Edition of Rees's *Cyclopaedia*."] Handwritten, 15 pp., with notes in pencil pertaining to the American Edition.

_____. [1987]. ["Index of Plates in Volume 1 of the American Edition of Rees's *Cyclopaedia*."] Handwritten, 1 p.

_____. 1987. "Rees's *Cyclopaedia*, American Edition, Published notes from the *Port Folio, The North American Review*, and other published sources." Typewritten in part, 3 pp. 17 June 1987.

_____. [1987]. "Rees's *Cyclopaedia*, Library holdings as indicated in the National Union Catalog; London edition; American edition." Typewritten, 2 pp. Manuscript.

_____. [1987]. [List of] "American Printers (Publishers or Distributors)." Handwritten, 1 p. + notes, 5 pp.

---

[50] June Z. Fullmer and Emanuel D. Rudolph. 1987. "Creation of the Readers' Guide to Rees's *Cyclopaedia*." Grant proposal submitted 30 October 1987 to the National Endowment for the Humanities, Washington, D.C. Typewritten, 106 pp.

Fig. I.8. Emanuel D. Rudolph's set of Rees's *Cyclopaedia*, in the Rare Book Room, Herbarium, The Ohio State University.

(Taken 4 December 1995, by Ronald L. Stuckey, in the collection of RLS)

Table I. 2.
Publication Dates of the American Edition of Rees's *Cyclopaedia* Taken from Contemporary Sources,[a] by Rudolph

Volume	Date	Reference Periodical
Announcement	10 Aug 1805	*Port Folio* 5(31): 245. 1805.
Further Announcement	17 Aug 1805	*Port Folio* 5(31): 252. 1805.
Vol. 1, 1st half	3 May 1806	*Port Folio* n. ser. 1(17): 267–268. 1806.
Vol. 2, 2nd half	"in press"	
Vol. 6, 1st half	31 Oct 1807	*Port Folio* n. ser. 4(18): 281–282. 1807.
Vol. 36, 2nd half	Mar 1818	*North American Review* 6: 432. 1818.
Vol. 37, 2nd half=part 74	May 1818	*Port Folio* 5th ser. 5: 224. 1818.
Vol. 38	Sep 1818	*North American Review* 7: 436. 1818.
Vol. 49[=39], part 1	May 1819	*Port Folio* 5th ser. 7: 428. 1819.
Vol. 39	Sep 1820	*Port Folio* 10: 255. 1820.
Publication completed	year 1822	Scharf and Westcott (1884)[b]
Full set donated to the American Philosophical Society	18 Jan 1822	Minutes of the meeting for 18 Jan 1822, published *Proceedings of the American Philosophical Society* 22: 508. 1885.

[a] Emanuel D. Rudolph. 1987. Rees's *Cyclopaedia*, American Edition. Published notes from *Port Folio*, *North American Review*, and other published sources. [Manuscript, Typewritten in part, 3 pp.]. Signed Date, 17 June 1987.

[b] J. Thomas Scharf and Thompson Westcott. 1884. History of Philadelphia 1609–1884. L. H. Everts & Co., Philadelphia. 3 vols. (2,399 pp.).

Exhibiting Botanical Illustrations (1988–1992)

In Rudolph's early years at The Ohio State University, when he was teaching in the course General Botany, all instructors at one or more times were asked to assemble in a hallway showcase a display on a botanical theme. These displays were made to complement topics taught in the course. Rudolph enjoyed this activity, having been involved in a similar way at the Brooklyn Children's Museum. He always had plenty of display materials for many topics and themes. On 18 July 1968 during his early years of studying the history of science, Rudolph[51] wrote to George Lawrence (1910–1978): "I do have a strong interest in botanical illustrations and thus would appreciate your exhibit catalogues." From this time on, Rudolph seriously began building a collection of botanical prints and other memorabilia, both for their beauty and educational value. Displaying and discussing examples from his collection of botanical illustrations were always significant items on the agenda when his class in the history of biology and others came to his home.

Rudolph's opportunity for a public exhibition of his own collected botanical illustrations came in 1988. At the request of the Friends of the Chadwick Arboretum on the campus of The Ohio State University, he was invited by its Director David A. Voyles and Development Officer Janet M. Oberliesen to exhibit his prints on the theme of "500 Years of Botanical Illustration c. 1481–1981." This all-day event, held 12 May 1988 at the Arboretum, was part of the second annual Chadwick Garden tour. After a luncheon, Rudolph presented a short lecture on his botanical exhibition. For the 8-page program, printed on purple-colored paper, Rudolph[52] wrote an "Introduction" and provided an annotated list of the 40 illustrations that were on display. These prints depicted various means of production, as follows: Eight woodcuts from herbals, 20 engravings, three lithographs, two etchings, two color halftones, and one each identified as nature printed, pencil sketch, color lithograph, linoleum block. One was not labeled. Familiar names among the illustrators were Leonhart Fuchs, Abraham Munting, George D. Ehret, Mark Catesby, William Curtis, Jacob Bigelow, Isaac Sprague, and Pierre Joseph Redouté. The last one named, Redouté (1759–1840), a popular, affluent French artist and teacher of art honored for his work, is considered the greatest of all known painters of flowering plants in the history of botanical art.[53] This exhibition was the only public showing of Rudolph's botanical print collection. Since his death, the exhibit has been reconstructed and displayed at the Columbus Museum of Art and the OSU Faculty Club.

---

[51] Emanuel D. Rudolph. 1968. Letter of 18 July to George H. M. Lawrence, The Rachel McMasters Miller Hunt Botanical Library, Pittsburgh, Pennsylvania. 1 p.

[52] Emanuel D. Rudolph [and others]. [1988]. [Program Brochure]. 2nd Annual Chadwick Arboretum Garden Tour; "500 Years of Botanical Illustrations c. 1481–1981." [Chadwick Arboretum, The Ohio State University, Columbus]. 8 pp.

[53] William T. Stearn. 1990. Pierre-Joseph Redouté, pp. 24–25. *In Botanical Masters: Plant Portraits by Contemporary Artists*. Prentice Hall Editions, New York. 160 pp.

Rudolph did not write or publish any papers on his botanical print collection. His copy of the program for the Chadwick Arboretum exhibition contains extensive marginal comments accompanied with notes written on small yellow sheets of tablet paper. No outline or manuscript exists of his after-lunch presentation. His one-paragraph "Introduction" is a concise summary of botanical illustration history during 500 years, from 1481–1981. Being worthy of preservation, it is slightly edited and reprinted here:

> The reasons for illustrating plants have changed over the five hundred years since the late fifteenth century. The earliest illustrations in printed books were meant to give the plants' essence of those used as medicines. Thus, the pictures in such books, or herbals, showed only the most diagnostic features, often crudely drawn. With greater interest in the accurate identification of medicinal plants in the sixteenth century, herbalists hired artists to draw particular plants to show all of their characteristics. Often these illustrations were copies, and many of the herbals of the seventeenth century still used earlier woodcuts. The new interest in gardens and decorative flowers led to production of books with flower illustrations, often of plants brought from distant places. In time, with increase of wealth and interest in gardens, these books of ornamental plants, or florilegia, became more and more elaborate, attracting better illustrators. Although the early books were sometimes hand-colored, using the limited pigments available, now more colors were available and the hand-coloring became more realistic and artistic. Exploring for plants new to science led to the publication of books in which the floral details were shown so that botanists could place the plants into proper classification. At the same time, colored plant illustrations became more available because of the publication of many gardening magazines. In recent times, color printing, either by lithography or halftones, makes fine botanical illustrations readily available to all.

Rudolph and Stuckey assisted OSU rare book librarian Geoffrey D. Smith in the preparation of one other exhibition. The "Plants of the Northwest Territory"[54] were featured in a display of prints and books in the Philip Sills Exhibit Hall at the Main Library of The Ohio State University. This exhibit was placed in conjunction with the annual meeting of the Council on Botanical and Horticultural Libraries, whose members Rudolph invited to the campus from 20–22 May 1992, just one month prior to his death.

When Rudolph's students in his class in the history of biology, along with several other friends, met for the last time in his home on 15 April 1992, he announced that his collection of botanical illustrations was to be a gift to OSU's Chadwick

[54] Geoffrey D. Smith. 1992. [Program Brochure of] The Philip Sills Exhibit Hall. Plants of the Northwest Territory: An Exhibit in Conjunction with the Meeting of the Council on Botanical and Horticultural Librarians, May 18 through 24 July 1992. The Philip Sills Exhibit Hall, The Ohio State University Libraries, Columbus. 12 pp.

Arboretum. This announcement was acknowledged in a letter to him of 8 May from Arboretum Development Officer Janet M. Oberliesen.[55] Its intent was honored by executor Stuckey at the time of the settlement of Rudolph's estate, although this gift was not specified in his will. All of his botanical prints were donated to the Chadwick Arboretum, with the exception of a few prints. These exceptions were those of Isaac Sprague, retained for display and research purposes in the Rudolph Memorial Library Room at the OSU Herbarium in the Museum of Biological Diversity. The illustrations of mint plants, that were collected and framed by Ann Rudolph (1934–1991), were requested by library personnel for the Rare Books and Manuscripts Collection. These prints are now displayed on a wall of the James Thurber Reading Room of University Libraries, The Ohio State University.

Summation and Prologue to Paper Introductions

Emanuel D. Rudolph's research contributions in the history of botany are like the many rehearsals of a theatrical production that has not reached the final performance. As reviewed in this general introduction, his studies represent the various acts and scenes, from building a sizeable book resource collection and completing those writings on the history of lichenology, the OSU Botany Department, and Isaac Sprague, to making presentations at scientific meetings. Not all of his writings were published, and consequently manuscripts have remained in all stages of development, including those worthy projects that lack manuscripts.

Meetings sponsored by professional societies, in which Rudolph was an active participant, are viewed as the backdrop and stage where his efforts have played a significant role in this theater of botanical history. Rehearsals are never final nor complete, and so was the situation with Rudolph's research projects. Thirty-eight of his papers, printed here, expand on topics briefly discussed earlier, and others introduce and examine new topics. This memorial volume is the presentation of his "concluding performance" for the botanical audience.

---

[55] Janet M. Oberliesen. 1992. Letter of 8 May to Emanuel D. Rudolph, Department of Plant Biology, The Ohio State University, Columbus, Ohio. 1 p.

GUIDE TO THE RUDOLPH PAPERS

By Ronald L. Stuckey

and

William R. Burk

Guide to the Rudolph Papers

Ronald L. Stuckey and William R. Burk

The 38 papers in this memorial volume are based on Emanuel D. Rudolph's handwritten and typewritten manuscripts, notes, abstracts, and short selections or summaries from his publications.

Titles and Numbering System

Each of the 38 papers is identified with a title and number. Most of the titles that Rudolph gave to his manuscripts are retained in the papers printed here; however, the editors have provided titles for those papers created from his notes, or portions of his publications. The number of each paper is followed in parentheses by a year or a range of years. Only one year indicates the year of that Rudolph publication and a range of years represents the period that Rudolph actively worked on that project which was not completed.

Rudolph's 38 papers are topically grouped into eight parts: Botany in Textbooks **(Part One)**, Botany in Children's Books **(Part Two)**, Botanical Teaching **(Part Three)**, Botanical Educators **(Part Four)**, Botanical Illustration **(Part Five)**, Women in American Botany **(Part Six)**, Writing Shaw's Garden History **(Part Seven)**, and Writing Botanical History **(Part Eight)**. The eight papers on "The Relationships between Science and Religion" **(Appendix)** are talks that Rudolph gave from 1955 to 1960 in the Houghton Memorial Chapel on the campus of Wellesley College, Massachusetts.

Documentation of Information Sources

The editors have made every effort to provide appropriate documentation of the sources for Rudolph's unpublished works and selections from his publications. Three kinds of documentation are used.

1. **Footnotes:** References cited in the text of introductions are represented by Arabic numbers and conform to conventional footnoting practice. They identify sources of information used in writing the general introduction and the introductions to the individual papers or groups of papers.

2. **References Cited:** The list of references at the completion of each paper's text is alphabetically arranged by author and placed below the heading "References Cited." At the left side of each citation is an Arabic number that represents a superscript in the text of the paper where that reference is used. The superscript numbers are not in numerical order in the text, thus allowing the references cited to be arranged in one alphabetical sequence by author.

3. **Publication Notes:** These notations follow the "References Cited" and represent the various states of development of Rudolph's writings in the history of botany. These sources are identified as manuscripts, abstracts, notes, and selected extracts from his published papers. When any of these sources was used by him in a presentation at a scientific meeting, the date, name, and location of the meeting are noted either in the papers' Introductions or in the Publication Notes. A citation is provided for each of these sources of information. If information from one or more of these sources is incorporated into the text of the printed paper, a lower case letter is placed at the beginning of each citation. These lower case letters are placed as superscripts at appropriate locations in the introductions and in the text of a paper as references to the sources used.

Terminology in Introductions and Publication Notes

The word **paper(s)** has two usages: (1) Rudolph's papers as published in journals or books, which are also referred to as publications, and (2) the 38 selected numbered papers that are printed in this book. A **manuscript** is any item that Rudolph prepared for (1) a **presentation** delivered at a scientific meeting, or (2) a **publication** in a journal or in preparation for a future published journal paper or chapter in a book. His manuscripts are **handwritten** or **typewritten.** The latter are sometimes identified in the text as a **typescript**. **Notes** are short items or quoted passages that Rudolph copied mostly from journal publications and books. These items had not been developed into manuscripts. **Abstracts** are short, usually one-paragraph, summaries of a manuscript or a published paper. Abstracts are prepared to accompany presentations at scientific meetings and distributed to those attending the meetings, or published in the organization's meeting program, journal, or special publication.

Location of Unpublished Materials

Rudolph's original letters, notes, abstracts, and manuscripts pertaining to the papers printed here have been collected into one master file. The file is arranged according to the numbered papers and deposited in the University Archives at The Ohio State University. A second file, including the complete set of photographs printed in the book, is in the Hunt Institute for Botanical Documentation, Pittsburgh, Pennsylvania.

List of Abbreviations

b.	born	no(s).	number(s)
ca.	circa	opp.	opposite
Co.	Company	p(p).	page(s)
ed.	edition	pt.	part
fl.	flourished	pl(s).	plate(s)
frontis.	frontispiece	publ.	publication
illust.	illustration(s)	r.	right
Inc.	Incorporated	ser.	series
l.	left	suppl.	supplement
misc.	miscellaneous	tab.	table
n. ser.	new series	vol(s).	volume(s)

THE RUDOLPH PAPERS

With Introductions
By Ronald L. Stuckey

Part One

BOTANY IN TEXTBOOKS

Emanuel D. Rudolph's research interests and investigative approaches into the fascinating theme of the popularization of botany and science were unique. Many appropriate topics with this theme are rarely studied by historians of science. The paucity of these studies, according to Rudolph, is largely because the raw materials for analyses are frequently overlooked.[1] The necessary books are not in the mainstream of the archival scientific literature.

Rudolph could engage in studies on the popularization of botany because he had the raw materials in his extensive library. Not only did he have a massive collection of children's natural history books, but he also had acquired numerous volumes of schoolbooks, catechisms and chapbooks of botany, and textbooks of natural science, biology, and botany. A large collection of biographies of scientists further strengthened his resources for this research. Rudolph's published papers focused on information synthesized from children's natural history literature, popular biographical magazines, the study of illustrations of plants, the role of women in both non-professional and professional nineteenth century botany, the contributions of other non-professionals to the development of biology and botany, and the contributions of botanical artists and practicing physicians.

Four of Rudolph's papers on the popularization of botany are placed in this group (**Part One**). They are Schleiden's contribution (**Paper No. 1**), the origin and development of American botanical textbooks (**Paper No. 2**), Asa Gray's botanical textbooks (**Paper No. 3**), and lichens in botanical textbooks (**Paper No. 4**). The only paper among these four topics on which Rudolph published was his, "Lichens in US Introductory Botany Textbooks, 1836–1986" (1988).

---

[1] Emanuel D. Rudolph. 1989. Emanuel D. Rudolph, 1 p. *In* Research Interests of the Faculty, College of Biological Sciences. [The Ohio State University, Columbus]. Typewritten.

Introduction No. 1

Schleiden's Contribution to Popularization (1965)

Books popularizing botany became readily available to Americans during the nineteenth century. The most numerous of these books were textbooks of botany written for all ages and all levels of education, from the primary grades through the secondary grades, as well as for college and university instruction. The word "schoolbook" is used here for those books at the lower educational levels, while the term "textbook," although applicable to most of these books, is usually restricted for use in the secondary grade levels and for instruction in colleges and universities. With the publication of schoolbooks and textbooks, the population at large, both professional and non-professional, had increased exposure to botanical science. These books were a major element in the popularization of botany.

Overlooked books often provide appropriate subjects for historical investigations. Rudolph's first manuscript in the history of botany dealt with an overlooked book on the popularization of botany. The topic concerned Matthias Jacob Schleiden's *Die Pflanze und ihr Leben* (1848), the English translation of the title being *The Plant, A Biography*. It is a very popular book of public lectures in the German language on a very general and broad approach to botany.

In 1853, Alphonso Wood (1810–1881) took an English version of Schleiden's book and edited it for an American audience. Wood's edition, published in New York and Cincinnati, was of particular interest to Rudolph, because its title, *Poetry of the Vegetable World*, . . . was different, evidently causing it to be overlooked by American botanists (Fig. 1.1). Wood, a teacher of botany in Ohio at the Cleveland Female Seminary in 1853, moved a year later to College Hill, near Cincinnati, where he was a professor and later president in the Ohio Female College.[1] He was also a popularizer of botany, whose reputation was established by his manual and textbook of botany, called "Class-book."[1,2,3] His books were designed to compete with the early editions of Asa Gray's textbooks of botany. Schleiden (1804–1881),

[1] O. R. Willis. 1881. Alphonso Wood. *Bulletin of the Torrey Botanical Club* 8: 53–56.

[2] Willard W. Eggleston. 1927. The botanical publications of Alphonso Wood. [U.S. Department of Agriculture]. *Agricultural Library Notes* 2: 95–100.

[3] Charles J. Lyon. 1939. A 50 edition best seller. Alphonso Wood, class of 1834, wrote botanical class books that reached a circulation of 800,000 copies. *Dartmouth Alumni Magazine* 31(March): 18, 81–83.

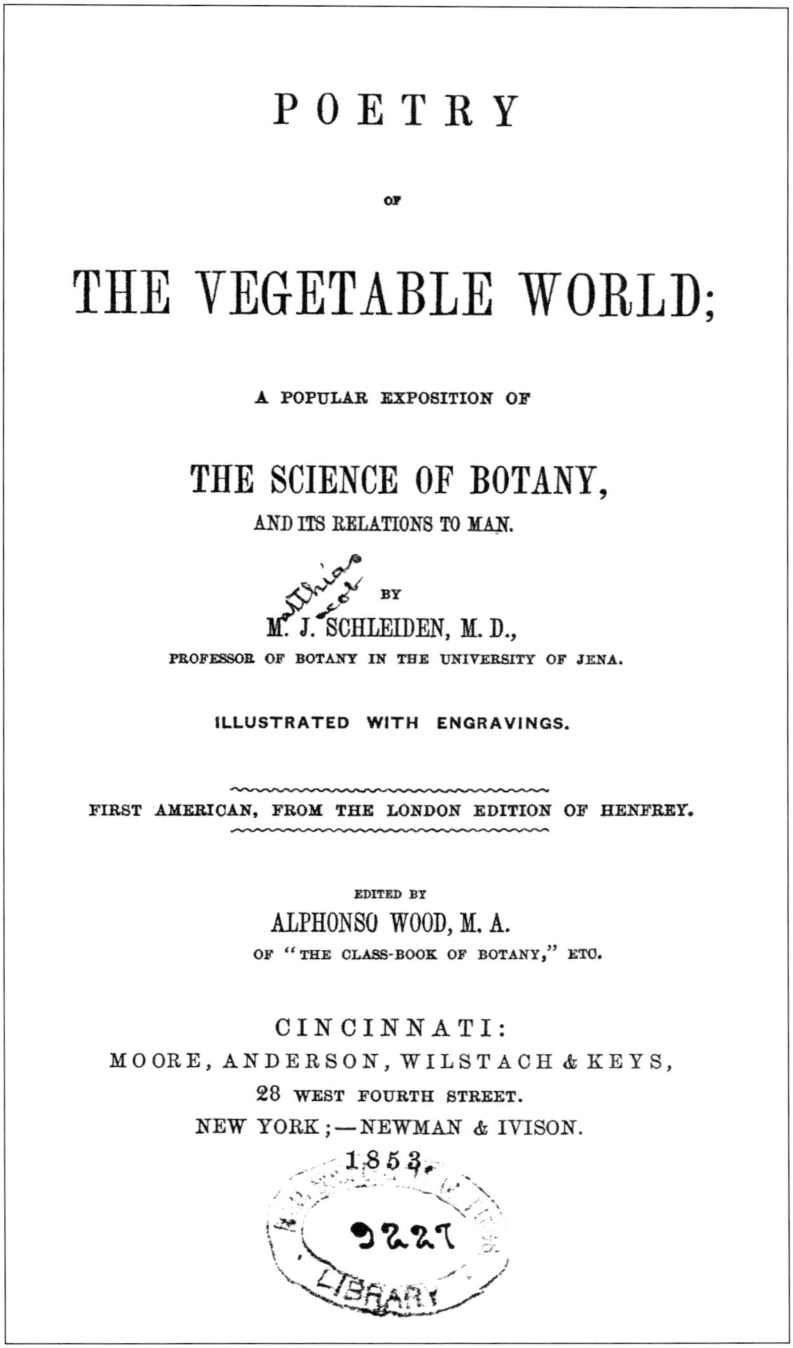

POETRY

OF

THE VEGETABLE WORLD;

A POPULAR EXPOSITION OF

THE SCIENCE OF BOTANY,

AND ITS RELATIONS TO MAN.

BY

M. J. SCHLEIDEN, M. D.,

PROFESSOR OF BOTANY IN THE UNIVERSITY OF JENA.

ILLUSTRATED WITH ENGRAVINGS.

FIRST AMERICAN, FROM THE LONDON EDITION OF HENFREY.

EDITED BY

ALPHONSO WOOD, M. A.

OF "THE CLASS-BOOK OF BOTANY," ETC.

CINCINNATI:

MOORE, ANDERSON, WILSTACH & KEYS,

28 WEST FOURTH STREET.

NEW YORK;—NEWMAN & IVISON.

1853.

Fig. 1.1. Title page from *Poetry of the Vegetable World*, edited by Alphonso Wood (1853).

(From the copy owned by Ronald L. Stuckey)

Fig. 1.2. Rudolph's copies of Wood's *Class-book,* Rare Book Room, OSU Herbarium Library.

(Taken 4 December 1995, by Ronald L. Stuckey; in the collection of RLS)

Fig. 1.3. Alphonso Wood (1810–1881).

(From Willis, *Bulletin of the Torrey Botanical Club* 8: 54. 1881)

the co-founder with Theodor Schwann (1810–1882) of the cell theory in biology, was important for the development of botany as an experimental science. According to Rudolph, however, he frequently was not appreciated for his writings as a popularizer of botany.

Rudolph first presented his study on Schleiden's edited textbook on 23 April 1965 to the Plant Sciences Section at the Annual Meeting of The Ohio Academy of Science held at Ohio University in Athens. The topic garnered attention by The Ohio State University News and Information Service, and was the subject of a press release[4] on that same day. This news release, the first of only two that the University ever issued on Rudolph's studies in the history of botany, quoted him in part as follows:

> It is interesting and significant to compare the English and American editions and note how Wood edited for the American audience. His negative comments on the evolutionary implications are particularly relevant. At times [Wood] simplified the sentence structure, added editorial notes, and in two instances suppressed theological views of the author. . . . Wood added marginal titles to almost every paragraph and countered Schleiden's views with footnotes.

Rudolph presented this subject a second time, later that year on 16 August at the first-held paper session of the Historical Section of the Botanical Society of America at the University of Illinois, Champaign-Urbana. Five years later on 18 April 1970

---

[4] Anonymous. 1965. [Emanuel D. Rudolph speaks on Schleiden's little known American edition of his *Die Pflanze und ihr Leben*]. Ohio State University News and Information Service, Columbus. Typewritten, 2 pp.

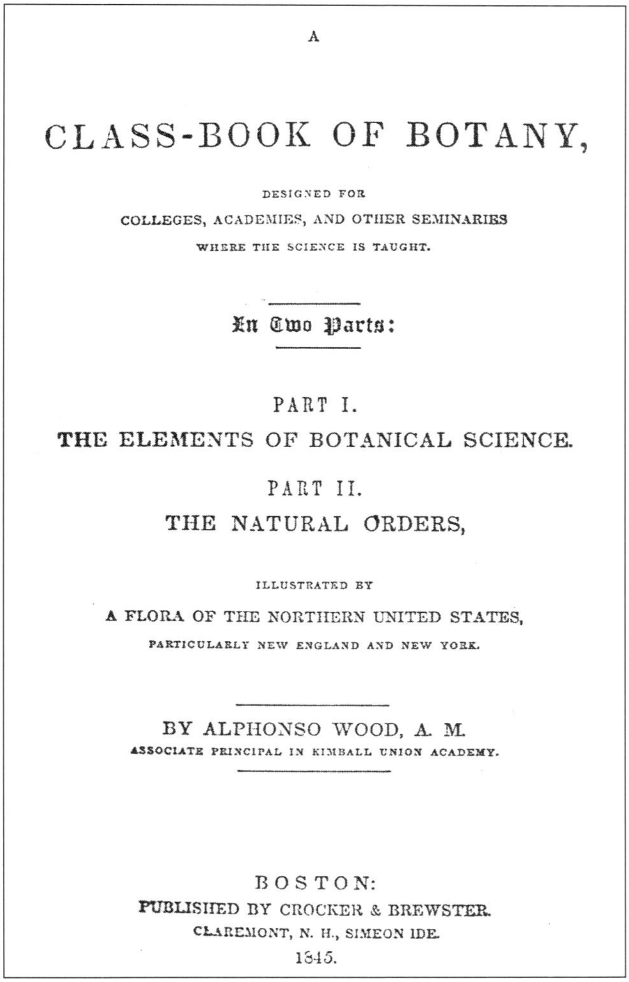
A

CLASS-BOOK OF BOTANY,

DESIGNED FOR
COLLEGES, ACADEMIES, AND OTHER SEMINARIES
WHERE THE SCIENCE IS TAUGHT.

In Two Parts:

PART I.
THE ELEMENTS OF BOTANICAL SCIENCE.

PART II.
THE NATURAL ORDERS,

ILLUSTRATED BY
A FLORA OF THE NORTHERN UNITED STATES,
PARTICULARLY NEW ENGLAND AND NEW YORK.

BY ALPHONSO WOOD, A. M.
ASSOCIATE PRINCIPAL IN KIMBALL UNION ACADEMY.

BOSTON:
PUBLISHED BY CROCKER & BREWSTER.
CLAREMONT, N. H., SIMEON IDE.
1845.

Fig. 1.4. Title page from the first edition of *Class-book of Botany*, by Alphonso Wood (1845).

(From Rudolph's copy in the Rare Book Room, Herbarium, The Ohio State University)

Rudolph spoke again on Schleiden as a lawyer, doctor, botanist, and anthropologist at the Annual Meeting of the Ohio Academy of Medical History in Granville.

Rudolph's manuscript[a] on Schleiden's overlooked American edition was never published. He apparently believed that additional information was necessary as stated in his letter of 1 June 1970 to George H. M. Lawrence (1910–1978), Director of the Hunt Botanical Library. Rudolph wrote:[5] "I should spend some time in Germany if I ever want to do something substantial with Schleiden." **(Paper No. 1)**.

[5] Emanuel D. Rudolph. 1970. Letter of 1 June 1970 to George H. M. Lawrence, The Rachel McMasters Miller Hunt Botanical Library, Pittsburgh, Pennsylvania. 1 p.

Matthias J. Schleiden, Popularizer of Botany

Matthias J. Schleiden (1804–1881)

Paper No. 1 (1965)

Matthias J. Schleiden's Contribution to the Popularization of Botany, Die Pflanze und ihr Leben *(1848)*

Matthias Jacob Schleiden (1804–1881), professor of botany at the University of Jena, now best known as one of the founders of the cell theory in biology, published a series of public lectures as a book, *Die Pflanze und ihr Leben: Populäre Vorträge* (1848).[7] In the same year an English translation made by Arthur Henfrey (1819–1859) appeared in London with the title, *The Plant; A Biography: In a Series of Popular Lectures* (1848).[8] Schleiden's introduction contained the following admonitions:

> The following Lectures have been composed during the course of the last eight years, at the instance of an intelligent and cultivated circle, free from 'the dust of schools,' and were by no means intended for publication; . . . My chief aim was, in fact, the satisfaction of what may be called a class-vanity. A large proportion of the uninitiated, even among the educated classes, are still in the habit of regarding the Botanist as a dealer in barbarous Latin names, a man who plucks flowers, names them, dries and wraps them in paper, and whose whole wisdom is expended in the determination and classification of this ingeniously collected hay. This portrait of the Botanist was, alas! once true, but it pains me to observe, that now, when it bears resemblance to so few, it is still held fast to by very many persons; and I have sought, therefore, in the present discourses, to bring within the sphere of general comprehension, the more important problems of the real science of Botany, to point out how closely it is connected with almost all of the most abstruse branches of philosophy and natural science, and to show how almost every fact or larger group of facts tends, as well in Botany as in every other branch of human activity, to suggest the most earnest and weighty questions, and to carry mankind forward beyond the possessions of sense, to the anticipation of the spirit.

Born 1804 in Hamburg, Matthias Schleiden[1,2,3,6] was the son of a physician. He earned a degree in jurisprudence from Heidelberg in 1827 and practiced law for a short time. Having found the legal profession not to his liking, he initially studied medicine and then botany at Göttingen and later at Berlin with almost a religious fervor. His aim was to purge the science of botany of its philosophical and deductive trappings, especially those of the Natur Philosophie advocates. The English

---

Frontispiece Credits: Matthias J. Schleiden (1804–1881). (From M. Möbius. 1904. *Matthias Jacob Schleiden.* Verlag von Wilhelm Engelmann, Leipzig; in the Rudolph photograph collection)

translation of his book, *Principles of Scientific Botany; or, Botany as an Inductive Science* (1849),[9] outlined the science in a clear and logical way. Although itself not completely free from philosophizing, it served as the model for future college textbooks of botany. Schleiden was a creative thinker and synthesizer whose enthusiasm frequently went further than his facts.

Soon after Schleiden began writing scientific papers and books, he began to present popular lectures, and the publication of his *Die Pflanze und ihr Leben* (1848)[7] is the result of these efforts. He continued to popularize botany and science by lecturing and producing other popular writings, including *Studien* (1855),[11] *Das Meer* (1867),[12] *Für Baum und Wald* (1870),[13] *Die Rose* (1873)[14] and *Das Salz* (1875).[15] After 1862, he relinquished botanical investigations for anthropology, history, and social studies and published in these fields before his death in 1881.

The Plant; A Biography has twelve chapters or lectures which range from consideration of the eye and the microscope to the structure of plants, their nutrition, propagation and uses, to the weather, as well as the aesthetics of the vegetable world. The book according to the author can be divided into four parts: the plant itself, the relations of the individual plants to each other, the relation of plants to the world, and the relation of man to plants. He believed that these topics covered the branches of botany which may be called theoretical, systematic, geographical, and applied. In both the German and English editions colored lithographed plates amplify the text, as do the charming black-and-white wood-engraved vignettes which appear at the head of each chapter.

The appeal of this popular work for the general reader is attested to by its five subsequent German editions, two English, two Dutch, one French, one Russian, and one Swedish all in translation. This work also had an American edition, which is not cited in the standard bibliographies of Schleiden probably because of its different title, *Poetry of the Vegetable World; A Popular Exposition of the Science of Botany, and its Relation to Man*.[10] This edition was edited by Alphonso Wood from the London edition of Henfrey and published in Cincinnati by Moore, Anderson, Wilstach and Keys and in New York by Newman and Ivison. The book lacks an author's name on the spine, and no indication of its original title is on the title page. Wood's name on the title page is as large as Schleiden's.

The editor, Alphonso Wood (1810–1881) is best known to many botanists as the author of a very popular high-school level textbook, *A Class-book of Botany*,[18] first published in 1845 and carried through "50 editions."[4,5,16] His book was very popular because it made plant identification easy and had a didactic approach to the subject of botany. He was a graduate of Dartmouth College and later taught and was principal at various academies and finally was professor of botany at the New York College of Pharmacy.[17] He tried to encourage Asa Gray to write a high-school textbook and failing, wrote his own. That he was a well-informed botanical thinker is indicated in the preface of his *Class-book*:[18]

> The time has arrived when Botany should no longer be presented to the learner encumbered with the puerile misconceptions and barren facts of the old school, but as a System of Nature, raised by recent researches to the dignity and rank of a science founded upon the principles of inductive philosophy . . . . That theory of the floral structure which refers each organ to the principle of the leaf, long since propounded in Germany by the poet Goethe, and recently admitted, by authors generally, to be coincident with facts, is adopted, of course, in the present work.

It is not surprising that Wood was impressed with the writings of Schleiden and must have read his first work translated into English, *The Plant; A Biography*.[8] The reason he edited an American edition is indicated in his introduction:

> In the production of the present admirable work he [Schleiden] has conferred a great favor upon every lover of Nature. By its peculiar character, it meets an actual want in literature which has long been felt. There is, indeed, no lack of floras and text-books for the schools—works which pursue the science of Botany to its minutest details. But such treatises are necessarily dry, unimaginative—regarding plants and flowers chiefly as 'materials for a herbarium,' or, on the other hand, as food for animals. But the Vegetable World has a higher significance than either the education of man's intellect, or even the maintenance of animal life . . . . Herein lies the charm of the present work. While its author has everywhere exhibited the principles of science in the most perspicuous and attractive style, he has also contrived to blend with them the imaginative and the spiritual, and thus to render his work the exponent of the relations of the plant to the human soul.

Wood wrote that in bringing the volume before the American public, using the English translation of Henfrey (1848),[8] he has at times simplified the sentence structure. He also added editorial notes, and in two instances suppressed theological views of the author "which were more liable to be construed as offensive or irreverent in this country than in that of the writer."[10]

In the American edition the lithographed plates are mostly uncolored, and the book lacks a frontispiece or vignettes. Wood has added marginal titles to almost every paragraph. The two suppressed theological passages are concerned with a question about a Biblical quotation and a criticism of the English Church for Irish exploitation and literal Biblical interpretation.

The editorial comments, made in footnotes, which question Schleiden's ideas about the evolution of plants, are of the most interest. Schleiden[10] wrote:

> If now we recall to mind the sketches above given of the successive epochs of vegetation, we see that the vegetable world begins in water, under the simplest forms, and in that very family in which the whole plant is represented by a single cell, most frequently, in the present time. In the succeeding periods the other groups are added to this, making their appearance in a series which corresponds through continually higher organization . . . . Thus the stemless Cryptogamia are followed by those provided with stems and leaves. Then the Gymnospores (Conifers and Cycadeae) enter upon the field; to these succeed the Monocotyledons, and lastly appear the Dicotyledons. Imperfect as are the documents we have obtained, and little as we have yet deciphered of them, yet in no period do we find the appearance of a wholly new creation, but the organic beings are always added gradually; . . . . This view, that the whole fullness of the vegetable world has been gradually developed out of a single cell and its descendants, by gradual formation of varieties, which became stereotyped into species, and then, in like manner, became the producers of new forms, is at least quite as possible as any other and is perhaps more probable and correspondent than any other, since it carries back the Absolutely Inexplicable, namely, the production of an Organic Being, into the very narrowest limits which can be imagined.

Alphonso Wood[10] in his footnote reply to this concept says:

> One can scarcely believe our author to be serious here, in proposing to account for the origin of plants by so wild an hypothesis as this. Indeed, he professes to be only 'dreaming,' while stating it. And yet his readers cannot avoid the conclusion, from the amount of argumentation bestowed, that in the author's estimation, this theory was more than a dream. It appears to be merely a modification of the famous 'development theory' of Lamarck, which has been so often refuted by the ablest writers. Our author here applies that theory only to the species of the vegetable kingdom. But if it be admitted that all plants, from the Moss to the Magnolia, have sprung from one common origin, viz. the Protococcus, then it is impossible to deny that the Animal Kingdom, with man at its head, had a similar development; and we must hence regard the primitive Monad as our great grand progenitor, and thence trace our descent down through shellfish, Saurians, Dolphins, and lastly Monkeys! And all this absurdity for what? It is merely to relieve the Almighty Creator from the fatigue of having created so many species, —nay, from creating at all! since that simple primogenetive cell might easily have been struck out of the seafoam by electricity! Alas, for the darkness of German Theology! A.W.

Wood no doubt expressed the American opinion of the times. In retrospect, it is of note that the details of Schleiden's cell theory were faulty and the details of Wood's evolutionary idea were vague. However, both ideas have become part of current botanical thinking. No one can deny that Schleiden's ideas were ahead of his time. One cannot say as much for his American editor, Alphonso Wood.

References Cited

1. [Anonymous]. 1883. Sketch of Matthias Jacob Schleiden. *Popular Science Monthly* 22: 256–262.

2. B[ar]y, A[nton] d[e]. 1881. M. J. Schleiden. *Botanische Zeitung* 39: 519–520.

3. Behrens, Wilhelm. 1881. Matthias Jakob Schleiden. *Botanisches Centralblatt* 7: 150–156, 183–190.

4. Lyon, Charles J. 1939. A 50 edition best seller. Alphonso Wood, class of 1834, wrote botanical class books that reached a circulation of 800,000 copies. *Dartmouth Alumni Magazine* 31(March): 18, 81–83.

5. _____. 1945. Centennial of Wood's "Class-book of Botany." *Science* 101: 484–486.

6. Möbius, M. 1904. *Matthias Jacob Schleiden zu seinem 100. Geburtstage.* Wilhelm Engelmann, Leipzig. 106 pp.

7. Schleiden, Matthias Jacob. 1848. *Die Pflanze und ihr Leben: Populäre Vorträge.* Wilhelm Engelmann, Leipzig. v, 329 pp.

8. _____. 1848. *The Plant; A Biography: In a Series of Popular Lectures.* Translated by Arthur Henfrey. Hippolyte Bailliere, London; J. B. Bailliere, New York. viii, 365 pp. (Quote, pp. 1–2).

9. _____. 1849. *Principles of Scientific Botany; or, Botany as an Inductive Science.* Translated by Edwin Lankester. Longman, Brown, Green, and Longmans, London. viii, 616 pp.

10. _____. 1853. *Poetry of the Vegetable World; A Popular Exposition of the Science of Botany, and its Relation to Man.* First American [ed.], from the London edition of Henfrey. Edited by Alphonso Wood. Moore, Anderson, Wilstach & Keys, Cincinnati; Newman & Ivison, New York. xxvii, 29–360 pp. (Quotes, pp. iii–iv, v, 291–292, 292–293).

11. _____. 1855. *Studien: Populäre Vorträge.* Mit einer Ansicht der Porta del popolo in Rom. Wilhelm Engelmann, Leipzig. viii, 318 pp.

12. _____. 1867. *Das Meer.* A. Sacco Nachfolger, Berlin. xii, 709 pp.

13. _____. 1870. *Für Baum und Wald: Eine Schutzschrift an Fachmänner und Laien gerichtet.* Wilhelm Engelmann, Leipzig. viii, 144 pp.

14. _____. 1873. *Die Rose: Geschichte und Symbolik in ethnographischer und kulturhistorischer Beziehung.* Ein Versuch von M. J. Schleiden. Wilhelm Engelmann, Leipzig. ix, 322 pp.

15. _____. 1875. *Das Salz: Seine Geschichte, seine Symbolik und seine Bedeutung im Menschenleben*. Eine monographische Skizze. Wilhelm Engelmann, Leipzig. viii, 236 pp.

16. Went, Frits. 1942. Centenary of Schleiden's textbook of botany: Have aims of botany changed? *Chronica Botanica* 7: [145]–148.

17. Willis, O. R. 1881. Alphonso Wood. *Bulletin of the Torrey Botanical Club* 8: 53–56. [Abstract of a "Biographical Sketch of Prof. Wood," read by Dr. O. R. Willis before the Torrey Botanical Club, 8th March].

18. Wood, Alphonso. 1845. *A Class-book of Botany, Designed for Colleges, Academies, and Other Seminaries Where the Science Is Taught*. In Two Parts: Part I. The Elements of Botanical Science. Part II. The Natural Orders, Illustrated by a Flora of the Northern United States, Particularly New England and New York. Crocker & Brewster, Boston; Simeon Ide, Claremont, N[ew] H[ampshire]. II Parts: (I: 124 pp.; II: ii, 3–474 pp.). (Quote, Part I, p. [3]).

Publication Notes

Revised and edited from a manuscript[a] read 23 April 1965 at the Annual Meeting of The Ohio Academy of Science, Ohio University, Athens; also read 16 August 1965 at the meeting of the Historical Section, Botanical Society of America, University of Illinois, Champaign-Urbana.

[a] [1965]. "Matthias Jacob Schleiden's Contribution to the Popularization of Botany, *Die Pflanze und ihr Leben*, with notes about a little known American edition and a bibliographic treatment of the editions." Handwritten, 11 pp.; typewritten, 9 pp.; references, handwritten, 2 pp. [Manuscript]. A second version exists with a shorter title, "M.J. Schleiden's Contribution to the Popularization of Botany, *Die Pflanze und ihr Leben*." Typewritten, 6 pp.; references, [1] p. [Manuscript].

[1965]. "Schleiden—American Edition Compared with English Edition: Lists of Substitute Words and Changes in Phraseology." Handwritten, 13 pp., 9 pp.; typewritten, 7 pp. [Manuscript].

1965. A generally unknown Cincinnati edition of Schleiden's *The Plant, A Biography*, edited by Alphonso Wood, p. 3. *In* Abstracts of Papers Presented before Section B, Plant Sciences, The Ohio Academy of Science Meeting at Ohio University, [Athens], 22–24 April 1965. Mimeographed. [Abstract].

1965. M. J. Schleiden's contribution to the popularization of botany, *Die Pflanze und ihr Leben*. *American Journal of Botany* 52: 631. [Abstract].

Introduction No. 2

American Botanical Textbooks (1970–1976)

Regrettably Rudolph did not publish any of his manuscripts on the origin and development of American schoolbooks on natural history and textbooks of botany that he read at four meetings. Three partially completed manuscripts[d,e,f] with extensive notes and quotations survive from the meetings. Each of them is marked with the title of the presentation. No manuscript is known for the presentation of 24 August 1990 on "Nineteenth Century American Schoolbooks and Textbooks of Botany." Rudolph's four presentations, spanning the years 1970 to 1976, are as follows:

> "Nineteenth Century American Schoolbooks and Textbooks of Botany." Presented 24 August 1970 to the Historical Section, Botanical Society of America, Indiana University, Bloomington.
>
> "The Origin and Development of the Botanical Textbook and Its American Aspects."[d] Presented 6 March 1971 to The Ohio Academy of Science, Battelle Memorial Institute, Columbus.
>
> "The Botanical Textbook in Nineteenth-century America As a Reflection of Botanical and Cultural Trends."[e] Presented 11 April 1975 to the Midwest Junto for the History of Science, Kansas State University, Manhattan.
>
> "The Americanization of Germanic Botany, 1865–1915."[f] Presented 1 June 1976 to the Historical Section, Botanical Society of America, Tulane University, New Orleans.

In Rudolph's partially written manuscript on "The Origin and Development of the Botanical Textbook"[d] presented in 1971, he defined a textbook in general, and a botanical textbook, in particular, and sketched reasons for studying them, especially those written in the nineteenth century. He noted that textbooks (1) represent a summary of knowledge of the science at the time-periods when they were written; (2) show the trends, aims, and changes of the subject at a particular time in history; (3) reveal relationships to the popularization of the subject at a time in history when a large audience was developed and increased in numbers; (4) provide evidence for the importance of science as a subject to be taught in schools and colleges; (5) document the content, format, and style developed by German botanists, whose textbook features were transferred to America in the programs and books referred to as the "New Botany;" and (6) make available sources of information that might be supportable for studying possible revolutions in the history of science as proposed by Thomas S. Kuhn (1962).[1]

Rudolph's insight into botanical textbooks is revealed in his partial manuscript, "The Botanical Textbook in Nineteenth-Century America as a Reflection of Botanical and Cultural Trends"[e] read in 1975. In the introduction, he wrote:

> There are changes evident in the botanical textbooks *published* in the United States during the nineteenth century that reflect profound shifts in botanical science. These shifts are the basic model of what a botanist does and how botanists are trained. The early model was British, or perhaps French, of a physician concerned with descriptive and taxonomic botany largely self-trained; the later model was German, of a university professor concerned with experiments and evolutionary botany, trained at a university in a non-medical program.
>
> In the American context, there are unique cultural factors that influenced these general world-wide shifts and that prepared the botanists of the 20th century to assume world leadership in certain aspects of botany. One way of understanding the American development is to look at textbooks which should provide an insight into the "normal" science of the time. . . . textbooks are windows to the posture that the science takes when being presented to the initiate.
>
> Some of the uniquely American cultural forces that influenced 19th century botany, in addition to the world-wide trends toward professionalization, specialization, nationalistic identification, and shifts toward evolutionary emphasis, are the following:
>
> 1. Development of "irregular" medical sects in the U[nited] S[tates] with strong botanical interests.
>
> 2. Development and increases of strong secondary public education, instruction, particularly as a result of the uses of the Treasury surpluses of 1834.
>
> 3. Founding and development of public agricultural colleges as a result of the Morrill Act.
>
> 4. Strong textbook publishing competition in mid-century and the final unification of the major publishing companies in the American Book Company in 1890.

---

[1] Thomas S. Kuhn. 1962. *The Structure of Scientific Revolutions*. University of Chicago Press, Chicago, Illinois. xv, 172 pp.

Fig. 2.1. A bookcase of classical botanical references and textbooks in the living room of the Arcadia Avenue Rudolph home.

(Taken August 1992, by Ronald L. Stuckey; in the collection of RLS)

Information about the history of schoolbooks and textbooks is extremely massive and Rudolph would have needed much time to synthesize and write a comprehensive manuscript of some length for publication. His own resources were numerous, consisting of many different schoolbooks on natural history, textbooks of biology and botany, and a file cabinet with extensive notes documenting the history of secondary and collegiate science education, histories of biology and botany textbooks, and methods of teaching science. The resources for future substantive studies of the history of these kinds of textbooks are available from his resources at The Ohio State University. The paper printed here has been assembled by combining and editing Rudolph's abstracts,[b,c] partial manuscripts,[d,e,f] and notes[e,f] from his presentations on the subject. Stuckey wrote additional paragraphs. **(Paper No. 2)**.

Botany Textbooks in the Rudolph Home

Botanical textbooks that were on opposite walls of the second floor hallway of the Rudolph Arcadia Avenue home, Columbus

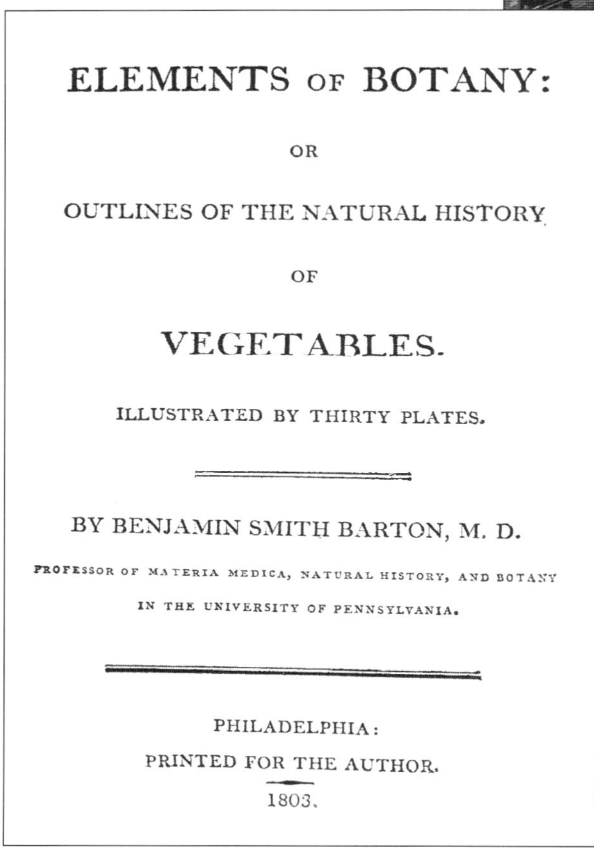

ELEMENTS OF BOTANY:

OR

OUTLINES OF THE NATURAL HISTORY

OF

VEGETABLES.

ILLUSTRATED BY THIRTY PLATES.

BY BENJAMIN SMITH BARTON, M. D.

PROFESSOR OF MATERIA MEDICA, NATURAL HISTORY, AND BOTANY

IN THE UNIVERSITY OF PENNSYLVANIA.

PHILADELPHIA:

PRINTED FOR THE AUTHOR.

1803.

Title page from the first American botanical textbook, *Elements of Botany,* by Benjamin Smith Barton (1803)

Paper No. 2 (1970–1976)

A Summary of the Origin and Development of American Botany Textbooks

A "true textbook is one especially prepared for the use of pupil and teacher in a school or class, presenting a course of study in a single subject, or closely related subjects."[1] Textbooks, especially those of wide use, are by their nature repositories of generally accepted knowledge in any discipline at a particular period.[d]

Textbooks are useful for the historian of science because they present in a concise way the generally held paradigms of a period. The botanical textbook is here defined as a work that introduces students to current aspects of plant science, rather than a manual for the identification of particular plants. The origins of textbooks are to be found in works such as Linnaeus' *Philosophia Botanica* (1751). The Linnaean model persisted throughout the eighteenth century and stressed gross structure and history as a tool for classification (Appendix 2.1, part I). This model, which was used in books in England and on the European Continent, was gradually replaced by works having greater emphases on plant function and geography.[c]

Benjamin Smith Barton's *Elements of Botany* (1803) published at Philadelphia by the author is considered to be the first American textbook of botany and contains portions relating strictly to American plants. Further, it was illustrated by thirty plates mostly drawn by William Bartram. Barton used a considerable amount of material from foreign texts, usually, but not always, giving credit. Amos Eaton (1776–1842), who taught in Troy, New York, in the early part of the nineteenth century, was a very influential teacher. He wrote a popular *Manual of Botany* (1817) which followed with eight editions and they together with his *Botanical Exercises* (1820) were used as textbooks. In reality they were not textbooks, but rather introductions to the native flora. Eaton directly inspired a number of later textbook writers including Jane K. Welsh (1819), Almira Hart Lincoln Phelps (1829), Laura Johnson (1834), and Asa Gray (1836, 1842, 1857, 1887)[a] (Appendix 2.1, part II). Other American students of botany early in the nineteenth century were using British texts or continental works that had been translated by Englishmen. Among them

---

Frontispiece Credits: American Botany Textbooks in the Rudolph Home. Botanical textbooks at the Rudolph Arcadia Avenue home. (Photographs by A. E. Spreitzer, August 1992; in the collection of RLS) Title page of *Elements of Botany*. (From a copy in the Herbarium Library of The Ohio State University)

Title page from the first American botanical textbook, *Elements of Botany* by Benjamin Smith Barton (1803). (In the Rudolph photograph collection)

was *An Introduction to Physiological and Systematical Botany* by James Edward Smith (1814). Some of these British works in their American editions introduced the Natural System of Antoine de Jussieu (1748–1836) and contained lists of American plant genera arranged by that system, for example, James Edward Smith's *A Grammar of Botany* (American ed., 1822), and John Lindley's *Introduction to the Natural System of Botany* (American ed. with an Appendix by John Torrey, New York 1831)[a,b] (Appendix 2.1, part II).

The publication of Matthias Jacob Schleiden's *Grundzüge der wissenschaftlichen Botanik* (1842–1843) (*Principles of Scientific Botany*, English ed. 1849) introduced a new and influential model that stressed physiology and microscopic anatomy (Appendix 2.1, part III). In the latter part of the nineteenth century, Schleiden's model was further modified by German botanists to include aspects of evolutionary thinking that stressed plant life-cycles and developmental anatomy.[c]

The textbook author Douglas Houghton Campbell (1859–1953),[2] wrote that the greatest advancement in the rise and development of botany took place in Germany during the middle half of the nineteenth century. Here in 1868, according to Campbell, the greatest of all textbooks, *Lehrbuch der Botanik*, was published by Julius von Sachs (1832–1897). It was an epitome of the results of the researches of outstanding German investigators (Fig. 2.2). No mere compendium of what had been accomplished to the time of its publication, it was a storehouse of new and important facts based upon the original investigations of the author. Although Sachs was regarded primarily as a physiologist, his textbook covered the whole range of botany including many new contributions to morphology and taxonomy. Its translation in 1875 into English was undoubtedly the most important factor in bringing about a change in attitude toward botanical teaching in both England and America, where the contributions to morphology and physiology were little known.

Previous to the 1870s, very few American botanists had studied in Germany. Early in the next decade, William G. Farlow (1844–1919) went to Europe to study under Anton H. de Bary (1831–1888) in Germany and Jean B. E. Bornet (1828–1911) in France. Farlow was interested especially in the morphology and taxonomy of the algae and fungi, and upon his return to America he was appointed to a professorship in cryptogamic botany which was established in 1878 at Harvard University. Farlow (1913)[4] described how he discovered Sachs' textbook.

> In the laboratory [of Anton de Bary] I noticed that the students seemed to refer frequently to a book of which I had never seen a copy or even heard. The book was Sachs's 'Lehrbuch' second edition, 1870. I bought the book and was perfectly amazed. I had never dreamed that botany covered so large a field. The 'Lehrbuch' was an admirable summary of what was known of all departments of botany up to that date, well written and excellently illustrated. . . . I know no single book which has had so great an influence in shaping the course of modern botany as Sachs's 'Lehrbuch.' It may be that the facts there given were generally known in Germany, but they were not

LEHRBUCH

DER

BOTANIK

NACH DEM

GEGENWÄRTIGEN STAND DER WISSENSCHAFT

BEARBEITET

VON

DR. JULIUS SACHS,

ORDENTL. PROFESSOR DER BOTANIK IN FREIBURG IM BR.

M. Hesselink

MIT 358 ABBILDUNGEN IN HOLZSCHNITT.

LEIPZIG,

VERLAG VON WILHELM ENGELMANN.

1868.

Fig. 2.2. Title page from *Lehrbuch der Botanik*, by Julius Sachs (1868), with photograph of Julius Sachs (1832–1897).

(From Rudolph's copy in the Herbarium Library, The Ohio State University; photograph from Noll, *Botanical Gazette* 25: opp. p. 1. 1898)

known in other countries. On returning home by way of England in 1874, I showed my copy of Sachs to several English botanists and it was evident that it was quite new to them. It was certainly unknown in America. If imitation is the sincerest flattery, the value of Sachs's 'Lehrbuch' was quickly recognized, for, using it as a model or basis, there soon appeared a large number of really excellent text-books in various languages in which one recognized Sachs translated, Sachs condensed, Sachs diluted, Sachs trimmed to suit local demands.

One of these new American textbooks based on Sachs' text-book was prepared by Charles E. Bessey (1845–1915), whose *Botany for High Schools and Colleges* (1880) attained a well-deserved success. The book had a great influence on the trend of both research investigation and botanical teaching, as it served to introduce the new botany to a host of eager American students.[2] John M. Coulter (1851–1928)[3] wrote that "the real segregation of botany from animal biology in university courses began with the appearance of Bessey's Botany (1880), which was in effect the introduction of Sachs' *Lehrbuch* into American laboratories."

Gradually the proportion of books by American authors increased, and although writers continued to use European models, by the end of the century most of the books used in schools were of American origin. Some of the mid-century works, namely those of Asa Gray (1810–1888), Alphonso Wood (1810–1881), and Almira Hart Lincoln Phelps (1793–1884), were widely used and repeatedly reprinted and revised (Appendix 2.1, part II). The rise of a nationalistic feeling in the United States can be seen in the production of strictly American books for American schools and colleges.[a,b] By examining 390 unique printings of botanical textbooks published in the United States during the nineteenth century with reference to the author's home country, it can be seen that the percentage of American authors increased remarkably as the century progressed (Appendix 2.2 and 2.3). In the first quarter of the century, only about one-third of the books were authored by American writers, but the percentage increased to 90% during the second half of the century. These data demonstrate that a major shift in botanical textbook writing transpired in the United States during the nineteenth century.[e]

Analysis also shows a shift from emphasis upon systematic botany and gross morphology to greater emphasis upon cytology, comparative morphology, and physiology, as well as a shift from British and French to German models. Although these general trends can be determined, many complexities occur in the history of the content of American textbooks of botany during the nineteenth century.[a,b]

The twentieth century textbook of botany has generally followed the late nineteenth century German model. Some notable exceptions, for example, are the ecologically oriented textbooks by the University of Chicago botanists, used at the beginning of the twentieth century (Appendix 2.1, part IV). Some recent textbooks have introduced concepts derived from studies on the molecular level of organization in plants (Appendix 2.1, part V).

References Cited

1. American Textbook Publishers Institute. 1949. *Textbooks in Education. A Report from The American Textbook Publishers Institute to Its Membership, Its Friends, and Any Others . . . .* The American Textbook Publishers Institute, New York City. xi, 139 pp. (Quote, p. 19).

2. Campbell, Douglas Houghton. 1933. The Sachs Text-book and Its Influence on the Development of Botany in America. *Bulletin of the Torrey Botanical Club* 60: 331–333.

3. C[oulter], J[ohn] M. 1911. Botany, pp. 425–429. *In* Paul Monroe (ed.), *A Cyclopedia of Education*, Vol. 1. Macmillan Co., New York. (Quote, p. 426).

4. Farlow, W[illiam] G. 1913. The change from the old to the new botany in the United States. *Science, n. ser.* 37: [79]–86. (Quote, pp. 83–84).

Publication Notes

Compiled and edited from abstracts,[a,b,c] manuscripts,[d,e,f] and notes[e,f] of papers read at meetings identified in the introduction to this paper and in citations below. Stuckey wrote paragraphs six, seven, and eight. Stuckey and Burk prepared Appendix 2.1 from notes left in a similar arrangement by Rudolph.

[undated]. "[An Abbreviated Alphabetical List of] 236 Botany Texts." Typewritten & handwritten, 3 pp. [Manuscript].

[a] [1970]. "Nineteenth Century American Schoolbooks and Textbooks of Botany." Read 24 August 1970 at Bloomington, Indiana. Typewritten, 2 pp. [Abstract, long version, not published]. [Manuscript not located].

[b] 1970. Nineteenth century American schoolbooks and textbooks of botany. *American Journal of Botany* 57(6, pt. 2): 768. [Abstract, short version].

[c] 1971. The origin and development of the botanical textbook and its American aspects, p. 6. *In* Abstracts of Contributed Papers Presented before The Ohio Academy of Science 1971 History of Science Meeting, March 6, 1971, Battelle Memorial Institute, 505 King Avenue, Columbus, Ohio. [Typewritten, Mimeographed]. [Abstract].

[d] 1971. "The Origin and Development of the Botanical Textbook and Its American Aspects." Read 6 March 1971 at Columbus, Ohio. Handwritten notes listing titles of textbooks with notations about them, 7, 1 pp. [Manuscript].

[e] [1975]. "The Botanical Textbook in Nineteenth-century America as a Reflection of Botanical and Cultural Trends." Read 11 April 1975 at Manhattan, Kansas. Handwritten, 6 pp. + 1 table. 63 pp. [Manuscript text and notes].

[f] 1976. "The Americanization of German Botany." Read 1 June 1976 at New Orleans, Louisiana. Handwritten outline with notes or quotations from various textbooks of botany, 12 pp. Additional notes from textbooks of botany, 51 pp. [Manuscript].

1976. The Americanization of Germanic botany, 1865–1915, p. 19. *In* Abstracts of Papers to be Presented at the Meetings of the Botanical Society of America and Certain Affiliated Groups at Tulane University, New Orleans, 30 May–4 June 1976. Botanical Society of America, Lawrence, Kansas. [Abstract].

APPENDIX 2.1.
SELECTED TITLES OF BOTANY TEXTBOOKS GROUPED ACCORDING TO THEIR SIMILARITIES AND DEVELOPMENTAL HISTORY,[e,f] BY RUDOLPH AND STUCKEY

Part I. Eighteenth Century Linnaean Model Textbooks.

A.) Carl Linnaeus. 1751. *Philosophia Botanica; In Qua Explicantur Fundamenta Botanica* . . . . Apud Godofr. Kiesewetter, Stockholmiae. 362 pp.

B.) James Lee. 1760. *An Introduction to Botany . . . Extracted from the Works of Linnaeus* . . . . J. and R. Tonson, London. xvi, 320 pp.

C.) Hugh Rose. 1775. *The Elements of Botany . . . Being a Translation of the Philosophia Botanica, and Other Treatises of . . . Linnaeus* . . . . T. Cadell; M. Hingeston, London. xii, 472 pp.

D.) Jean Jacques Rousseau. 1785. *Letters on the Elements of Botany. Addressed to a Lady.* Translation into English, with notes, and twenty-four additional letters, fully explaining the system of Linnaeus. By Thomas Martyn. B. White and Son, London. xxiii, 503 pp. (For bibliographical background see Dale E. Johnson. *Huntia* 6: 79–82. 1985).

E.) John Thompson. 1798. *Botany Displayed, Being a Complete and Compendious Elucidation of Botany, According to the System of Linnaeus.* John Thompson and A. Nunes, London. 10 pp. + 12 pls.

Part II. Nineteenth Century American Textbooks Based on the Linnaean Model or Books that Introduced the Natural System of Classification into America.

A.) Benjamin Smith Barton. 1803. *Elements of Botany: or Outlines of the Natural History of Vegetables.* Illustrated by thirty plates. Printed for the Author, Philadelphia. 2 Vols. (1: xii, 302 pp. + 1 pl.; 2: 168, 38 pp. + 29 pls.).

B.) James Edward Smith. 1814. *An Introduction to Physiological and Systematical Botany.* 1st American, from the 2nd English ed., with notes by Jacob Bigelow. Bradford and Read, Boston; Anthony Finley, Philadelphia. xii, 415 pp.

APPENDIX 2.1. (CONTINUED)

C.) A[mos] Eaton. 1817. *A Manual of Botany for the Northern States, ...* By the Members of the Botanical Class in Williams' College (Mass.), from a Manuscript System Compiled by the Author of Richard's Botanical Dictionary. Websters and Skinners, Albany, New York. vi, 164 pp. [The 8th ed., entitled North American Botany, was written with John Wright].

D.) A[mos] Eaton. 1820. *Botanical Exercises, Including Directions, Rules and Descriptions, Calculated to Aid Pupils in the Analysis of Plants with a Labelling Catalogue, for the Assistance of Teachers*. Websters and Skinners, Albany, N[ew] Y[ork]. 171 pp.

E.) James Edward Smith. 1822. *A Grammar of Botany*. American ed. James V. Seaman, New-York. xvi, [1], 18–284 pp.

F.) Almira Hart Lincoln [Phelps]. 1829. *Familiar Lectures on Botany*. H. and F. J. Huntington, Hartford; G. & C. & H. Carvill, New York; Richardson & Lord, Boston. x, [11]–335, [1], 4 pp. [At least 39 different printings and two major revisions (1845, 1852) were issued through 1869].

G.) John Lindley. 1831. *An Introduction to the Natural System of Botany: or, A Systematic View of the Organization, Natural Affinities, and Geographical Distribution of the Whole Vegetable Kingdom* . . . 1st American ed., with an appendix by John Torrey. G. & C. & H. Carvill, New York. lxxix, 393 pp.

H.) Laura Johnson. 1834. *Botanical Teacher for North America: In Which Are Described the Indigenous and Common Exotic Plants, Growing North of the Gulf of Mexico*. Steele, Albany, New York. 268 pp. [A second edition, modernized for 1840, to accompany Elias Gates' 8th edition of North American Botany, was published by Elias Gates, Troy, New York. 268 pp.].

I.) Asa Gray. 1842. *The Botanical Text-book for Colleges, Schools, and Private Students: Comprising Part I. An Introduction to Structural and Physiological Botany. Part II. The Principles of Systematic Botany* . . . . Wiley and Putnam, New-York; Little and Brown, Boston. xii, 413 pp.

J.) Asa Gray. 1857. *First Lessons in Botany and Vegetable Physiology*. Ivison & Phinney and G. P. Putnam & Co., New York. xii, 236 pp.

K.) Asa Gray. 1887. *The Elements of Botany for Beginners and for Schools*. Revised ed. American Book Co., New York, Cincinnati, Chicago. viii, 226 pp.

L.) Alphonso Wood. 1845. *A Class-book of Botany, Designed for Colleges, Academies, and Other Seminaries Where the Science Is Taught.* In Two Parts: Part I. The Elements of Botanical Science. Part II. The Natural Orders, Illustrated by a Flora of the Northern United States, Particularly New England and New York. Crocker & Brewster, Boston; Simeon Ide, Claremont, N[ew] H[ampshire]. II Parts: (I: 124 pp.; II: ii, 3–474 pp.). [Over 60 editions were issued, the last one posthumously in 1881].

APPENDIX 2.1. (CONTINUED)

Part III. Nineteenth Century German Model Textbooks Made Available in the English Language.

A.) Matthias Jacob Schleiden. 1842–1843. *Grundzüge der wissenschaftlichen Botanik*. Verlag von Wilhelm Engelmann, Leipzig. 2 Vols. (1: xxvi, 289 pp.; 2: xviii, 564 pp.). English translation, 1849. *Principles of Scientific Botany; or, Botany as an Inductive Science*. Translated by Edwin Lankester. Longman, Brown, Green, and Longmans, London. viii, 616 pp.

B.) Julius Sachs. 1868. *Lehrbuch der Botanik*. Verlag von Wilhelm Engelmann, Leipzig. xii, 632 pp. English translation, 1875. *Text-book of Botany. Morphological and Physiological*. Translated and annotated by Alfred W. Bennett, assisted by W. T. Thiselton Dyer. At the Clarendon Press, Oxford [England]. 858 pp.

C.) E[duard] Strasburger [and others]. 1884. *Das botanische Practicum*. G. Fischer, Jena. xxxvi, 664 pp. English translation, 1887. *Handbook of Practical Botany*. Edited from the German by W. Hillhouse. Revised by the author and with many additional notes by author and editor. S. Sonnenschein, Lowrey & Co., London. xxiv, 425 pp.

D.) Anton DeBary. 1877. *Vergleichende Anatomie der Vegetationsorgane der Phanerogamen und Farne*. *In* W. Hofmeister (ed.), *Handbuch der physiologischen Botanik*, Band 3. W. Engelmann, Leipzig. xvi, 663 pp. English translation, 1884. *Comparative Anatomy of the Vegetative Organs of the Phanerogams and Ferns*. Translated and annotated by F. O. Bower and D. H. Scott. At the Clarendon Press, Oxford, [England]. xvi, 659 pp.

E.) Charles E. Bessey. 1880. *Botany for High Schools and Colleges*. [2nd ed., 1881; 3rd ed. revised, 1883.] *American Science Series*. Henry Holt and Co., New York. vi, 611 pp.

F.) W[illiam] A. Kellerman. 1883. *The Elements of Botany* . . . . John E. Potter & Co., Philadelphia, New York, Boston, Chicago. x, 11–360 pp.

G.) Douglas Houghton Campbell. 1890. *Elements of Structural and Systematic Botany, for High Schools and Elementary College Courses*. Published by Ginn & Co., Boston. ix, 253 pp.

H.) Volney M. Spalding. 1893. *Guide to the Study of Common Plants. An Introduction to Botany*. D. C. Heath & Co., Publishers, Boston. xxiii, 246 pp.

Part IV. Late Nineteenth and Early Twentieth Century United States Model Textbooks with an Ecological and/or Natural History Orientation.

A.) George Francis Atkinson. 1898. *Elementary Botany*. Henry Holt and Co., New York. xxiii, 444 pp.

Appendix 2.1. (Continued)

B.) George Francis Atkinson. 1900. *Lessons in Botany*. Henry Holt and Co., New York. xv, 365 pp.

C.) George Francis Atkinson. 1905. *A College Text-book of Botany*. 2nd ed., revised. Henry Holt and Co., New York. xvi, 737 pp.

D.) Charles Reid Barnes. 1898. *Plant Life Considered with Special Reference to Form and Function*. Henry Holt & Co., New York. x, 428 pp.

E.) John M. Coulter. 1899. *Plant Relations. A First Book of Botany.* Twentieth Century Text-books. D. Appleton and Co., New York. vii, 264 pp.

F.) John M. Coulter. 1900. *Plant Structures. A Second Book of Botany*. Twentieth Century Text-books. D. Appleton and Co., New York. ix, 348 pp.

G.) John M. Coulter. 1900. *Plant Studies. An Elementary Botany*. Twentieth Century Text-books. D. Appleton and Co., New York. ix, 392 pp.

H.) John M. Coulter. 1905. *A Text-book of Botany for Secondary Schools*. Twentieth Century Text-books. D. Appleton and Co., New York. vii, 365 pp.

I.) John Merle Coulter, Charles Reid Barnes, and Henry Chandler Cowles. 1910. *A Textbook of Botany for Colleges and Universities.* Vol. I. *Morphology and Physiology*. American Book Co., New York, Cincinnati, Chicago. viii, 484 pp. + index (pp. A–L).

J.) John Merle Coulter, Charles Reid Barnes, and Henry Chandler Cowles. 1911. *A Textbook of Botany for Colleges and Universities.* Vol. II. *Ecology.* American Book Co., New York, Cincinnati, Chicago. pp. [i]–x, 485–964 + index (pp. a–q).

K.) L[iberty] H[yde] Bailey. 1897. *Lessons with Plants*. Suggestions for Seeing and Interpreting Some of the Common Forms of Vegetation. The Macmillan Co., New York. xxxi, 491 pp.

L.) L[iberty] H[yde] Bailey. 1900. *Botany. An Elementary Text for Schools*. The Macmillan Co., New York. xiv, 355 pp.

M.) Douglas Houghton Campbell. 1910. *A University Text-book of Botany.* 2nd ed.—Revised and Corrected. The Macmillan Co., New York. xv, 579pp.

Part V. Twentieth Century Textbooks That Introduce Concepts from Studies on the Molecular Level of Organization Published before 1977 and Selected by Ronald L. Stuckey (as Professor Rudolph listed no examples in his notes).

A.) Arthur Cronquist. 1961. *Introductory Botany*. Harper & Row, Publishers, New York and Evanston, [Illinois]. ix, 902 pp.

APPENDIX 2.1. (CONTINUED)

B.) Victor A. Greulach and Edison Adams. 1962. *Plants. An Introduction to Modern Botany*. John Wiley & Sons, Inc., New York, London. xvi, 557 pp.

C.) Peter H. Raven and Helena Curtis. 1970. *Biology of Plants*. Worth Publishers, Inc., New York. xi, 706 pp.

D.) Knut Norstog and Robert W. Long. 1976. *Plant Biology*. W. B. Saunders Co., Philadelphia, London, Toronto. vi, 585 pp.

E.) The following textbook and its later editions show the changes from a traditional first (1941) edition textbook to revised editions introducing molecular concepts in the fourth (1963) and fifth (1972) editions.

1. Harry J. Fuller. 1941. *The Plant World. A Text in College Botany*. Henry Holt and Co., New York. xi, 592 pp.

2. Harry J. Fuller. 1951. *The Plant World. A Text in College Botany.* Revised ed. Henry Holt and Co., New York. xiv, 769 pp.

3. Harry J. Fuller. 1955. *The Plant World. A Text in College Botany.* 3rd ed. Henry Holt and Co., New York. vii, 584 pp.

4. Harry J. Fuller and Zane B. Carothers. 1963. *The Plant World. A Text in College Botany.* 4th ed. Holt, Rinehart and Winston, Inc., New York. ix, 564 pp.

5. Harry J. Fuller, Zane B. Carothers, Willard W. Payne, and Margaret K. Balbach. 1972. *The Plant World.* 5th ed. Holt, Rinehart and Winston, Inc., New York. xv, 553 pp.

Appendix 2.2.
Numbers of Botany Textbooks in the United States (1800–1899), Based on 390 Unique Printings,[e] by Rudolph

Quarter of the Century	Total Number of Textbooks	Percentage of Authors American	Numbers of Authors American	English	French	German
First Quarter (1800–1824)	32	31%	10	22	0	0
Second Quarter (1825–1849)	87	79%	69	16	2	0
Third Quarter (1850–1874)	104	94%	98	2	4	0
Fourth Quarter (1875–1899)	167	90%	150	11	0	6
Totals	390	94%	327	51	6	6

Appendix 2.3.
Numbers of Textbooks Published at Selected Year Intervals between 1800 and 1910,[e] by Rudolph

Years	Numbers of Textbooks: Totals per 10-Year Intervals	Totals per 50-Year Intervals
1750–1799		90
1800–1809	75	
1810–1819	105	
1820–1829	150	
1830–1839	270	
1840–1849	300	900
1850–1859	353	
1860–1869	360	
1870–1879	525	
1880–1889	300	
1890–1899	270	1808
1900–1909	225	

Introduction No. 3

Asa Gray's Botanical Textbooks (1970–1976)

As a component to Rudolph's study of botanical textbooks, he had compiled extensive notes[a] on the 11 different editions of the textbooks written by Asa Gray (1810–1888). Published during the second half of the nineteenth century, Gray's books were considered by his contemporaries as the best models for current botanical knowledge covering the entire scope of the subject as known in his day. Gray wrote them at a level of comprehension and complexity different from most textbooks, in order that they would be available and useful to individuals of all age-levels of knowledge. When one sees the various titles and dates of publications, the bibliographic complexity and confusion of this subject become apparent.

Stuckey (1997)[1] has clarified the bibliographic situation in his paper on the merits of Gray's botanical textbooks by using information from (1) the prefaces of Gray's books and his correspondence with botanical friends, and from (2) commentaries from other writers in published reviews of his textbooks. Based on their titles, contents, and purposes, Gray's textbooks may be placed into three categories: those that (1) provide the essential basic botanical elements, written for beginning students in schools, published in 1836, 1857, and 1887 (Fig. 3.1); (2) have the word "textbook" in the title, written for students at the secondary and collegiate levels, published in six rewritten editions, 1842, 1845, 1850, 1853, 1858, and 1879 (Figs. 3.2, 3.3); and (3) have a morphological-physiological orientation written for use in elementary schools, published in 1858 and 1872 (Appendices 3.1 and 3.2). Gray's textbooks were original contributions that organized botany into a more useful science and fulfilled the educational needs of the time. They were in demand for use in schools and by the public in general, but after Gray's death, no one continued preparing revisions, and thus both their use and usefulness ceased.

How Rudolph would have written a paper on Asa Gray's textbooks is not known, and no manuscript can be reconstructed adequately from his notes. He did, however, prepare a list of titles, each with the year of its original printing, along with quotations about some of the books from Gray's own writings and published reviews by his contemporaries. Rudolph was an advocate for the search of book reviews useful for writing critical analyses of the importance of books. These notes that Rudolph

[1] Ronald L. Stuckey. 1997. Commentaries on Asa Gray's botanical textbooks by Gray and his contemporaries (1836–1887). *Sida* 17: 769–802.

assembled served initially as the basis for the publication by Stuckey.[1] A chronological list of the first editions of Gray's botanical textbooks has been formulated from Rudolph's notes. As a resource for the study of these textbooks, Burk and Stuckey have added a list of the known published reviews.[b] **(Paper No. 3)**.

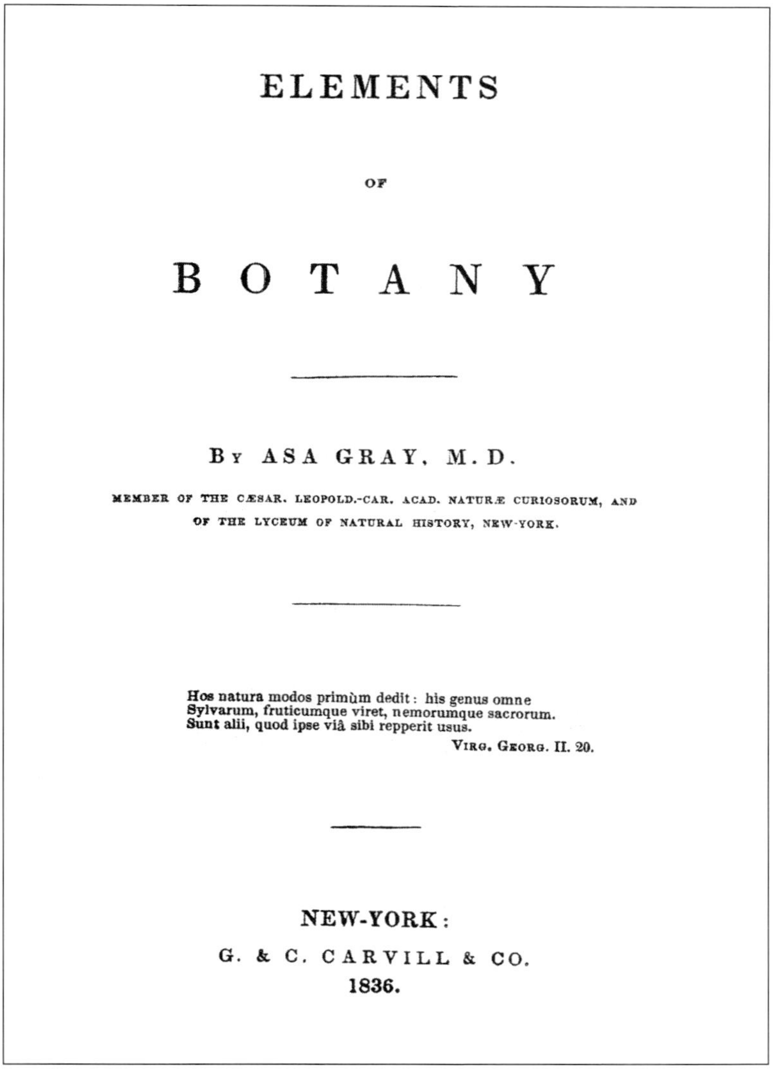
ELEMENTS

OF

BOTANY

By ASA GRAY, M.D.

MEMBER OF THE CÆSAR. LEOPOLD.-CAR. ACAD. NATURÆ CURIOSORUM, AND OF THE LYCEUM OF NATURAL HISTORY, NEW-YORK.

Hos natura modos primùm dedit: his genus omne
Sylvarum, fruticumque viret, nemorumque sacrorum.
Sunt alii, quod ipse viâ sibi repperit usus.
VIRG. GEORG. II. 20.

NEW-YORK:
G. & C. CARVILL & CO.
1836.

Fig. 3.1. Title page from *Elements of Botany*, by Asa Gray (1836).

(From a copy in the Herbarium Library, The Ohio State University)

Asa Gray and His Botany Textbooks

Asa Gray (1810–1888)

Line illustration of Gray's lily, from *How Plants Grow*, by Asa Gray (1858)

Paper No. 3 (1970–1976)

Sources of Reviews for Asa Gray's Botany Textbooks

Asa Gray (1810–1888), considered the foremost nineteenth century botanist in the United States, authored 11 different editions of his textbooks of botany, designated as among the best at that time in the United States. This paper[a] is a list of the first edition printings of Gray's textbooks in the order in which they were written and published (Appendix 3.1).[b] With the exception of the fourth-listed textbook (1853), accompanying each title is one or more references to published contemporary reviews, provided as a resource for the study of these textbooks. A total of 31 reviews from three languages are cited. Of Gray's textbooks, Joseph C. Arthur (1895)[1] of Purdue University, in his vice-presidential address of 29 August 1895 to Section G, Botany, of the American Association for the Advancement of Science spoke as follows:

> Botany, as a substantial part of the curriculum, cannot be said to have received recognized standing in the American educational system until the time of Asa Gray. In the latter part of the decade of the thirties his first text-book, the 'Elements of Botany,' appeared, and in the decade following the 'Text-book for Colleges' and the 'Manual,' all of which works showed a true appreciation of the best features of the science and the needs of the time. They were so well conceived, and so much in demand, that new editions rapidly succeeded one another; and to the present day they hold a high place in the estimation of botanical teachers. These works possessed a specially potent element of virility in being the expression of knowledge at first hand, the words of the master. In so far as inspiration was drawn from foreign sources it came chiefly from French and English scholars, of whom De Candolle the eldest and Robert Brown were the representatives.

British botanist, Joseph D. Hooker (1888),[3] writing after Gray's death, stated that

> . . . the *Elements of Botany* (1836) was the first of a series of editions of a work that has been for nearly half a century the text-book of schools and colleges throughout the United States, and the later issues of which have been generally recommended by the botanical professors of the United Kingdom as the best of its class.

Frontispiece Credits: Asa Gray and his Botany Textbooks. Asa Gray (1810–1888). (Original photograph source unknown; in the Rudolph photograph collection)

Line illustration on page [1] of Asa Gray's lily. (From Asa Gray. 1858. *How Plants Grow*. Ivison, Blakeman, Taylor & Co., New York; permission to reproduce illustration courtesy of Rare Books and Manuscripts Library, OSU)

Gray's colleague in cryptogamic botany at Harvard University, William G. Farlow (1895),[2] praised Gray's textbooks in this note:

> The numerous editions of Gray's Text-Book made his name known in all the higher schools in the country, ... He left a great nation in which, very largely through his exertions, the value of botany had become generally recognized and in which a crowd of young workers had arisen anxious to carry out the good work . . . .

References Cited

1. Arthur, J. C. 1895. Development of vegetable physiology. *Science, n. ser.* 2: 360–373. [Also published in *Proceedings of the American Association for the Advancement of Science* 44: [163]–184. 1896]. (Quotes, pp. 362, 370).

2. Farlow, William G. 1895. Memoir of Asa Gray, 1810–1888. Read before the National Academy, April 17, 1889. *Biographical Memoirs, National Academy of Sciences.* 3: 161–175. (Quote, pp. 173–175).

3. H[ooker], J[oseph] D. 1888. Professor Asa Gray. *Nature* 37: 375–377. (Quote, p. 376).

Publication Notes

Compiled from notes[a] of a paper read 11 April 1975 at the Midwest Junto for the History of Science, Kansas State University, Manhattan, and from a list of the titles of Asa Gray's textbooks of botany. Burk and Stuckey (1996) added a list of the known published reviews for these books.[b]

[a] [1975]. "The Botanical Textbook in Nineteenth-century America as a Reflection of Botanical and Cultural Trends." Handwritten, 6 pp. + 1 Table. Titles and Quotations from books, and related information, including items on Asa Gray's textbooks. 63 pp. [Manuscript and notes].

[b] William R. Burk and Ronald L. Stuckey. 1996. "Asa Gray's Botanical Textbooks Listed with Citations of Published Book Reviews." [Appendix 3.1].

Appendix 3.1.
Asa Gray's Botany Textbooks Listed with Citations of Published Book Reviews,[b] by Burk and Stuckey

1836. *Elements of Botany*. G. & C. Carvill & Co., New-York. xiv, 428 pp. (Fig. 3.1). *American Gardener's Magazine* 2: 421–424. 1836. By "D." [Andrew Jackson Downing].

American Journal of Science 30: 399. 1836.

1842. *The Botanical Text-book for Colleges, Schools, and Private Students: Comprising Part I. An Introduction to Structural and Physiological Botany. Part II. The Principles of Systematic Botany; With an Account of the Chief Natural Families of the Vegetable Kingdom, and Notices of the Principal Officinal or Otherwise Useful Plants.* Illustrated with numerous engravings on wood. Wiley and Putnam, New-York; Little and Brown, Boston. xii, [13]–413 pp.

American Journal of Science and Arts 43: 388–389. 1842. By "W.D." [= William Darlington].

Annals and Magazine of Natural History, including Zoology, Botany and Geology 10: 352–354. 1842.

Botanische Zeitung 1: 50–51. 1843. By "S-l." = [D. F. L. von Schlechtendal].

Gardener's Magazine 2(3rd ser.): 634. 1842. J. C. Loudon, ed.

North American Review 56: 192–207. 1843. [George Barrell Emerson].

The London Journal of Botany 1: 636. 1842. By "W. J. H." = [William J. Hooker].

1845. *The Botanical Text-book, for Colleges, Schools, and Private Students: Comprising Part I. An Introduction to Structural and Physiological Botany. Part II. The Principles of Systematic Botany; With an Account of the Chief Natural Families of the Vegetable Kingdom, and Notices of the Principal Useful Plants.* 2nd ed. Illustrated with more than a thousand engravings on wood. Wiley and Putnam, New-York. 509 pp.

Botanische Zeitung 4: 98–100. 1846. By "S-l." = [D. F. L. von Schlechtendal].

North American Review 61: 254–258. 1845. [George Barrell Emerson].

North American Review 67: 180–182. 1848. [John Carey].

1850. *The Botanical Text-book, an Introduction to Scientific Botany, Both Structural and Systematic. For Colleges, Schools, and Private Students.* 3rd ed., rewritten and enlarged. Illustrated with twelve hundred engravings on wood. George P. Putnam, New York. xii, [13]–520 pp. (Fig. 3.3).

Botanische Zeitung 9: 157–158. 1851. By "S-l." = [D. F. L. von Schlechtendal].

APPENDIX 3.1. (CONTINUED)

1853. *The Botanical Text-book, an Introduction to Scientific Botany, Both Structural and Systematic. For Colleges, Schools, and Private Students.* 4th ed. Illustrated with twelve hundred engravings on wood. George P. Putnam & Co., New York. xii, [13]–528 pp. [No book reviews known].

1857. *First Lessons in Botany and Vegetable Physiology, Illustrated by over 360 wood engravings, from original drawings, by Isaac Sprague. To which is added a copious glossary, or dictionary of botanical terms.* Ivison & Phinney and G. P. Putnam & Co., New York; S. C. Griggs & Co., Chicago. xii, 236 pp.

American Journal of Science and Arts 73: 439–440. 1857. By "T. C. P." = [Thomas Conrad Porter].

Botanische Zeitung 16: 151. 1858. By "S-l." = [D. F. L. von Schlechtendal].

Journal of Botany & Kew Garden Miscellany 9: 154–160. 1857.

North American Review 87: 322. 1858.

1858. *Introduction to Structural and Systematic Botany, and Vegetable Physiology, Being a Fifth and Revised Edition of The Botanical Text-book.* Illustrated with over thirteen hundred woodcuts. 5th ed. Ivison and Phinney, New York; S. C. Griggs & Co., Chicago. xii, [13]–555 pp. [title page bears date 1858; verso of title page bears copyright 1857].

North American Review 86: 587. 1858.

North American Review 87: 326–327. 1858. [Isabella James].

1858. *How Plants Grow, a Simple Introduction to Structural Botany. With a Popular Flora, or an Arrangement and Description of Common Plants, Both Wild and Cultivated. Botany for Young People and Common Schools.* Illustrated by 500 wood engravings. Ivison, Blakeman, Taylor, & Co., New York. 233 pp.

American Journal of Science and Arts 76(2nd ser., vol. 26): 139–140. 1858.

Bulletin de la Société Botanique de France 5: 655–656. 1858.

North American Review 87: 322–326. 1858. [Isabella James].

1872. *How Plants Behave: How They Move, Climb, Employ Insects To Work for Them, &c. Botany for Young People. Part II.* Ivison, Blakeman, Taylor, and Co., New York and Chicago. viii, [9]–46 pp.

American Naturalist 6: 475–477. 1872. By "E. C. B." = [Rev. E. C. Bolles].

Bulletin of the Torrey Botanical Club 3: 35. 1872.

Appendix 3.1. (Continued)

1879. *Structural Botany, or Organography on the Basis of Morphology. To Which Is Added the Principles of Taxonomy and Phytography, and a Glossary of Botanical Terms. Gray's Botanical Text-book (sixth edition) Vol. I.* Ivison, Blakeman, Taylor and Co., New York and Chicago. xii, 442 pp.

American Journal of Science and Arts 118: 73–76. 1879. By "G. L. G." = [George Lincoln Goodale].

Bulletin de la Société Botanique de France 26: 165. 1879.

Bulletin of the Torrey Botanical Club 6: [317]–320; 328–329. 1879.

Journal of Botany British and Foreign, new ser. 8: 253–254. 1879. By "H. T." = [Henry Trimen].

American Naturalist 14: 870. 1880. Charles E. Bessey.

1887. *The Elements of Botany for Beginners and for Schools. Gray's Lessons in Botany.* Revised ed. American Book Co., New York, Cincinnati, Chicago. viii, 226 pp.

American Journal of Science and Arts 134: 495. 1887. By "G. L. G." = [George Lincoln Goodale].

American Naturalist 22: 46–47. 1888. Charles E. Bessey.

Botanical Gazette 13: 41. 1888. [editors, John M. Coulter, Charles R. Barnes, J. C. Arthur].

Journal of Botany British and Foreign 26: 92–93. 1888. [James Britten, editor].

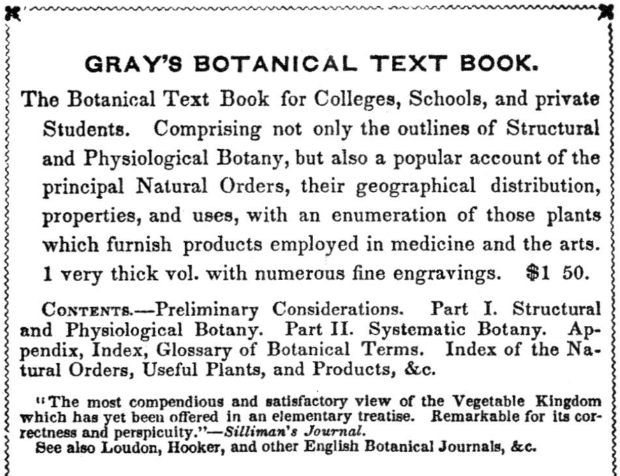

GRAY'S BOTANICAL TEXT BOOK.

The Botanical Text Book for Colleges, Schools, and private Students. Comprising not only the outlines of Structural and Physiological Botany, but also a popular account of the principal Natural Orders, their geographical distribution, properties, and uses, with an enumeration of those plants which furnish products employed in medicine and the arts. 1 very thick vol. with numerous fine engravings. $1 50.

Contents.—Preliminary Considerations. Part I. Structural and Physiological Botany. Part II. Systematic Botany. Appendix, Index, Glossary of Botanical Terms. Index of the Natural Orders, Useful Plants, and Products, &c.

"The most compendious and satisfactory view of the Vegetable Kingdom which has yet been offered in an elementary treatise. Remarkable for its correctness and perspicuity."—*Silliman's Journal.*
See also Loudon, Hooker, and other English Botanical Journals, &c.

Fig. 3.2. Advertisement for Asa Gray's second edition of his *The Botanical Text-book* (1845).

(From Rudolph's copy in the Herbarium Library, The Ohio State University)

APPENDIX 3.2.
SUMMARY OF SELECTED FEATURES OF ASA GRAY'S BOTANICAL TEXTBOOKS,[1] BY STUCKEY

Abbreviated Title of Book	Year on Title Page	Year of Copyright	Copyright Owner	Number of Additional Printings
Group 1. *Elements and Lessons*:				
Elements	1836	1836	G. & C. Carvill & Co.	none
First Lessons	1857	1857	George P. Putnam & Co.	17
Elements		1887	Asa Gray	3
Group 2. "True" *Textbooks*:				
Textbook ed. 1.	1842	1842	Wiley & Putnam	none
Textbook ed. 2.	1845	1842	Wiley & Putnam	none
Textbook ed. 3.	1850	1850	George P. Putnam	none
Textbook ed. 4.	1853	1853	George P. Putnam & Co.	none
Textbook ed. 5.	1858	1857	Ivison & Phinney	15
Textbook ed. 6.		1879	Asa Gray	5
Group 3. "Botany for Young People:"				
How Plants Grow		1858	Ivison & Phinney	25
How Plants Behave		1872 1900	Asa Gray Jane L. Gray	4

[1] Ronald L. Stuckey. 1997. Commentaries on Asa Gray's botanical textbooks by Gray and his contemporaries (1836–1887). *Sida* 17: 769–802. (From p. 772).

[2] Figures are numbered by two methods:

[a] Each figure contains one or more individual illustrations, each designated with the letters a, b, c, etc.

[b] Each individual illustration is given a figure number.

Date of Preface	Number of Illustrations[2]	Illustrator	Changes from Previous Edition
April 1836	125[a]		_____
1 January 1857	367[b]	Isaac Sprague	_____
March 1887	589[b]	Isaac Sprague	"new and much revised"
No Preface	78[a]	Miss Agnes Mitchell; Mr. J. J. Butler	_____
March 1845	1,045[b]	additional illustrations by Isaac Sprague	portions"entirely rewritten"
April 1850	1,207[b]	additional illustrations by Isaac Sprague	"rewritten and enlarged"
March 1853	1,207[b]	same as edition 3	"corrections and minor alterations"
September 1857	1,336[b]	Isaac Sprague	"almost entirely rewritten"
10 April 1879	695[b]		"entirely rewritten"
No Preface	519[b]	Isaac Sprague	_____
2 February 1872	40[b]	Isaac Sprague	_____

THE

BOTANICAL TEXT-BOOK,

AN

INTRODUCTION TO SCIENTIFIC BOTANY,

BOTH STRUCTURAL AND SYSTEMATIC.

FOR COLLEGES, SCHOOLS, AND PRIVATE STUDENTS.

THIRD EDITION,

REWRITTEN AND ENLARGED.

Illustrated with twelve hundred Engravings on Wood.

BY ASA GRAY, M. D.,

FISHER PROFESSOR OF NATURAL HISTORY IN HARVARD UNIVERSITY.

NEW YORK:

GEORGE P. PUTNAM.

1850.

Fig. 3.3. Title page from *The Botanical Text-book, An Introduction to Scientific Botany.* Third edition, by Asa Gray (1850).

(From Rudolph's copy in the Herbarium Library, The Ohio State University)

Introduction No. 4

Lichens in Botanical Textbooks (1986, 1988)

Sixteen years after his first presentation in 1970 on the history of botanical textbooks, Rudolph prepared an analysis of lichens which are usually discussed in most textbooks of botany (Fig. 4.1). This study, "Coverage of Lichens in American Introductory Botany Textbooks 1860–1985," was presented on 12 August 1986 at the meeting of the Historical Section of the Botanical Society of America, at the University of Massachusetts in Amherst. Rudolph (1988) published a full discussion and analysis of this topic in *BioScience*. The edited meeting abstract[b] and selected statements from his introduction and analysis in *BioScience*[c] are reproduced here. **(Paper No. 4)**.

Fig. 4.1. Emanuel D. Rudolph studying lichens in his office at The Ohio State University.

(Taken 1969, by G. Willey, OSU Research Foundation; in the collection of RLS)

Lichens in Botany Textbooks

Emanuel D. Rudolph in his OSU office studying a lichen, 1977

Table 1. Botanical textbooks surveyed.

1836	A. Gray. *Elements of Botany.*
*1845	A. Gray. *The Botanical Text-book for Colleges, Schools, and Private Students.* 2nd ed. (No lichens in 1st ed. 1842; same information with some deletions to 5th ed. 1876.)
1853	H. Coultas. *The Principles of Botany, as Exemplified in the Cryptogamia.*
†1868	A. Wood. *Class-book of Botany: Being Outlines of the Structure, Physiology, and Classification of Plants.* (Same in editions to 1882.)
†1880	C. E. Bessey. *Botany for High Schools and Colleges.* (Same in editions to 1892.)
1883	W. A. Kellerman. *The Elements of Botany.*
†1883	G. Macloskie. *Elementary Botany.* (Same in 1887 edition.)
†1884	C. E. Bessey. *The Essentials of Botany* (Same in editions to 1912.)
†1889	E. S. Bastin. *College Botany.* (Same in editions to 1918.)
1889	A. Chambers-Ketchum. *Botany for Academies and Colleges.*
1890	D. H. Campbell. *Elements of Structural and Systematic Botany, for High Schools and Elementary College Courses.*
1896	M. Westermair. *A Compendium of General Botany.*
†1897	W. A. Kellerman. *A Text-book of Elementary Botany.* (Same in editions to 1902.)
†1898	G. F. Atkinson. *Elementary Botany.* (Same in editions to 1902 as *College Text-book of Botany.*)
†1898	C. R. Barnes. *Plant Life with Special Reference to Form and Function.* (Same in 1900 edition as *Outlines of Plant Life.*)
1898	C. H. Clark. *A Laboratory Manual in Practical Botany.*
1899	D. H. Campbell. *Lectures on the Evolution of Plants.*
*1899	H. H. Rusby and E. J. Smith. *Morphology and Histology of Plants.* (With some omissions in H. H. Rusby. *A Manual of Structural and Descriptive Botany.* 1924.)
†1900	L. H. Bailey. *Botany: An Elementary Text for Schools.* (Same in editions to 1910.)
†1900	J. M. Coulter. *Plants: A Text-book of Botany.* (Same in editions to 1909.)
‡1901	J. Y. Bergen. *Foundations of Botany.* (Additions in his *Essentials of Botany,* 1908.)
1901	R. G. Leavitt. *Outlines of Botany for the High School Laboratory and Classroom.*
1904	O. W. Caldwell. *Handbook of Plant Morphology.*
1905	J. M. Coulter. *A Text-book of Botany for Secondary Schools.*
†1906	J. Y. Bergen and B. M. Davis. *Principles of Botany.* (Same in printings to 1934.)
1907	F. E. Clements. *Plant Physiology and Ecology.*
*1907	C. C. Curtis. *Nature and Development of Plants.* (Same with slight modification in editions to 1918.)
†1910	G. F. Atkinson. *Botany for High Schools.* (Same in editions to 1912.)
1910	D. H. Campbell. *A University Text-book of Botany.* 2nd ed.
†1910	J. M. Coulter, C. R. Barnes, and H. C. Cowles. *A Textbook of Botany for Colleges and Universities.* (Same in editions to 1930.)
1911	E. F. Andrews. *A Practical Course in Botany.*
1913	J. G. Coulter. *Plant Life and Plant Uses.*
1913	F. L. Sargent. *Plants and Their Uses: An Introduction to Botany.*
1914	J. Y. Bergen and O. W. Caldwell. *Introduction to Botany.*

Portion of Table 1, listing botanical textbooks surveyed in lichen study by Rudolph (1988)

Paper No. 4 (1986, 1988)

Lichens in American Introductory Botany Textbooks 1836–1986

Lichenologists, over the years, have had significant problems in explaining some basic aspects of lichen biology. The questions that follow are difficult to answer with simple statements: What is the nature of the relationship between fungi and algae in lichens; can the components also be free-living; and do lichen scientific names belong to the whole lichen or to its individual symbionts? By reviewing the treatments of lichens in 93 American introductory botany textbooks published between 1836 and 1986, a case study was conducted of how textbook writers treat a complex and sometimes controversial biological topic. The selected textbooks were evaluated concerning these questions in relation to the available standard American introductory lichenology books. A list of these textbooks is in Rudolph (Table 1, 1988),[c] and a portion of that table for the years 1836 to 1914 appears in the frontispiece to this paper. Those textbooks which have had several editions and only minor changes were considered as one title.[a,b]

Most textbook writers do not consider many of the issues involved in the understanding of lichens, and they fail to use information about lichens available in contemporary lichenology books. Most authors repeat the same information in revision after revision, sometimes spanning many years, even when the available information has changed significantly. Because lichens present a challenge in terms of clear-cut explanations, textbook authors have regarded them as difficult to treat, and as a result they have presented the information poorly.[a,b]

Lichens take their forms from the growth of fungal symbionts, which enclose the algal autotrophic components. In the late 19th century independent culturing of the symbionts demonstrated their composite form. As a result, lichens have generally been considered in textbooks as either fungi or algae, or sometimes in a section on symbiosis.[c]

Four books published before 1880 considered lichens an independent group of green plants. The other textbooks considered them composite plants consisting of a fungus and an alga. Of these books, 40% stated that the fungus is an ascomycete, and fewer than 10% wrote that it also might be a basidiomycete. The description of

---

Frontispiece Credits: Emanuel D. Rudolph in his OSU office studying a lichen, 1977. (Original photograph source unknown; in the Rudolph photograph collection)

Portion of Table 1, listing botanical textbooks surveyed in lichen study. (From Rudolph, *BioScience* 38: 414. 1988, permission to reproduce portion of table, courtesy of American Institute of Biological Sciences ©1988)

the algae in only about 50% of the books indicated that they may be either green algae or blue-green bacteria. Although introductory lichen texts for over 80 years, as in the example (Fig. 4.2) by Schneider (1897),[2] have reported that the same lichen thallus can contain both a green alga and blue-green bacteria, general botany texts have not reported this fact. Other textbooks have reported that the fungal component provided the algal component with moisture (50%), beneficial minerals (30%), protection from physical injury (25%), and supplementary carbon dioxide (5%). Only 40% of the texts considered whether the algal component can be free-living. With regard to naming a lichen, only four recent textbooks followed the 1954 *International Code of Botanical Nomenclature*[1] which states that the name refers only to the fungal component.[c]

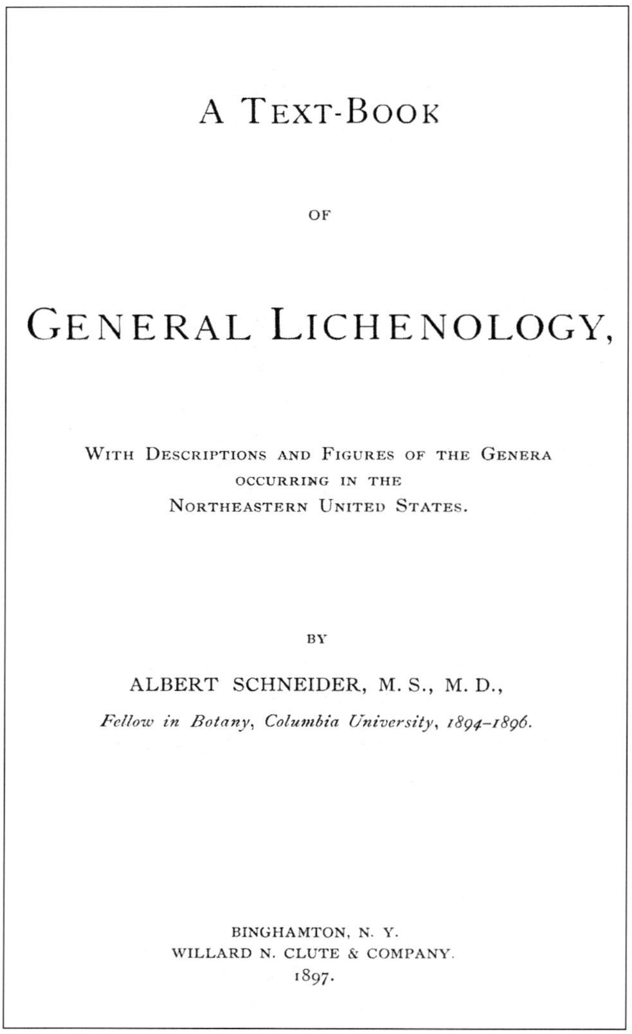

A TEXT-BOOK

OF

GENERAL LICHENOLOGY,

WITH DESCRIPTIONS AND FIGURES OF THE GENERA
OCCURRING IN THE
NORTHEASTERN UNITED STATES.

BY

ALBERT SCHNEIDER, M. S., M. D.,

Fellow in Botany, Columbia University, 1894–1896.

BINGHAMTON, N. Y.
WILLARD N. CLUTE & COMPANY.
1897.

Fig. 4.2. Title page from *A Textbook of General Lichenology*, by Albert Schneider (1897).

(From Rudolph's copy in the Herbarium Library, The Ohio State University)

Many authors who prepare revisions or new editions of their textbooks merely repeat the same information about lichens, in some instances for 30–40 years. New editions or revisions of textbooks by these authors are out-of-date with respect to lichens at the time of printing. The lessons that can be learned from this case history study are: (1) controversial issues are difficult for a simple pedagogical presentation preferred by some authors, (2) introductory textbooks often give fragmentary or erroneous information about lichens, (3) introductory textbook authors rarely cite experimental evidence, and (4) most authors are conservative about revising their textbooks, resulting in outdated information being repeated in one or more revisions.[c]

References Cited

1. Lanjouw, J. [and others] (eds.). 1956. *International Code of Botanical Nomenclature Adopted by the Eighth International Botanical Congress, Paris, July 1954. Regnum Vegetabile*, vol. 8. International Bureau for Plant Taxonomy and Nomenclature of the International Association for Plant Taxonomy, Utrecht, Netherlands. 338 pp. [See Article 13, Note 4 on p. 14].

2. Schneider, Albert. 1897. *A Text-book of General Lichenology, with Descriptions and Figures of the Genera Occurring in the Northeastern United States.* Willard N. Clute & Co., Binghamton, N[ew] Y[ork]. xvii, 230 pp. + 76 pls.

Publication Notes

Compiled and edited from the manuscript[a] and abstract[b] of the paper read 12 August 1986 at the meeting of the Historical Section, Botanical Society of America, University of Massachusetts, Amherst; and excerpts from the same paper[c] published in *BioScience* 38: 412–416. 1988.

[a] [1986]. "Coverage of Lichens in American Introductory Botany Textbooks 1836–1986." Typewritten, 6 pp. + 1 handwritten p. [Manuscript].

[b] 1986. Coverage of lichens in American introductory botany textbooks 1860–1985. *American Journal of Botany* 73: 691. [Abstract].

[1986]. "Coverage of Lichens in American Introductory Botany Textbooks, 1836–1986." Typewritten, 15 pp.; also earlier drafts 1 and 2. [Manuscript]. [Published in *BioScience* 38: 412–416. 1988].

[1986]. "Tratamiento de los Liquenes en Libros Norteamericanos de Introduccion a la Botanica entre los Anos 1836–1986." Read 3 July 1986 at the IV Congreso Latinoamericano de Botanica, Medellin, Colombia, South America. Typewritten in Spanish, 4 pp.; handwritten translation in Spanish from the English, 12 pp. [Manuscript].

[c] 1988. Lichens in US introductory botany textbooks, 1836–1986. *BioScience* 38: 412–416.

APPENDIX 4.1.
A LIST OF EMANUEL D. RUDOLPH'S PUBLISHED JOURNAL CONTRIBUTIONS TO LICHENOLOGY, BY BURK AND STUCKEY

1953. The role of lichens in soil formation and plant succession. *Ecology* 34: 805–807. (with R. Cooper)

1953. A contribution to the lichen flora of Arizona and New Mexico. *Annals of the Missouri Botanical Garden* 40: 63–72.

1955. Lichenological notes on the flora of the Antarctic Continent and the subantarctic islands. I–IV. *Annals of the Missouri Botanical Garden* 42: 131–149. (with Carroll W. Dodge)

1956. An analysis of variation in a variable population of *Cladonia*. *Evolution* 10: 147–156. (with Edgar Anderson)

1963. Vegetation of Hallett Station area, Victoria Land, Antarctica. *Ecology* 44: 585–586.

1964. Ecology and floristic investigations of Antarctic lichens. *Bulletin of the U.S. Antarctic Project Officer* 5(10): 64–65.

1965. Antarctic lichens and vascular plants: Their significance. *BioScience* 15: 285–287.

1966. Ecology and floristic investigations of lichens. *Antarctic Journal of the United States* 1: 218.

1966. Electron microscope studies of lichen reproductive structures in *Physcia aipolia*. *Mycologia* 58: 786–796. (with Robert M. Giesy)

1967. Biological survey of Marie Byrd Land. *Antarctic Journal of the United States* 2: 96–97.

1970. Conserving the Antarctic terrestrial ecosystem. *Biological Conservation* 3: 52–54.

1971. Water relations in living, dead, and cellulose models of the lichen *Umbilicaria papulosa*. *The Bryologist* 74: 444–450. (with Ray E. Showman)

1974. The role of lichens and associated fungi in the chemical weathering of rock. *Mycologia* 66: 648–660. (with Michael E. Williams)

1975. Lichen accumulation of some heavy metals from acidic surface substrates of coal mine ecosystems in southeastern Ohio. *The Ohio Journal of Science* 75: 113–117. (with James D. Lawrey)

Part Two

BOTANY IN CHILDREN'S BOOKS

Sometime before 1959, while Emanuel D. Rudolph[1] was teaching at Wellesley College in Massachusetts, he began building a collection of natural history books written for children. A colleague there, Mary Eleanor Prentiss (1893–1974), while preparing for retirement, presented him with a little book entitled, *The Child's Botany* (1831), by Samuel Goodrich. With that acquisition, followed by the addition of other children's natural history books, Rudolph soon became devoted to building a library of children's literature that grew to over eight thousand volumes. His collection consisted of chap books, shape books, pop-up books, toy books, animal alphabet books, and plant books.

About 10 years after Rudolph acquired his first book on botany for children, he began making presentations on this topic at scientific meetings. Between 1966 and 1977, three of these presentations were published and are reprinted here (**Paper Nos. 5, 7, 8**). Rudolph remained active in this endeavor until about 1985, when he presented his last paper on the topic. Four other manuscripts in typescript are also printed here as new contributions (**Paper Nos. 6, 9, 10, 11**).

---

[1] Emanuel D. Rudolph. 1979. Children's scientific books. *Friends Line. Newsletter of the Friends of the Libraries of The Ohio State University*. [Columbus]. No. 16: 4–5. February.

Introduction Nos. 5–11

Botany in Children's Books (1968–1983)

Rudolph[1] gave four reasons for studying children's literature on biological subjects. They were "to understand the: a) general biological thinking of an age, b) attitudes toward educating children, c) social class distinctions, and d) popularization of science." These were the themes that he often considered in writing about children's books. His major discovery involved how information about plants and animals was taught to children during the nineteenth and twentieth centuries. These discoveries led him into one of his main research themes, the history of the popularization of science involving the disciplines of natural history, biology, and botany. He published three papers as a result of studies with his children's book collection.[2,3,4] Several studies were left as abstracts and in manuscripts, and other topics were in the research phase.

Following publication of the three above-mentioned papers during the 1970s and as an effort to continue research in the popularization of science using children's books, Rudolph[5] received during 1986 and 1987, some financial support from a Research Challenge Program of the Office of Research and Graduate Studies at The Ohio State University. These funds were to be used for travel to other libraries to continue his investigations of American children's books. Beyond the resources in his own library, he sought information on plants and animals that were discussed in children's books elsewhere in the United States. His aim was to determine what information was available and potentially useful in stimulating the interest of

---

[1] Emanuel D. Rudolph. 1978. Letter of 15 April to Sandra Herbert, Department of History, University of Maryland, Baltimore County. 1 p.

[2] Emanuel D. Rudolph. 1973. Botany in American and British chapbooks before 1860. *Plant Science Bulletin* 19: 34–36. [Reprinted in *Morris Arboretum Bulletin* 25: 60–62, 1974, with the title: Botany in British and American chapbooks before 1860].

[3] Emanuel D. Rudolph. 1973. How it developed that botany was the science thought most suitable for Victorian young ladies. *Children's Literature* 2: 92–97.

[4] Emanuel D. Rudolph. 1978. Learning botany by rote, the way of the nineteenth century catechisms. *Plant Science Bulletin* 24: 39–40.

[5] Emanuel D. Rudolph. 1986. Organisms in 19th and 20th Century Children's Books: An Analysis. Application for Grant-in-Aid from Research Challenge Program Funds. Office of Research and Graduate Studies, The Ohio State University, [Columbus]. Typewritten, 4 pp.

children. Rudolph believed this information was useful in the education of children, who could become professional biologists or informed citizens with an understanding of biological problems.

The libraries that Rudolph visited were The American Antiquarian Society in Worcester, Massachusetts, during August 1986 and the Osborne and Lillian Smith Collection of the Toronto Public Library on 28 October 1986. The latter library, according to Rudolph, has one of the largest and strongest collections of children's literature in North America. Here Rudolph[6] studied books to determine how plants and animals were presented in instructional books for children. He was, however, unable to locate children's laboratory manuals or guidebooks dealing with the study of plants. The Rare Book Collection and Special Children's Book Division of the Free Library in Philadelphia were examined on 7–9 May 1987 and the Rare Book and Special Collection Division of the Library of Congress, Washington, D.C. was used on 25–26 June 1987. From data obtained during these library studies, and from other libraries, Rudolph[7] intended to revise and complete several of his manuscripts pertaining to children's natural history books and the popularization of science for young people. No additional work appears to have been accomplished on these topics, the major reason probably being that Rudolph was in his last year as chairman of the Department of Botany. The following year, in September of 1988, he was invited and accepted the invitation to write a scientific history of the Missouri Botanical Garden. This project occupied most of his available time until his death.

In his Report of Research Travel to Libraries in 1987, Rudolph[7] wrote that he intended to complete a journal article, probably on "What Animal Is That?: The Animals in Nineteenth Century American Children's Books Compared to Those of the Twentieth Century." He further stated his intention to write a proposal to a granting agency "for the preparation of a more extensive book using these findings. This book will examine the ways in which plants and animals have been presented to children over the years and the importance of this information for making children potential biologists or well informed adults cognizant of the contemporary understanding of natural history." The contents of this proposed book are not known, but among his manuscripts on this topic, however, is an outline of what appears to be a table of contents for such a book. This outline "Approaches for the Young in Understanding the Biotic World; An Historical Appraisal in America 1700–1950" is reproduced verbatim as follows:

[6] Emanuel D. Rudolph. 1986. Organisms in 19th and 20th Century Children's Books: An Analysis. Interim Report of 2 December 1986, Office of Research and Graduate Studies, The Ohio State University, Columbus. Typewritten, 2 pp.

[7] Emanuel D. Rudolph. 1987. Organisms in 19th and 20th Century Children's Books: An Analysis. Report of 7 July 1987, Office of Research and Graduate Studies, The Ohio State University, Columbus. Typewritten, 2 pp.

Contents for Proposed Book on Natural History

I. Introduction.

II. The Publications and Their Uses.

III. Nature Illustrates the Wisdom and Design of God.

IV. Studying Nature is Good for the Mind and Body.

V. Nature as Adventure.

VI. Man Uses Nature for His Purposes.

VIII. Learning about Nature is Sound Educational Practice.

IX. Biology is an Important Science to be Studied.

Presentations Made on Children's Books at Scientific Meetings (1968–1985)

From 1968 to 1985, Rudolph presented thirteen papers on botany and related subjects about American children's natural history books. He determined that the format and manner of presentation of the information, in children's books before 1900, as well as the book's content itself, were heavily influenced by attitudes toward the education of children. Books published especially for children were not common until the latter part of the eighteenth century. In the United States, these early works were reprinted from or modeled on English or European books. Botany was taught first within the context of moral lessons, in which the scientific information was modified for entertainment and educational purposes. Rudolph's thirteen meeting presentations, arranged chronologically, are:

"Botany in American Children's Books: The Period before 1900." Presented 10 February 1968 to the History of Science Meeting, The Ohio Academy of Science, Battelle Memorial Institute Auditorium, Columbus. (Paper No. 6).

"Botany in American Children's Books before 1900." Presented 3 September 1968 to the Historical Section, Botanical Society of America, The Ohio State University, Columbus. (Paper No. 6).

"Natural History Chapbooks: Throwbacks to Pliny and the *Physiologus* as Well as Windows to 18th Century Biology." Presented 8 April 1972 to the History of Science Meeting, The Ohio Academy of Science, The Ohio Historical Center, Columbus. (Paper No. 9).

"Botany in English and American Chapbooks." Presented 30 August 1972 to the Historical Section, Botanical Society of America, University of Minnesota, Minneapolis-St. Paul. (Paper No. 7).

"How Biological Information Was Presented to the Young, 1750–1910: A Beginning of an Understanding of the Books Used." Presented 7 April 1973 to the Joint Atlantic Seminar in the History of Biology, Boston University Law School, Boston. (Paper No. 6).

"Botany by Rote: Early 19th Century Catechisms for Teaching about Plants." Presented 23 August 1977 to the Historical Section, Botanical Society of America, Michigan State University, East Lansing. (Paper No. 8).

"What Nineteenth Century Children Were Taught about Physiology." Presented 29 April 1978 to The Ohio Academy of Medical History, Howard Dittrick Museum of Historical Medicine, Cleveland.

"What Animal Is That?: The Animals in Nineteenth Century American Children's Books Compared with Those of the Twentieth Century." Presented 28 October 1978 to the Midwest Junto for History of Science, University of Wisconsin, Madison.

"A 150 Year History of the Botanist as a Career Model for American Children." Presented 14 August 1979 to the Historical Section, Botanical Society of America, Oklahoma State University, Stillwater. (Paper No. 11).

"Personifying Plants from Linnaeus to the Present." Presented 10 August 1982 to the Historical Section, Botanical Society of America, Pennsylvania State University, University Park. (Paper No. 10).

"A Brief History of Plant Personification." Presented 23 April 1983 to the Annual Meeting, The Ohio Academy of Science, Bowling Green State University, Bowling Green, Ohio. (Paper No. 10).

"How Children Became Introduced to the Biological World: One Approach to the History of Biology." Presented 1 May 1985 to the Department of Anatomy Seminar, The Ohio State University, Ohio.

"Books on Natural History for Children." Presented 23 September 1985 to the children's literature class, Marietta College, Marietta, Ohio.

Rudolph wrote more manuscripts about children's books than any other topic in this memorial volume (**Part II**). Seven of these manuscripts have been developed into papers printed here (**Paper Nos. 5–11**).

Introduction to Papers on Children's Books

How Rudolph[1] began his children's book collection, as discussed earlier, was the subject of an article[a] in the *Friends Line* newsletter. This account, written by Rudolph, was based on a talk he had given to the Friends of the Libraries at The Ohio State University. That article, with editorial modifications, is reprinted here. (**Paper No. 5**).

Rudolph's first research endeavor on children's books, "Botany in American Children's Books . . . before 1900,"[a] was read in Columbus on 10 February 1968. If this research had been published, it would have represented an important phase of his study on the popularization of botany. Among his manuscript materials are outlines of the paper and many handwritten quotations copied from books. In a comprehensive 4-page typewritten outline,[c] he (1) asked the question why study children's books, (2) listed several problems concerned with his research, (3) summarized attitudes towards children's education and books, and (4) listed examples of books, including some with appropriate quotations.

As reconstructed here the first paragraph for **Paper No. 6** is the abstract,[e] "How Biological Information Was Presented to the Young, 1750–1910: A Beginning of an Understanding of the Books Used." Rudolph prepared this abstract and a manuscript for a symposium, sponsored by the "Joint Atlantic Seminar in the History of Biology," and organized by Diana L. Hall of Boston University and Everett Mendelsohn of Harvard University, held 7 April 1973 at the Boston University Law School. The Joint Atlantic Seminar in the History of Biology represented an itinerant annual program of papers devoted to the history of biology hosted by different institutions along the Atlantic seaboard. The abstract appeared in the mimeographed meeting program, and was not otherwise made available. This symposium was the first one in which Rudolph was invited to participate on a topic in the history of biology.

The second paragraph of **Paper No. 6**, "Botany in American Children's Books before 1900," is taken from his abstract[a] that has the same title. He presented this study twice, first on 10 February 1968, at The Ohio Academy of Science History of Science meeting, and second, later that same year on 3 September, at the annual meeting of the Historical Section of the Botanical Society of America held in Columbus. The abstract[b] for the latter presentation contains two additional sentences, and accordingly, it was chosen for printing here. The paper concludes with Rudolph's four-page typewritten outline[c] edited for printing here. (**Paper No. 6**).

Abstracts for nine of Rudolph's 13 presentations on botany and related subjects in children's books are in existence. Five of these abstracts are for presentations given at the annual meetings of the Historical Section of the Botanical Society of America and were published by Rudolph (1968, 1972, 1977, 1979, 1982), whereas the remaining four are typewritten and mimeographed. Only two of his studies were published: one on chapbooks, "Botany in American and British Chapbooks before 1860," and one on catechisms, "Learning Botany by Rote, the Way of the Nineteenth Century Catechisms." They appeared in the *Plant Science Bulletin*, (1973, 1978), and are reprinted here. **(Paper No. 7 on Chapbooks; and Paper No. 8 on Catechisms)**.

Three studies were left as typewritten manuscripts. One of them, "Natural History in Nineteenth Century Chapbooks for Children,"[a] was submitted for publication, but was returned as inappropriate for the journal. This manuscript may have been developed from his presentation on 8 April 1972 at the Annual Meeting of the Ohio Academy of Medical History, listed in the program as "Natural History Chapbooks: Throwbacks to Pliny and the *Physiologus* as Well as Windows to 18th Century Biology."[b] The manuscript[a] has been modified and printed here. **(Paper No. 9)**.

Rudolph's study[a] on "Personifying Plants from Linnaeus to the Present" represents the second of his manuscripts relating to children's natural history books. This contribution is printed here. **(Paper No. 10)**.

The third manuscript of six pages, "The Botanist as a Career Model for American Children, a 150 Year History,"[a] is a study that demonstrates how biographical accounts of botanists and scientists in general can be used to influence children to appreciate and understand scientists, or even to become one later in life. This manuscript,[a] presented 14 August 1979 at the annual meeting of the Historical Section, Botanical Society of America in Stillwater, Oklahoma, is printed here. **(Paper No. 11)**.

Children's Scientific Books

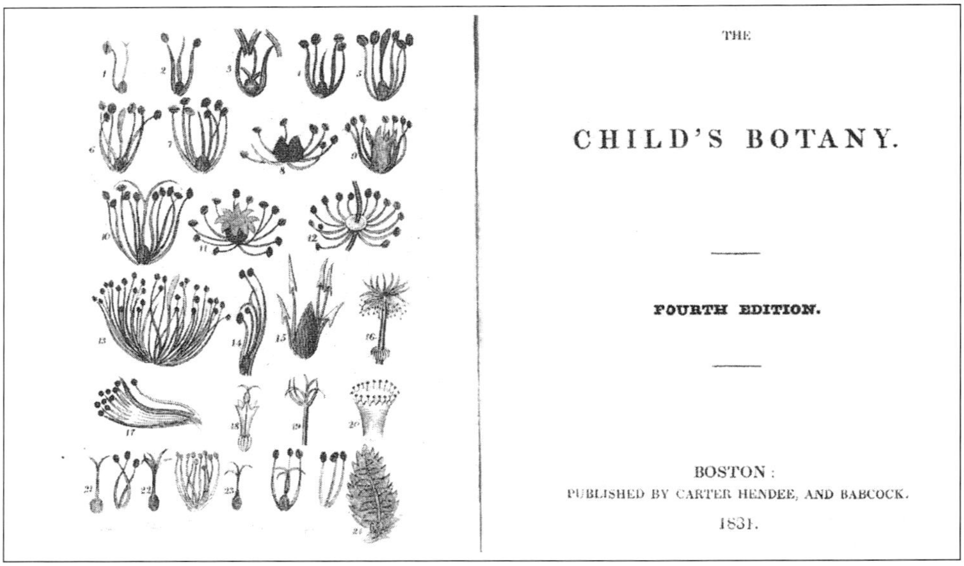

Frontispiece and title page from *The Child's Botany*, by Samuel G. Goodrich (1831)

Title page from *A Week Spent in a Glass Pond by the Great Water Beettel*, published by Emirk and Binger, London ([ca. 1880]) [Not Cited]

Paper No. 5 (1968)

Children's Scientific Books

I am fortunate to combine my joy of collecting books with my professional interest in the history of science by making a collection of natural history books for children. This pursuit began over twenty years ago, before 1959, with the gift of a colleague at Wellesley College, where I was then teaching. In cleaning house preparatory to retiring she came upon and presented to me a little book that had been passed on to her by an aunt. It is an unpretentious volume of 4-3/8" by 5-1/4" with an unlabelled calf spine and plain yellow paper over boards, entitled *The Child's Botany* (1831).[1] Although anonymous, the book bears on the verso of its title page the 1820 Massachusetts copyright of S[amuel] G[riswold] Goodrich (1793–1860), better known by his pen name of Peter Parley. He was the author of many popular children's books that appeared during the early part of the nineteenth century, and no doubt he also wrote this one.

The book's four plates and 103 pages of text introduce the young reader to the world of plant life. On every page parents are provided with footnoted questions meant to benefit children's inquisitive minds. Reading this charming little book, I was surprised to discover just how informative and interesting it was and how much it told me about the way science was distilled for youthful instruction. Having made this discovery, I was hooked on finding other children's books for comparison.

Over the years I have experienced much enjoyment in pursuing these volumes for children and learning about children's literature in general. My collection, now several thousand volumes, comprises many categories: chapbooks (cheap, small, paper-covered books of the 18th and 19th centuries), shape books, pop-up books and toy books, as well as more usual books of every size. All of them have in common the provision of instructional content. Animal fantasies and fairy tales are excluded; no Peter Rabbit, no Mickey Mouse. The collection encompasses a broad range of periods (although mostly 19th and 20th centuries), languages (although chiefly English), reading levels and original prices, because only thus can a complete picture emerge of the natural history literature available to all children.

I soon discovered some special aspects to collecting children's books. First, children do not take very good care of books, a fact that renders physical quality somewhat less important for children's book collectors than for collectors of other categories of books. Second, most children's books are illustrated and therefore attractive to

---

Frontispiece Credits: Children's Scientific Books. Two title pages and one frontispiece of children's books. (From Rudolph, *Friends Line. Newsletter of the Friends of the Libraries of The Ohio State University*, No. 16: 4–5. 1979; permission to reproduce illustrations courtesy of the Friends of the Libraries, OSU)

collectors of book illustrations. This factor added to a recent renewed interest in collecting children's books is making these books ever more difficult to find and to afford. All of these attributes are part of the fun of hunting for these unusual books.

In studying and using my collection, I have discovered how information about plants and animals has been taught to children over the years. The history of the popularization of science is a fascinating discipline to study. My own collection has helped me to better understand it and stimulated me to do further research in other children's book collections at the Free Library, Philadelphia; the Toronto Public Library; the Library of the American Antiquarian Society; the Library of Congress; and others. Book collecting provides me with excitement, adventure, intellectual stimulation, and pleasure. The excitement and adventure come from searching out books that will be integral to the collection, and the intellectual stimulation from studying books that open windows to readers' and writers' minds. The more books the more windows and the greater the illumination. Pleasure comes from seeing the collection grow and mature and from examining books, some beautiful and made with care, and each, whether beautiful or not, showing individuality and personality. My books are a never-ending source of pleasure. The illustrations are from books in my library (Frontisp. to Paper No. 5; Fig. 5.1).

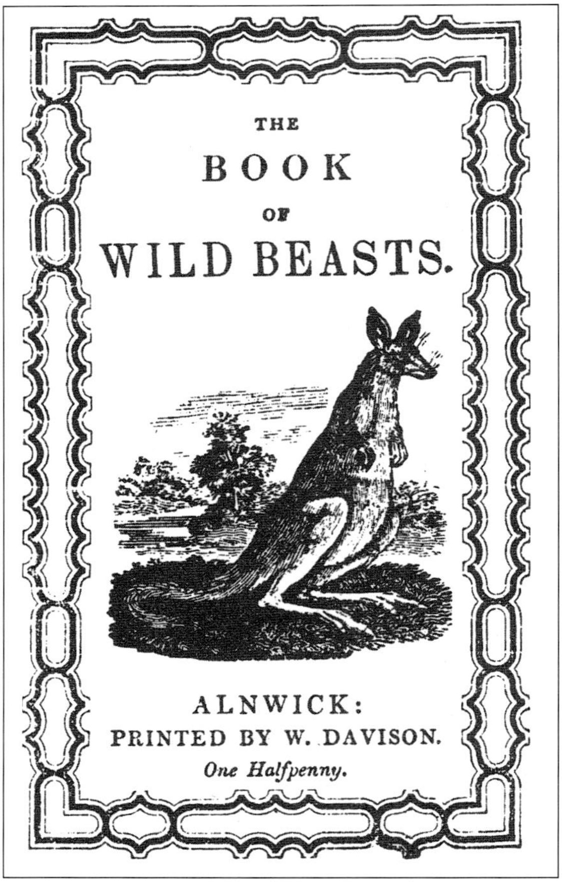

Fig. 5.1. Title page from *The Book of Wild Beasts*, by W. Davison, (ca. 1848). [Not Cited]

(From Rudolph, *Friends Line, Newsletter of the Friends of the Libraries* OSU No. 16: 4–5, 1979; permission to reproduce illustration courtesy of the Friends of the Libraries, OSU)

Reference Cited

1. Goodrich, S[amuel] G[riswold]. 1831. *The Child's Botany.* 4th ed. Carter, Hendee, and Babcock, Boston. 103 pp.

Publication Notes

Modified and edited from an article,[a] that Emanuel D. Rudolph was invited to write for the Friends of the Libraries of The Ohio State University. (Permission to reprint article courtesy of the Friends of the Libraries, OSU)

[a] 1979. Children's scientific books. *Friends Line. Newsletter of the Friends of the Libraries of The Ohio State University*. [Columbus]. No. 16: 4–5. February 1979. [Manuscript not located].

Appendix 5.1.
Recent Studies on Rudolph's Children's Books, by Stuckey

Since Rudolph's death in 1992, two students have conducted studies on his children's books. Undergraduate student Ethan Stock[a] has written a senior honors thesis on a topic using about 300 of Rudolph's children's books. He calculated from the Rudolph records "that from 1962 to 1980 he purchased on average four books per day, and from 1980 until his death an average of seven books per day." Stock further wrote:

> It is clear from such sources as advertising pages contained in the children's books that the Rudolph collection is an incomplete representation of the published children's scientific literature of Victorian England. However, the large number of titles in the collection (approximately 500 books from the period 1830–1890) and the manner in which the books were accumulated . . . make it likely that the collection is a fair and representative sample of the books which were publicly available in the nineteenth century.

John W. Frederick[b] wrote an evaluation of the pre-twentieth-century component of the 8,218 children's books in the collection. He determined that of 1,286 (16%) books in this component, 809 (63%) of them relate to natural history. The latter represents about 10% of his total children's book collection. The remaining pre-twentieth-century titles generally may be classified under "Readers" or "Conduct of Life."

[a] Ethan Stock. 1994. Themes of Order and Sources of Authority in Popularized Victorian Natural History 1844–1889. Senior Thesis. [Department of History, The Ohio State University, Columbus]. 62 pp.

[b] John W. Frederick. 1997. An Evaluation of Pre-twentieth-century Materials in the Rudolph Collection of Children's Science Books in The Ohio State University Libraries. M.L.S. Thesis, School of Library and Information Science, Kent State University, Kent, Ohio. 422 pp.

Botany in American Children's Books

Page [15] from *The First Book of Botany*, by Eliza A. Youmans (1881)

Frontispiece to *The American Forest*, by Uncle Philip (1840)

Paper No. 6 (1968, 1973)

Botany in American Children's Books before 1900

What is learned in youth influences the ways in which later observations are interpreted. With this in mind, and as a way of analyzing public understanding of biological materials and problems during the latter part of the eighteenth and nineteenth centuries, a study has been conducted of the literature prepared primarily for children and students. Literature varying from elusive chapbooks and gift books to the standard schoolbooks and textbooks written anonymously or by hack, popular, and scientific authors, are the major sources of information for analysis. During the one and a half centuries, the changes seem not to be so much in the general accuracy of the content in terms of the knowledge of the time, but rather in the differences in mode of presentation, scope of content, and philosophical position. Trends in terms of popularization and professionalization of science, nationalistic science, and scientific fashion clearly emerge from the study of these books. Some scientist-authors present unguarded opinions and philosophies in their works for a juvenile audience that cannot be located in their other writings.[d,e]

The manner and presentation, as well as the content, of children's books have been very much determined by the attitude toward the education of children and the format of books written for children. Books especially written for children were not common until the eighteenth century. In the United States the early works were mostly reprinted from or modeled on English or European books. Botany was early taught in the context of moral lessons; later it was presented in books in which the scientific information was modified for entertainment and educational purposes. In the second half of the 19th century some well known American botanists, for example Asa Gray (1810–1888), William J. Beal (1833–1924), and Liberty Hyde Bailey (1858–1954), wrote botanical works designed for children. Comparison is made of the types of botany included in books before and after the initiation of writings by botanists. It is proposed that in their works for children the botanists distilled their science and included only those aspects which they deemed most important. Thus children's botany books may prove to be a key to the better understanding of the history of botany.[a,b]

Frontispiece Credits: Botany in American Children's Books. Two pages of children's books. (From the Rudolph collection of children's books, Eliza A. Youmans. 1881. *The First Book of Botany*. D. Appleton & Co., New York; [Uncle Philip]. 1840. *The American Forest: Or, Uncle Philip's Conversations with the Children about the Trees of North America*. Harper and Brothers, New York; permission to reproduce these illustrations courtesy of Rare Books and Manuscripts Library, The Ohio State University)

I. Aspects of Studying Children's Botany Books[c]

A. Why Study Children's Books?

1. Youth become adults and may be scientists, and how they are educated is important.

2. What is important in science may be distilled in children's books.

3. The books show the adult's attitude toward children and their potentiality.

4. These books are interesting in themselves and amusing.

B. Problems:

1. Not very many botany books were written for children, particularly in the early period.

2. Few of these books are still left of those that were published.

3. Dating and authorship of these books frequently present problems.

C. Some Attitudes toward Children's Education and Books:

1. Few books especially for children were written before mid-eighteenth century.

2. Early books were primers, moral tales, or picture books.

3. John Newbery started publishing children's books in London 1744.

Circle of the Sciences. 1744. Printed for J. Newbery. 10 vols (Mathematics, Logic, Geography)

The History of Little Goody Two Shoes. 1766. Macmillan, New York City. xiv, 91 pp.

4. By early 19th century, the books were more of the amusement types.

Wm. Roscoe. 1809. *The Butterfly's Ball and the Grasshopper's Feast*. B. C. Buzby, Philadelphia. 16 pp.

Mrs. Reeve. 1809. *The Flowers at Court*. C. and R. Baldwin, London. 30 pp.

The Rose's Breakfast. 1808. J. Harris, London. 32 pp.

II. Examples of American Children's Books of Botany[c]

A. Many of the earliest of these books were European imports:

1. English:

a. Priscilla Wakefield. 1811. *An Introduction to Botany in Series of Familiar Letters with Illustrative Engravings*. 6th ed. Kimber and Conrad, Philadelphia. 179 pp.

b. [Sarah Waring]. 1829. *The Young Botanist, or a Sketch of the Life of Linnaeus Designed for Young Persons*. Boston, New-York. 126 pp.

c. [Jane Marcet]. 1830. *Conversations on Vegetable Physiology; Comprehending the Elements of Botany, with their Application to Agriculture*."Adapted to the use of Schools," by J. L. Blake. Crocher and Brewster, Boston. 372 pp.

2. German:
   a. J. F. Schreiber. [early 1800s]. *Kleine Naturgeschichte des Thier-, Pflanzen-,u. Mineralreichs für die Jugend*. Neue Auflage. Verlag von Schäffer und Koradi, Philadelphia. xvi, 204 pp.

3. French:
   a. [Arnaud] Berquin. 1841. *An Introduction to the Study of Nature, with Other Pieces*. Crissy and Markley, Philadelphia. 142 pp.

B. American works of the first half of the nineteenth century, with quotes from some of them:

1. [Anonymous]. 1826. *Winter Evenings' Conversations between a Father and His Children on the Works of God*. Revised. . . American Sunday School Union, Philadelphia. 200 pp.

"E[liza] - But what may be the reason why there are so many different colours in flowers, while the grass is universally green?
F[ather] - To this I might reply, that God gives to every plant the form and colour of the flower that pleases him; but chemists have said that all the different colours of flowers are produced by a very small portion of iron being diffused in the juice of the plant, that goes to the nourishing and perfection of the flower.
H[enry] - I do not understand that.
F[ather] - And I confess that it is somewhat strange. At any rate, we perceive that there is perfection of skill in giving to each its portion of colour to produce its precise effect." (Quote, p. 93).

2. [Samuel G. Goodrich] (Peter Parley). 1831. *The Child's Botany*. 4th ed. Carter, Hendee, and Babcock, Boston. 103 pp.

After discussing Linnaeus: "There are some eminent botanists in this country; and some of them have written books about the flowers which grow in America. One of the best of these, is that written by Dr. [Jacob] Bigelow, of Boston, describing the wild flowers of New-England. Two works, one by [Thomas] Nuttal [i.e. Nuttall] and one by [Amos] Eaton, are very good. Dr. [John] Torrey, of New-York is publishing a very large Botany of New-England and the Middle States, which will be very valuable. When you are old enough, you should study one of these books." (Quote, pp. 25–26).

[also wrote: *Peter Parley's Illustrations of the Vegetable Kingdom*. 1849].

3. Sarah Josepha Hale. 1832. *Flora's Interpreter; or, The American Book of Flowers and Sentiments*. Marsh, Caper and Lyon, Boston. 223 pp.

4. [Almira Hart Lincoln] Phelps. 1833. *Botany for Beginners: An Introduction to Mrs. Lincoln's Lectures on Botany.* 2nd ed. F. J. Huntington, Hartford, [Connecticut]. 212 pp. [At least 27 printings were issued, 1833–1891].

   "You have been so long engaged upon a certain set of studies, that I perceive they have become tiresome; I think of introducing a new study into school: to-morrow I shall give a lecture on Botany; you may bring with you all the wild lilies, (or all the violets, or any kind of common flower) that you can find in the fields." (Quote, pp. [v]–vi).

5 Joseph Breck. 1833. *The Young Florist; Or Conversation on the Culture of Flowers, and on Natural History*. Russell Odiorne and Co., Boston. 168 pp.

   M[argaret] - Here is the Coreopsis; was not this brought from the West?
   H[enry] - Yes; we are indebted to Mr. [Thomas] Nuttall for this, who found it on the prairies near the Rocky Mountains, . . ." (Quote, p. 113).

6. Comstock, J[ohn] L. 1835. *The Young Botanist: Being a Treatise on the Science, Prepared for the Use of Persons Just Commencing the Study of Plants*. Robinson, Pratt & Co., New-York. x, [9]–259 pp.

7. [Uncle Philip]. 1840. *The American Forest: Or, Uncle Philip's Conversations with the Children about the Trees of America. Boy's and Girl's Library*. No. XXII. Published by Harper & Brothers, New-York. 250 pp.

8. Anonymous. [ca. 1840]. *Book of Fruits*. Mahlon Day & Co., and Baker, Crane & Co., New-York. 30 pp.

C. American botanical works of the later half of the nineteenth century.
   1. Books that moralize or amuse.
      a. Mrs. Horace Mann. 1862. *The Flower People*. New ed. Lothrop, Lee and Shepard Co., Boston. 176 pp.

      b. Captain Mayne Reid. 1866. *The Plant Hunters, or Adventures among the Himalaya Mountains*. Ticknor and Fields, Boston. vi, 353 pp.

      c. Theodore Thinker." [Francis Channing Woodworth]. 1870. *First Lessons in Botany*. A. S. Barnes and Co., New York and Chicago. vi, [7]–141 pp.

      [Old fashioned, still follows Linnaean system; quotes Isaac Watts (1674–1748) early 18th century children's writer:

      "How doth the little busy bee
      Improve each shining hour.
      She gathers honey all the day
      From every opening flower"]. (Quote, p. 74).

2. Books that have a scientific approach; some by well known botanists:
   a. Asa Gray. 1858. *How Plants Grow, a Simple Introduction to Structural Botany, with a Popular Flora, or an Arrangement and Description of Common Plants, Both Wild and Cultivated. Botany for Young People.* Ivison and Phinney, New York; S. C. Griggs & Co., Chicago. 233 pp.

   b. Asa Gray. 1872. *How Plants Behave: How They Move, Climb, Employ Insects to Work for Them, &c. Botany for Young People.* Part II. American Book Co., New York, Cincinnati, Chicago. viii, 46 pp. (Fig.6.1).

Fig. 6.1. Book cover from *How Plants Behave*, by Asa Gray (1872).

(From the Rudolph collection of children's books; permission to reproduce this illustration courtesy of Rare Books and Manuscripts Library, The Ohio State University)

The familiar rhymes of Dr. Watts directed the attention of young people to the bee visiting the flower as a model of industry. With a slight change of a couplet, adapting it to our present knowledge and to the lesson of mutual helpfulness, we may read:

'How doth the little busy bee
Improve each shining hour,
While gathering honey day by day,
To fertilize each flower.'" (Quote, pp. 39–40).

c. Eliza A. Youmans. 1873. *The First Book of Botany. Designed to Cultivate the Observing Powers of Children.* New and enlarged ed., with 300 engravings. D. Appleton and Co., New York. xvii, 202 pp.

[First published in 1870 with 183 pp., this book was significantly revised (1871, 1883) and published with at least twelve printings, two in London, England].

d. Liberty Hyde Bailey. 1898. *First Lessons with Plants.* The Macmillan Co., New York, London. x, 117 pp.

e. W[illiam] J. Beal. 1898. *Seed Dispersal.* Ginn & Co., Publishers, Boston. vii, 90 pp. (Fig. 6.2).

"Fruits that sling their seeds are to be found in every neighborhood, and are first-class objects for the curious person to see and handle. Very fortunate is the girl or boy who is never fully satisfied with what he reads and sees pictured, but has a strong desire to learn how plants are made and how they behave." (Quote, pp. 59–60).

Publication Notes

Compiled and edited from abstracts,[a,b,e] manuscripts,[c,d] and notes[d] of papers read at meetings identified in the introduction to this paper and in the citations below.

[a] 1968. Botany in American children's books: the period before 1900, p. 3. *In* Abstracts of Papers Presented before The Ohio Academy of Science 1968 History of Science Meeting, February 10, 1968, Battelle Memorial Institute Auditorium, 505 King Avenue, Columbus. Mimeographed. [Abstract].

[b] 1968. Botany in American children's books before 1900. *American Journal of Botany* 55: 718. [Abstract].

[c] [1968]. "Botany in American Children's Books: The Period before 1900." An outline, typewritten, 4 pp. [Manuscript].

[d] [1973]. "How Biological Information Was Presented to the Young, 1750–1910: A Beginning of an Understanding of the Books Used." Typewritten and handwritten outline of notes, 6 pp. + numerous pages of quotations from books. Read 11 April 1973 at Boston, Massachusetts. [Manuscript].

[e] 1973. How biological information was presented to the young, 1750–1910. A beginning of an understanding of the books used, p. 2. *In* Joint Atlantic Seminar in the History of Biology, Saturday, April 7, 1973, Boston University Law School, Massachusetts. Mimeographed. [Abstract].

NATURE STUDY

THE JANE ANDREWS BOOKS.
By JANE ANDREWS.
The Seven Little Sisters. Revised Edition. With new full-page illustrations. For introduction, 50 cents.
Each and All. Revised Edition. With new full-page illustrations. For introduction, 50 cents.
Stories Mother Nature Told her Children. Revised Edition. With new full-page illustrations. For introduction, 50 cents.

STICKNEY'S STUDY AND STORY NATURE READERS.
By J. H. STICKNEY, author of the Stickney Readers.
Earth and Sky. A First Grade Nature Reader and Text-Book. For introduction, 30 cents.
Pets and Companions. For primary grades. For introduction, 30 cents.
Bird World. A Bird Book for Children. By J. H. STICKNEY, assisted by RALPH HOFFMANN. For introduction, 60 cents.

STRONG'S ALL THE YEAR ROUND.
By FRANCES L. STRONG of the Teachers' Training School, St. Paul, Minn.
Part I. Autumn. For introduction, 30 cents.
Part II. Winter. For introduction, 30 cents.
Part III. Spring. For introduction, 30 cents.

EDDY'S FRIENDS AND HELPERS.
Compiled by SARAH J. EDDY. For introduction, 60 cents.

LONG'S WAYS OF WOOD FOLK. (First Series.)
By WILLIAM J. LONG. For introduction, 50 cents.

MORLEY'S LITTLE WANDERERS.
By MARGARET WARNER MORLEY. For introduction, 30 cents.

WEED'S STORIES OF INSECT LIFE. (First Series.)
By CLARENCE M. WEED, Professor of Zoölogy and Entomology in the New Hampshire College of Agriculture and the Mechanic Arts. For introduction, 25 cents.

MURTFELDT AND WEED'S STORIES OF INSECT LIFE. (Second Series.)
By MARY E. MURTFELDT and CLARENCE MOORES WEED. For introduction, 30 cents.

WEED'S SEED-TRAVELLERS.
By CLARENCE M. WEED. For introduction, 25 cents.

BEAL'S SEED DISPERSAL.
By W. J. BEAL, Professor of Botany and Forestry in Michigan State Agricultural College. For introduction, 35 cents.

BURT'S LITTLE NATURE STUDIES FOR LITTLE PEOPLE.
From the Essays of JOHN BURROUGHS. Edited by MARY E. BURT.
Volume I. A Primer and a First Reader. For introduction, 25 cents.
Volume II. A Second Reader and a Third Reader. For introduction, 25 cents

BERGEN'S GLIMPSES AT THE PLANT WORLD.
By FANNY D. BERGEN. For introduction, 50 cents.

HALE'S LITTLE FLOWER PEOPLE.
By GERTRUDE ELISABETH HALE. For introduction, 40 cents.

GINN & COMPANY, Publishers,
Boston. New York. Chicago. Atlanta. Dallas. San Francisco. London.

Fig. 6.2. An advertisement for books on nature study.

(From W. J. Beal. 1900. *Seed Dispersal.* Ginn & Co., Boston; from a copy in the Herbarium Library, The Ohio State University)

American and British Chapbooks

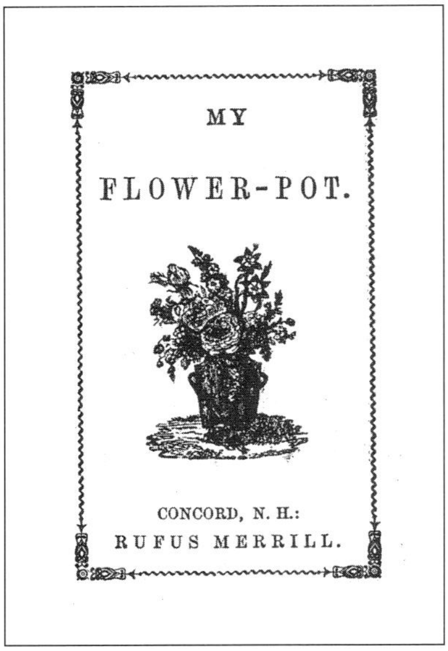

MY

FLOWER-POT.

CONCORD, N. H.:
RUFUS MERRILL.

Title page from *Grandmamma Easys Pretty Poetry about Trees, Fruit, and Flowers*, published by Brown, Bazin, and Co., Boston ([ca. 1860])

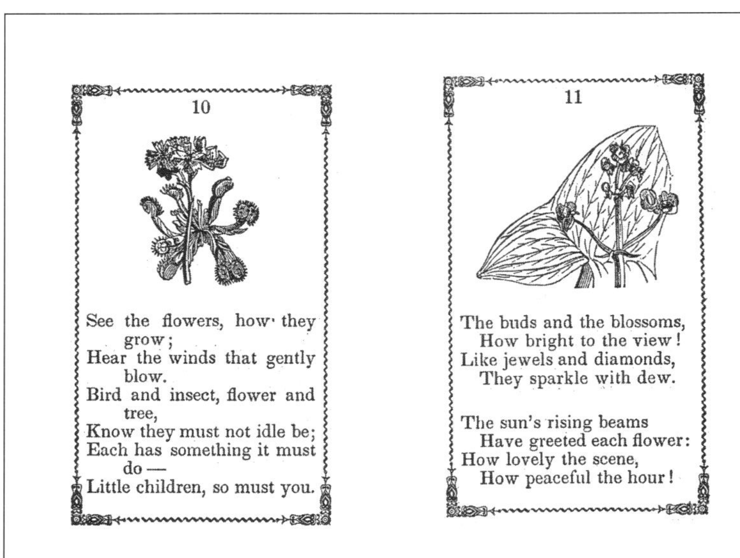

10

See the flowers, how they
grow;
Hear the winds that gently
blow.
Bird and insect, flower and
tree,
Know they must not idle be;
Each has something it must
do —
Little children, so must you.

11

The buds and the blossoms,
How bright to the view!
Like jewels and diamonds,
They sparkle with dew.

The sun's rising beams
Have greeted each flower:
How lovely the scene,
How peaceful the hour!

Title page (above left) and two text pages (below) from *My Flower-pot*, published by Rufus Merrill, Concord, New Hampshire ([ca. 1840])

Paper No. 7 (1972)

Botany in American and British Chapbooks before 1860

Very popular, cheaply produced, widely disseminated, anonymously authored, small books of the eighteenth and early nineteenth centuries termed "chapbooks" are a surprising place to find botanical information. One of the major reference works, *A Book about Chapbooks* (Weiss, 1942),[11] begins:

> Devils and angels, scoundrels and heroes, love and hate, murders, deathbed statements, witchcraft, riddles, tragedy, romance, song, jests, fairy tales, religion, hymns, speeches, executions, and all that goes to make up life, real and unreal, are reflected in the ephemeral chapbooks, that once circulated so freely and are now so scarce.

These books could be sold for a penny or two because they generally were made by folding a single sheet printed on both sides, with a paper cover or none, and were often copied and recopied by local printers. Chapbooks were sold by peddlers known as chapmen probably because they sold cheap or "chap" goods.

In addition to the adult fare referred to above, this book format was ideal for low-priced children's books before the period when many children's books were readily available. Less than five percent are devoted to natural history subjects for children judging from those I have seen or referred to in published works about chapbooks. Most of these are about animals in a format that has a small woodcut, often very crudely done, of an animal on the top of each page followed by a brief description. A very small number, estimated to be less than one percent, of the chapbooks consider plants. These works have become of interest to me in a study of the books that were used to initiate the young to biological information. They were a surprising place to find natural history writing.

As a group, chapbooks have a chatty and informal style that is generally lacking in the more formal books of the period. For example, in the *Book of Fruits* (ca. 1840);[2] with title page shown (Fig. 7.1), it says:

> Was there ever known a little boy or girl who did not jump [t]o take a fine Orange, Peach, or Apple, when offered? Or refuse a Lemon, Tamarind,

---

Frontispiece Credits: American and British Chapbooks. Title page of *Grandmamma Easys Pretty Poetry . . . .* (From Rudolph photographic collection; also in *Plant Science Bulletin* 19(3): 35. 1973; permission to reproduce illustration courtesy of the Botanical Society of America)

Title page and two text pages of *My Flower-pot*. (From the Rudolph collection of children's books; permission to reproduce these illustrations courtesy of Rare Books and Manuscripts Library, The Ohio State University)

> Pineapple, or a bunch of cherries? Very seldom, I dare say. Well, here is a short history of those fruits with some others that find there [sic] way to our numerous tempting fruit stands that are so profusely dressed off near the Fulton Market, or fine showy fruit stores in Broadway or Chatham-street. This little book will tell you many things about fruit that will be useful to you to know.

This work follows the format with cut and description for each fruit and has some unusual fruits described—mangostan, guava, and shaddock. Some chapbooks have a distinctively religious overtone, particularly those of later dates. For example, *The Fairy-craft of Nature* (ca. 1850),[4] written in dialogue form, contains the following conversation:

> Lucilla: 'Thank you, dear mamma, for your pretty tale.' Mother: 'Say rather, my history, for all the changes and all the misfortunes of Flax-seed have really happened as I have related them.' Lucilla: 'What! this little seed of flax that I had in my hand just now, may become first a pretty plant, then fine thread, cambric, lint, and even paper at last?' Mother: 'Yes, my love: at the command of the Great Being who creates all these wonders, the little seed bursts forth from the ground with its tiny green wings, which quickly expand into stems and leaves; these stems are soon adorned with pretty blue flowers; the flowers pass away . . . .

Another religious approach is in *My Flower-pot* (ca. 1840):[3]

> Say, Ma! did God make all the flowers/ That richly bloom to-day?/ And is it he that sends sweet showers/ To make them look so gay?/ Did he make all the mountains/ That rear their heads so high?/ And all the little fountains/ That glide so gently by?/ And does he care for children small?/ Say, Ma! does God love me?/ Has he the guardian care of all/ The various things we see?/ Yes! Yes! my child, he made them all,—Flowers, mountains, plants, and tree;/ No man so great, no child so small,/ That from his eye can flee!

In the category of strictly instructional chapbooks three common types appear: ABCs, object lessons, and catechisms. The ABC type uses the alphabet to instruct very young children in the letters and in words using them. *Grandmamma Easys Pretty Poetry About Trees, Fruit and Flowers* (ca. 1860)[5] uses plants from A to Z:

> The A stands first in Apple Tree,/ An Apricot so sweet;/ In Acorn, too, the A you'll see./ And Almonds, nice to eat./...The Z is seen in Hazel-tree/ On which nice nuts do grow;/ In Furze, also, the Z you'll see,/ And Zephyr winds that blow.

In *Set of Flowers, Alphabetically Arranged for Little Children*, by S. S. Wood, (1819)[12] each page has the name of the plant, a capital and small letter, with a picture, and very brief text from A—Aloe to W—Wall-Flower, with X, Y, Z on one page having no plants or pictures.

The object lesson books attempt to instruct by telling about familiar objects in common use. The *First Steps to Knowledge: Common Things* (1870)[9] contain the following questions and answers:

> What is cinnamon? A. The dried under bark of a tree. Q. Where are cinnamon trees found? A. In the island of Ceylon.

Quite a number of plant materials and products are considered in this work including fruits, vegetables, spices, and fibers. *Rural Scenes or a Peep into the Country for Children* (1818)[10] tells children that:

> Apples grow spontaneously in many woods of America, and are there so sweet, as to need no sugar, either in pies or puddings [and that] Walnut-tree wood was once in great request, and much used in clock-cases, looking-glass frames, drawers, and other household furniture; but now mahogany has driven the walnut-tree almost out of doors, as well as the solid oak.

These object lessons tell us not only about how plants were used but also about what people at that time had as their concerns.

The catechism is a format that required the memorization of information in set answers to questions. In *A Botanical Catechism: Containing Introductory Lessons for Students in Botany* (1819),[1] the book is addressed, "... to teachers as learners, preparatory to the use of 'Eaton's Manual of Botany.'"[6] It teaches the Linnaean System of classification as did Eaton. William Mavor's *Catechism of Botany* (1820)[8] gives the following questions and answers:

> Q. Is not the knowledge of plants very useful? A. Without some knowledge of plants, we should not be able to distinguish between such as are noxious and such as are good for food, or for the various purposes of life. . . . Q. How is the knowledge of plants to be acquired? A. By the study of botany, a science which formerly implied an acquaintance with the nature, uses, and cultivation of vegetables; but which, in its modern acceptation, is applied, in the first instance, at least, to the classification of plants; or that systematic arrangement by which, from general marks or characters, the botanist is enabled to refer them to their proper place in the system. Without this, all would be confusion.

The catechisms teach the definitions of plant parts and the classification of plants. C. Irving's *A Catechism of Botany* (1829)[7] also says:

> Q. What advantages are derived from the study of botany? A. The Study of this science as an amusement, is both healthful and pleasing, as it attracts us often into the country, and makes us acquainted with the wonderful works of nature: it also enlarges the mind, by implanting new and useful ideas, and fills it with the most exalted admiration of the great Creator of the Universe.

Who then would not want to study botany?

Chapbooks provide windows to understand the botanical information that was generally available to large numbers of adults and children during the eighteenth and early nineteenth centuries. They also provide materials for studying the differences between the botany known by the educated and uneducated at that time and how it was presented. In the latter part of the nineteenth century these little works were replaced by more elaborate books that then became available at modest cost. It is now difficult to find these chapbooks which were either worn out or thrown away. The search for them and of their contents is well worth the effort if one is interested in the history of ideas.

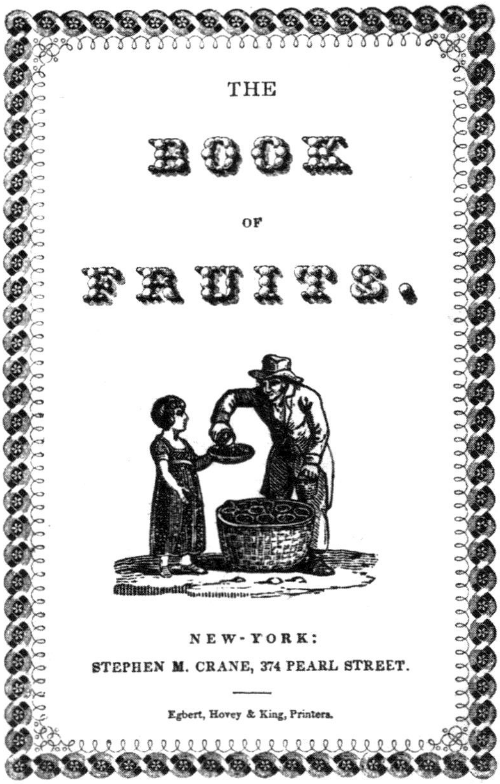

Fig. 7.1. Title page from *The Book of Fruits*, published by Stephen M. Crane, New York. [ca. 1840].

(In the Rudolph photograph collection; also in Rudolph, *Plant Science Bulletin* 19(3): 34. 1973; permission to reproduce illustration courtesy of the Botanical Society of America)

References Cited

1. A Lady. [Pseudonym for Jane Kilby Welsh]. 1819. *A Botanical Catechism: Containing Introductory Lessons for Students in Botany*. T. W. Shepard and Co., Northampton, [Massachusetts]. iv, [1], 6–34 pp. (Quote, p. [iii]).

2. [Anonymous]. [ca. 1840]. *Book of Fruits*. Mahlon Day & Co., and Baker, Crane & Co., New-York. 30 pp. (Quote, p. [4]).

3. [_____]. [ca. 1840]. *My Flower-pot*. Rufus Merrill, Concord, N[ew] H[ampshire]. 16 pp. (Quote, pp. 15–16).

4. [_____]. [ca. 1850]. *The Fairy-craft of Nature. Stories for Summer Days & Winter Nights, Second Series*. Groombridge and Sons, London. 46 pp. (Quote, p. 19).

5. [_____]. [ca. 1860]. *Grandmamma Easys Pretty Poetry about Trees, Fruit and Flowers*. Brown, Bazin and Co., Boston, Massachusetts. [8] pp. (Quotes, pp. [2], [8]).

6. Eaton, Amos. 1817. *A Manual of Botany for the Northern States, . . .* By the Members of the Botanical Class in Williams' College (Mass.), from a Manuscript System Compiled by the Author of Richard's *Botanical Dictionary*. Websters and Skinners, Albany, New York. vi, 164 pp.

7. Irving, C. 1829. *A Catechism of Botany; Containing a Description of the Most Familiar and Interesting Plants Arranged According to the Linnaean System. With an Appendix on the Formation of an Herbarium.* Adapted to the Use of Schools in the United States. 3rd American ed., improved and enlarged. Collins and Hannay, New-York. 82 pp. (Quote, p. [5]).

8. Mavor, William. 1820. *Catechism of Botany; or, An Easy Introduction to the Vegetable Kingdom: For the Use of Schools and Families*. Samuel Wood & Sons, New York; Samuel S. Wood & Co., Baltimore. iv, [5]–70 pp. (Quotes, pp. 7, 8).

9. Pinnock, George, Rev. 1870. *First Steps to Knowledge: Common Things. Part I.* T. J. Allman, London. 62 pp. [Original printing, 1854]. (Quote, p. 19).

10. [Taylor, Ann]. 1818. *Rural Scenes, or, A Peep into the Country for Children*. Printed for Darton, Harvey & Darton, London. 60 pp. (Quotes, pp. 18, 31).

11. Weiss, H. B. 1942. *A Book about Chapbooks: The People's Literature of Bygone Times*. [The Author], Trenton, N.J. v, 149 pp. (Quote, p. [1]).

12. Wood, S. S. 1819. *Set of Flowers, Alphabetically Arranged for Little Children*. Samuel Wood & Sons, New York; Samuel S. Wood & Co., Baltimore. 24 pp.

Publication Notes

Modified and edited from a paper published 1973 in the *Plant Science Bulletin*. The paper was read 30 August 1972 at the meeting of the Historical Section, Botanical Society of America, The University of Minnesota, Minneapolis-St. Paul. (Permission to reprint article courtesy of the Botanical Society of America)

1972. Botany in English and American chapbooks. *American Journal of Botany* 59: 674. [Abstract].

1973. Botany in American and British chapbooks before1860. *Plant Science Bulletin* 19: 34–36. [Reprinted in *Morris Arboretum Bulletin* 25: 60–62, 1974, with the title: Botany in British and American chapbooks before 1860]. [Manuscript not located].

Nineteenth Century Catechisms

Catechism at heart of early botany

Question: Why are the plants all of a green hue? Answer: The earth is full of God's goodness, and green is more pleasing to the eye than any other color.

Such was the information which the early 19th century botany student accepted and memorized, word for word, from his catechism on botany, says Emanuel D. Rudolph of the botany department.

Rudolph recently reported on early teaching methods in a paper before the historical section of the Botanical Society of American meeting at Michigan State University.

"This format of question and answer, based on the concept of purposefulness in nature, which the student learned without questioning validity, was an important pedagogic tool," said Rudolph.

"It influenced the thinking of several generations of botanists from the time J.F. Martinet introduced 'The Catechism of Nature for the Use of Children' in the last quarter of the 18th century."

Reprinted many times

Rudolph said Martinet's book, translated from the Dutch, was very popular and was reprinted many times well into the 19th century. It contained several chapters on plants oriented toward design in nature and bearing the influence of the early Swedish botanist, Carolus Linnaeus. Also popular during the period were books of conversations for teaching children about botany in the home.

Use of such catechisms as school books faded when textbooks more closely resembling today's began appearing in the 1830s, said Rudolph.

"The catechisms were not very scientific and did not develop a questioning attitude toward the teaching of botany until such authors as Amos Eaton of Rensselaer Polytechnic Institute broke with tradition by advocating learning by observation early in the last century."

Rudolph said it took a full generation of teachers to introduce the newer methods.

CATECHISM

OF

BOTANY;

OR,

AN EASY INTRODUCTION TO

THE

Vegetable Kingdom:

FOR

THE USE OF SCHOOLS AND FAMILIES.

BY WILLIAM MAVOR, L. L. D.

AUTHOR OF TRAVELS, AND MANY OTHER POPULAR WORKS.

New-York:

PUBLISHED BY SAMUEL WOOD & SONS,

NO. 261, PEARL-STREET;

And Samuel S. Wood & Co. No. 212, Market-street; Baltimore.

1820.

Their size was small

The catechisms were typically of modest size — approximately 4 by 5 inches — and contained sparsely illustrated texts of about 64 pages, he said.

"Today they are extremely rare because few have survived, so we don't know much about their numbers, although they were probably abundant during the period."

Rudolph said "it is interesting to look at the education of scientists who made significant contributions to botany during the last century and see how they overcame the tradition of the catechisms."

Probably most notable, he said, was Charles Darwin, the English naturalist and author of the monumental "Origin of the Species and the Ascent of Man," inspired by his voyages on the "Beagle."

"Darwin, in his autobiography, says he could accomplish what he did by looking at things for himself rather than accepting the established knowledge of his time," Rudolph said.

"He was a rather poor student who later attended Cambridge University where he learned from Henslow and Lyell, accompanying them in the field and making his own observations."

Portion of newspaper article about Rudolph's paper on *Nineteenth Century Catechisms.* Title page from *Catechism of Botany,* by William Mavor (1820)

Paper No. 8 (1977)

Botany by Rote, the Way of the Nineteenth Century Catechisms

How to provide the young with knowledge has been a problem for adults ever since the accumulation and transmission of information. One method that is old, yet is not so different from some computer instruction now thought very modern, is rote memorization. Such instruction was popular in the latter part of the eighteenth and early nineteenth centuries. A formulation of it was provided by little books called, for good reasons, "catechisms" of this and that (Fig. 8.1). They covered a wide range of subjects. (Appendix 8.1).

The intent of these works is made clear in the introduction to *Catechism of Botany* by the English educator William Mavor (1820):[4]

> As names and distinctions must be acquired, before any further progress can be made in the study, and as the tenacious memory of youth is peculiarly fitted for receiving and retaining them, a work of this kind cannot too early be put in their hands. It will be to Botany, what a grammar is to language—the first step, though not the ultimate end. In the last chapter, however, an attempt has been made to enlist curiosity in the service, by showing that Botany is not unproductive of advantage; and that it is not merely a science of names, as the ignorant and idle have wished to represent it.

Another catechism by C. Irving (1829)[2] says this about studying botany:

> Q.: What advantages are derived from the study of botany? A.: The study of this science as an amusement is both healthful and pleasing, as it attracts us often into the country, and makes us acquainted with the wonderful works of nature: it also enlarges the mind, by implanting new and useful ideas, and fills it with the most exalted admiration of the great Creator of the universe.

The pre-Darwinian, generally held belief of God's design in nature is very strong in these catechisms. Dr. Martinet (1790)[3] has his pupil ask, "I think you told me once that there are not two leaves alike, even on one tree?" and the tutor answers:

Frontispiece Credits: Nineteenth Century Catechisms. Portion of newspaper article about Rudolph's paper on "Nineteenth Century Catechisms." (From *OSU on Campus*, 25 August 1977; permission to reproduce article courtesy of *OSU on Campus*)

Title page of *Catechism of Botany*. (In the Rudolph photograph collection; also in *Plant Science Bulletin* 19(3): 35. 1973; permission to reproduce illustration courtesy of the Botanical Society of America)

> True, two things exactly alike, are not to be found in the whole world. So various are the works of God. The uppermost surface of the leaves is smooth or varnished, that the water may not lodge upon them, which would soon make them decay. The undermost is without that varnish, that it may imbibe the dew.

Even disagreeable flower-smells, the same pupil is told, have a purpose. They may possibly correct or absorb the noxious vapours. They may serve for food, to some animals and insects, or they may have some medicinal virtues."[3]

The prime emphasis of these volumes is the learning of parts of plants and plant classification. The reader is told by Mavor (1820)[4] that "modern" botany

> . . . is applied, in the first instance at least, to the classification of plants; or that systematic arrangement by which, from general marks or characters, the botanist is enabled to refer them to their proper place in the system. Without this, all would be confusion . . . a variety of systems have been invented and rejected; but that of Linnaeus, as the simplest and most certain, and which is founded on the parts of fructification, which will hereafter be explained, has gained, as it deserves general adoption, under the name of the sexual system.

Some of the information given about plant functions seems quaint or erroneous to us. For example, Dr. Martinet (1790)[3] has this to say about fruit production (Fig. 8.1):

> In many plants, the fruit proceeds from the same part where you see the blossom has been. But in the small nut, and the chestnut trees—in the Turkish and Indian corn, the fruit comes where the blossom has never been; otherwise almost all fruits are preceded by blossom. But the fig comes to perfection without a flower. And in melons, cucumbers, and some others, the fruit is seen before the flower.

In Irving (1829),[2] children learned that

> . . . the green colour of vegetables is supposed to arise from the oil which they contain, acted upon by the rays of the sun: it is also owing to their exposure to light, as may be seen by the common practice of blanching celery by covering it from the light; and by plants raised in darkness, which are of sickly white.

Plants are green because, according to Martinet (1790),[3] "Green is more pleasing and beneficial to the eye, than any other colour whatever." Pinnock (1831)[5] gives this answer to the question of the uses of leaves:

> Their particular uses in contributing to the growth of the plant are not very well known; but they absorb, and probably also give out moisture, and they

> form carbonic acid, by giving out a portion of charcoal, which unites with the oxygen of the air.

Children would learn as well about the uses of plants and some exciting facts about them. For example, in Irving (1829),[2] "the inhabitants of Quito prepare from [rubber] a kind of cloth, which they apply to the same purpose as our oil cloth." Or, that the venus fly trap plant has, according to Pinnock (1831):[5]

> Several leaves, each of which [is] divided into two lobes at the extremity, having long teeth on the margin like the antennae of insects, and armed within with six spines: these lie spread upon the ground round the stem, and are so irritable, that when a fly happens to light upon a leaf, it immediately folds up and crushes it to death.

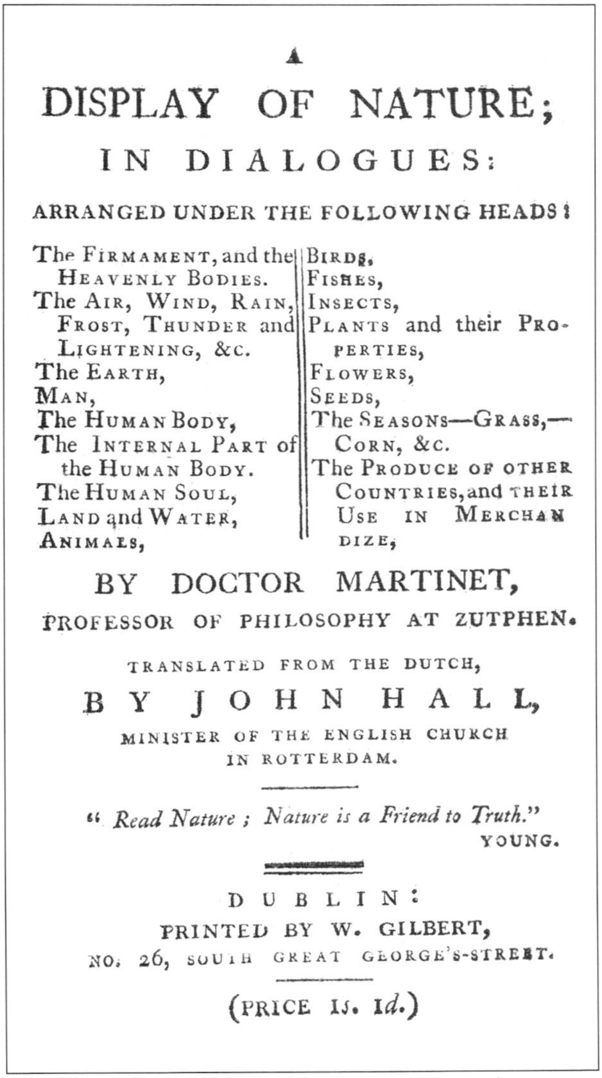

A

DISPLAY OF NATURE;

IN DIALOGUES:

ARRANGED UNDER THE FOLLOWING HEADS:

The Firmament, and the Heavenly Bodies.
The Air, Wind, Rain, Frost, Thunder and Lightening, &c.
The Earth,
Man,
The Human Body,
The Internal Part of the Human Body.
The Human Soul,
Land and Water,
Animals,
Birds,
Fishes,
Insects,
Plants and their Properties,
Flowers,
Seeds,
The Seasons—Grass,—Corn, &c.
The Produce of other Countries, and their Use in Merchandize,

BY DOCTOR MARTINET,

PROFESSOR OF PHILOSOPHY AT ZUTPHEN.

TRANSLATED FROM THE DUTCH,

BY JOHN HALL,

MINISTER OF THE ENGLISH CHURCH IN ROTTERDAM.

" *Read Nature ; Nature is a Friend to Truth.*"
YOUNG.

DUBLIN:

PRINTED BY W. GILBERT,

NO. 26, SOUTH GREAT GEORGE'S-STREET.

(PRICE 1s. 1d.)

Fig. 8.1. Title page from a catechism book, *A Display of Nature*;. . . by Doctor Martinet (1790).

(From the Rudolph collection of children's books; permission to reproduce this illustration courtesy of Rare Books and Manuscripts Library, The Ohio State University)

Historians of science believe that a set of beliefs can be termed "normal science," which is the framework accepted by investigators at any particular time. It is this normal science that is taught to and believed by the young. Only by changing these concepts can any basic changes be made in our scientific theories.

The world of the catechisms that has been described is a world very different from ours. It is a world in which big trees have small seeds, as stated by Martinet (1790),[3] and who also noted, "If they were larger they might injure us by the fall." It is a world before Darwin, who in his youth might well have learned his normal science from these catechisms or books much like them. He, and others like him, changed their normal science by overcoming their learning. Darwin said it well in his autobiography, as published by Barlow (1969):[1]

> As far as I can judge, I am not apt to follow blindly the lead of other men. I have steadily endeavoured to keep my mind free so as to give up any hypothesis, however much beloved (and I cannot resist forming one on every subject), as soon as facts are shown to be opposed to it.

References Cited

1. Barlow, Nora (ed.). 1969. *The Autobiography of Charles Darwin 1809–1882. With original omissions restored.* W. W. Norton and Co., New York. 253 pp. (Quote, p. 141).

2. Irving, C. 1829. *A Catechism of Botany; Containing a Description of the Most Familiar and Interesting Plants, Arranged According to the Linnaean System.* With an Appendix on the Formation of an Herbarium. Adapted to the Use of Schools in the United States. 3rd American ed., improved and enlarged. Collins and Hannay, New-York. 82 pp. (Quotes, pp. [5], 12–13, 71).

3. Martinet, Dr. [Johannes Florentius]. 1790. *A Display of Nature; in Dialogues.* Translated from the Dutch by John Hall. W. Gilbert, Dublin. 69 pp. (Quotes, pp. 52–53, 54, 57, 58).

4. Mavor, William. 1820. *Catechism of Botany; or, An Easy Introduction to the Vegetable Kingdom: For the Use of Schools and Families.* Samuel Wood & Sons, New York; Samuel S. Wood & Co., Baltimore. iv, [5]–70 pp. (Quotes, pp. [iii]–iv, 8).

5. [Pinnock, William]. 1831. *Pinnock's Catechism of the Elements of Botany.* A new & improved ed. Whittaker, Treacher & Co., London. 72 pp. (Quotes, pp. 20, 54).

Additional References Not Cited

Irving, C. [ca. 1849]. *A Catechism of Botany: Containing a Description of Some of the Most Familiar and Interesting Plants, Arranged According to the Linnaean System.* With an Appendix on the Formation of an Herbarium. A new ed. corrected and improved. Souter and Law, London. 72 pp.

Kerney, M. J. (ed.). 1853. *Irving's Catechism of Botany: Containing a Description of the Most Familiar and Interesting Plants, Arranged According to the Linnaean System.* With an Appendix on the Formation of an Herbarium. Adapted to the Use of Schools in the United States. 4th American ed., revised and improved by M. J. Kerney. J. Murphy and Co., Baltimore. 94 pp.

Martinet, Dr. [Johannes Florentius]. 1795. *The Catechism of Nature: For the Use of Children.* Translated from the Dutch, by J. Hall. David West, Boston. 99 pp.

Publication Notes

Modified and edited from a paper[a] published 1978 in the *Plant Science Bulletin.* The paper was read 23 August 1977 at the meeting of the Historical Section, Botanical Society of America, Michigan State University, East Lansing. (Permission to reprint article courtesy of the Botanical Society of America)

1977. "Learning Botany by Rote, the Way of the Nineteenth Century Catechisms." Typewritten, 6 pp.; handwritten, 20 pp. [Manuscript].

1977. Botany by rote; Early 19th century catechisms for teaching about plants. *Botanical Society of America, Miscellaneous Series, Publication* 154: 31. [Abstract].

[a] 1978. Learning botany by rote, the way of the nineteenth century catechisms. *Plant Science Bulletin* 24: 39–40.

Appendix 8.1.
List of Subjects in Catechisms,[a] by Rudolph

Agriculture, algebra, animated nature, anatomy, ancient geography, ancient history, architecture, arithmetic, astronomy, Biblical history, botany, British biography, chemistry, chronology, classical biography, conchology, drawing, duty to parents, electricity, English grammar, English history, English law, entomology, evidences of Christianity, first catechism, French grammar, Grecian antiquities, Greek grammar, health, Hebrew grammar, heraldry, history of England, history of Scotland, history of Ireland, history of France, history of Rome, history of Greece, history of America, history of the Jews, horticulture, hydronemics, ichthyology, Italian grammar, Jewish antiquities, land surveying, Latin grammar, logic, mechanics, medicine, mental philosophy, mineralogy, natural history, natural philosophy, natural theology, navigation, optics, ornithology, painting, perspective, poetry, religion, religious demonations, rhetoric, Roman antiquities, sacred geography, Spanish grammar, trades and commerce, universal history, use of the globes, and zoology.

Natural History Chapbooks

ANIMAL BIOGRAPHY:

OR

BOOK OF NATURAL HISTORY;

DESIGNED FOR THE

AMUSEMENT AND INSTRUCTION

OF

LITTLE FOLKS.

ILLUSTRATED BY TWENTY-ONE BEAUTIFUL WOOD ENGRAVINGS.

New-Haven.

PRINTED AND PUBLISHED BY S. BABCOCK.

1841.

Title page from *Animal Biography*, published by S. Babcock (1841)

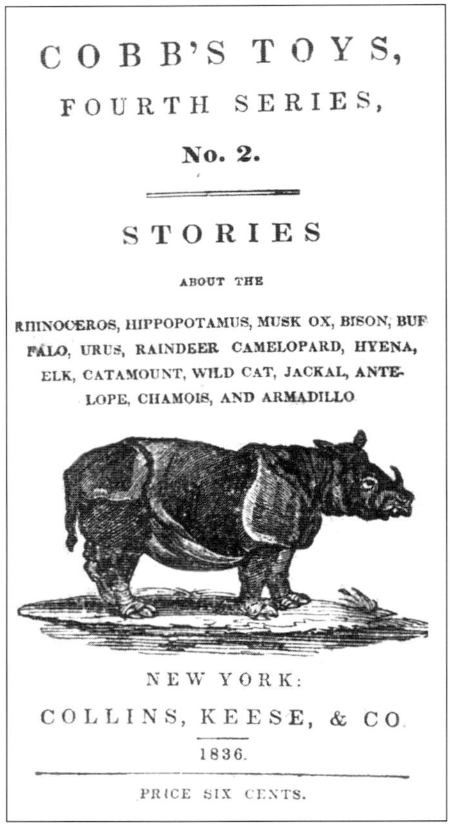
COBB'S TOYS,

FOURTH SERIES,

No. 2.

STORIES

ABOUT THE

RHINOCEROS, HIPPOPOTAMUS, MUSK OX, BISON, BUFFALO, URUS, RAINDEER CAMELOPARD, HYENA, ELK, CATAMOUNT, WILD CAT, JACKAL, ANTELOPE, CHAMOIS, AND ARMADILLO

NEW YORK:

COLLINS, KEESE, & CO.

1836.

PRICE SIX CENTS.

Title page from *Cobb's Toys, Fourth Series*, No. 2, by Lyman Cobb ([ca. 1836])

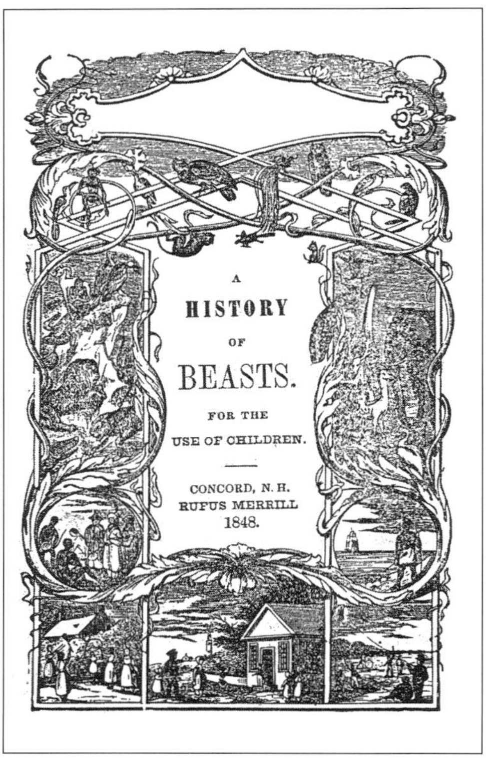
A

HISTORY

OF

BEASTS.

FOR THE

USE OF CHILDREN.

CONCORD, N. H.

RUFUS MERRILL

1848.

Title page from *A History of Beasts*, by Rufus Merrill, Concord, New Hampshire (1848) [Not Cited]

Paper No. 9 (1972)

Natural History in Nineteenth Century Chapbooks for Children

Chapbooks treating natural history subjects are rare and generally late in appearing. Weiss (1928),[23] in the only analysis of a natural history chapbook, notes that only two titles out of over 2,300 in the Harvard University Library catalogue of chapbooks[15] are ones on natural history. Most of the titles appeared in the first half of the nineteenth century. The majority of these are devoted to providing brief descriptions of animals which follow cuts illustrating them (Figs. 9.1, 9.2). Some have titles such as *A History of Beasts, A History of Birds*, or *A Book of Fishes*. Only a small fraction of these books considers plants. The readers learn in one[6] of them that:

> The questions which children, and even very little children too, put to those about them, by way of gathering knowledge of what they see or hear fully prove that the youthful mind is open, at a very early age, to receive instruction. Parents should avail themselves of this desire in their children, in order to bring them to be acquainted, by little and little, with everything that is proper for them to know.

Later in this work and also in Cobb (1836),[13] the reader is instructed that the works of nature are God's works and magnify his glory.

> Natural History is of great importance to children, particularly, as it is always the study of truth; and, besides nothing more interests, exalts, and improves their minds, especially when it is illustrated by appropriate pictures or cuts.

The pictures are of varying quality, but usually provide identifiable organisms. In a few cases the figures are wrong; for example, the sea-horse is illustrated by a nondescript quadruped in *Tommy Trip's History of Beasts*.[2] The most popular animals for inclusion are the elephant, tiger, and lion, which is surprising, as they are all exotic types. This interest in the exotic extends as well to the descriptive plant books, where tamarinds, mango-stems, and pomegranates are included in one,[8] and apple of Sodom, dragon's blood tree, Frankincense, and mandrake in another.[7] Knowledge for its own sake is good, we are told:[8]

> If there should be any part of it, or any words that you do not understand, ask your parents, or some kind friend to tell you. In this way, you will get

---

Frontispiece Credit: Natural History Chapbooks. Three title pages of children's books. (In the Rudolph photograph collection)

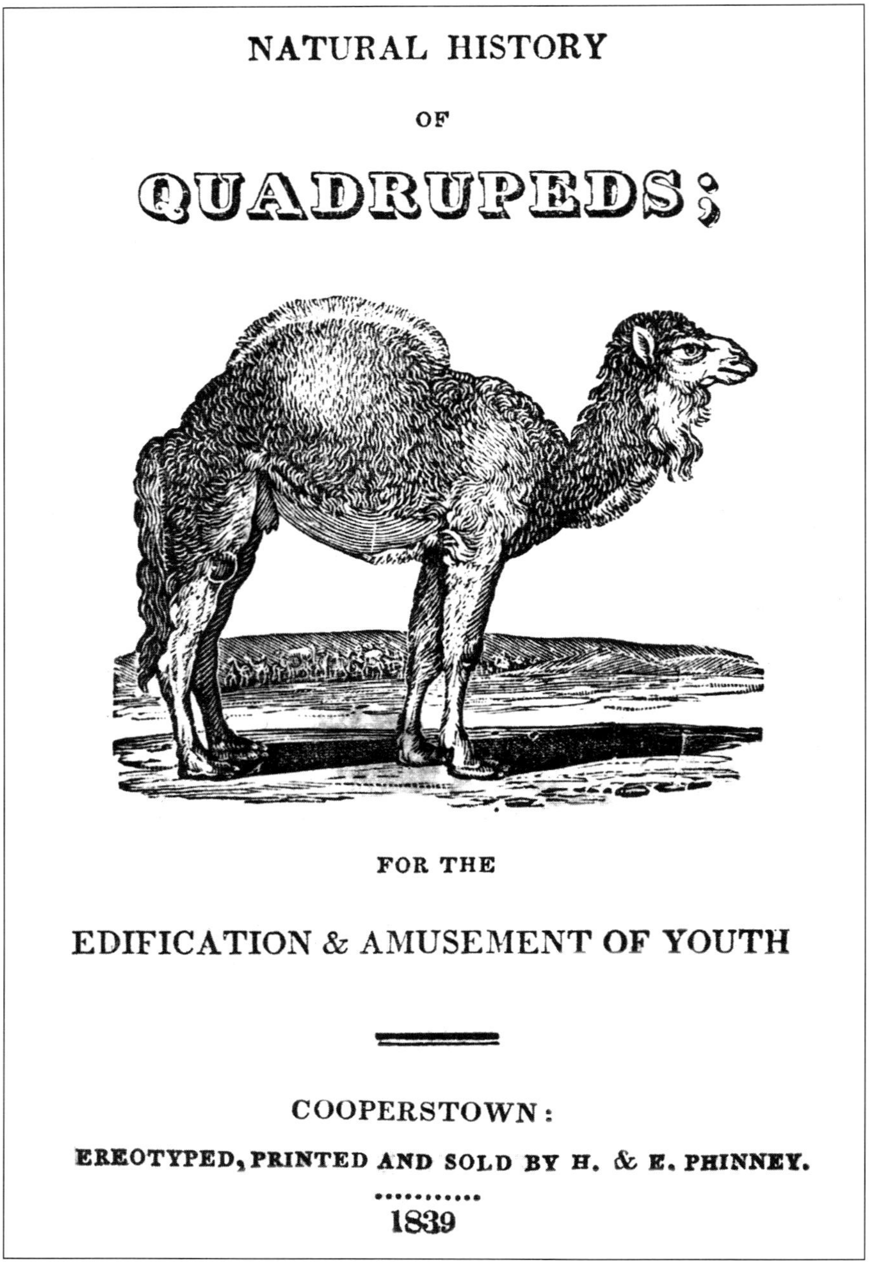
NATURAL HISTORY

OF

QUADRUPEDS;

FOR THE

EDIFICATION & AMUSEMENT OF YOUTH

COOPERSTOWN:

EREOTYPED, PRINTED AND SOLD BY H. & E. PHINNEY.

1839

Fig. 9.1. Title page from *Natural History of Quadrupeds*, printed and sold by H. & E. Phinney (1839). [Not Cited]

(In the Rudolph photograph collection)

> some valuable knowledge and be very likely to grow up useful men and women, a comfort to your parents and friends, and honored and respected by all who may know you.

And nature is a fine industrious example:[9]

> See the flowers, how they grow;
> Hear the winds that gently blow.
> Bird and insect, flower and tree,
> Know they must not idle be;
> Each has something it must do-
> Little children, so must you.

Thus, a strong moralistic tone comes from these books along with general descriptive natural history.

In addition to giving insight into the social attitudes of the time, natural history chapbooks provide windows for us to understand the biological knowledge taught to children in the early nineteenth century. If understanding "normal science" in a period is important to appreciate scientific change, according to Kuhn,[17] then these little books should give us clues to the essence of normal understanding of natural

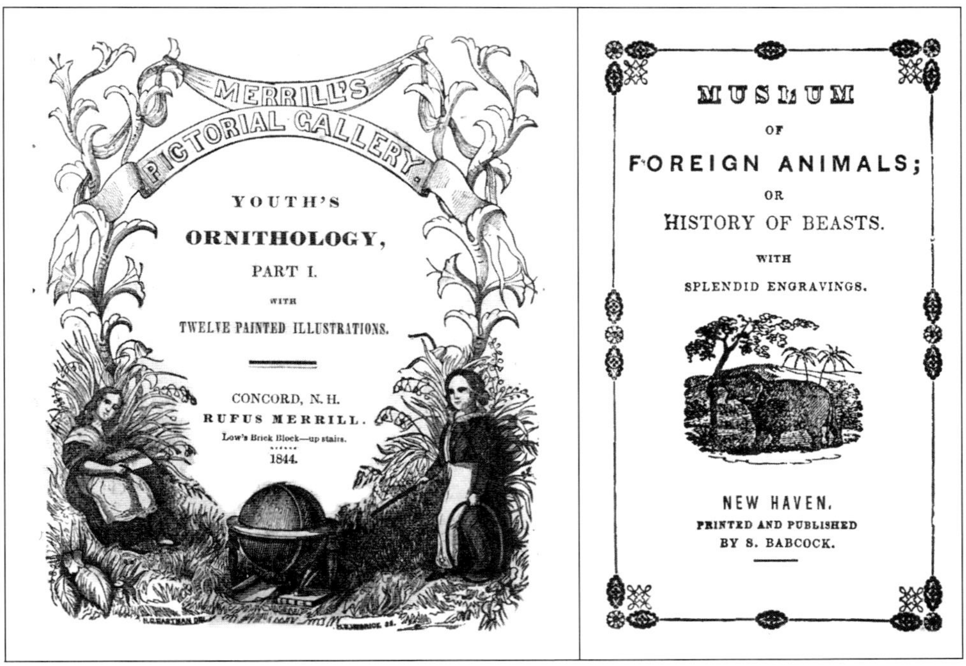

MERRILL'S PICTORIAL GALLERY.

YOUTH'S ORNITHOLOGY, PART I.

WITH TWELVE PAINTED ILLUSTRATIONS.

CONCORD, N. H.
RUFUS MERRILL.
Low's Brick Block—up stairs.
1844.

MUSEUM OF FOREIGN ANIMALS; OR HISTORY OF BEASTS.

WITH SPLENDID ENGRAVINGS.

NEW HAVEN.
PRINTED AND PUBLISHED BY S. BABCOCK.

Fig. 9.2. Title page from *Youth's Ornithology Part I,* published by Rufus Merrill ([ca. 1844]); title page from *Museum of Foreign Animals,* published by B. Babcock (without year).

(In the Rudolph photograph collection)

history at that time. Many of the chapbooks do provide accurate, if only brief, information about organisms. For instance Cobb (1835)[12] states:

> In this as well as in all the other Series nothing has been permitted to find a place which is false, unnatural, or unphilosophical, or any details of conversations among animals which *never* did, and which never *can* take place. It is believed that in the large field of Nature and Art, there are sufficient materials for descriptions and stories without launching into the field of Fiction and Falsehood, to find subjects which will be interesting to children.

Some chapbooks attempt to correct earlier misinformation. One[3] says of the ostrich that, "she does not leave them [eggs] to be hatched by the sun's heat, as has been a general tradition in every age." Another[4] says of the swan, "Willoughby seems to think the swan may live 300 years, but although this appears very doubtful, if not absolutely incredible, it is universally allowed that it reaches the age of 100." In another[11] the reader learns that the black-tail or mule deer, "was almost unknown to naturalists until Lewis and Clarke's [i.e. Clark's] expedition gave some information respecting it." However, traditional fabulous animal anecdotes, some dating from antiquity, find a place in chapbooks. For example, the reader is told in another book[10] that the marmots

> . . . cut the finest grass, others collect it together, and they carry it home upon a *cart*, made of one of their companions for the purpose. The Marmot which is used as the *cart,* lays himself down upon his back, extends his limbs, and allows himself to be loaded with hay; when done others of the party trail him along by the tail, taking great care not to turn over their curious *cart*.

In another reference[5] children learned that "eagles sometimes carry away children." Thus, one finds the accurate along with the inaccurate in these trifles for little children; overall they are better than one might expect from this sort of anonymous, much plagiarized book.

Chapbooks include some types that make the transition to the more formal school- and home-instruction books,[1,16,18,19,20,21,22] as well as to the finely illustrated entertaining works more generally available later. In studying the British and American chapbooks of the nineteenth century, one is examining the knowledge of the general public. As Harvey Darton (1958)[14] says of these works, "they were read and re-read and loved, . . . and were the books of the people of England." And it may be added, that many British ones were brought to and copied in America; others originated on this side of the Atlantic, so that they were also the books of the people of America, and their children.

Fig. 9.3. Title page from *The Child's Book of Beasts*, by J. Q. Preble, New York (without year). [Not Cited]

(In the Rudolph photograph collection)

References Cited

1. A Lady. [Pseudonym for Jane Kilby Welsh]. 1819. *A Botanical Catechism: Containing Introductory Lessons for Students in Botany*. T. W. Shepard and Co., Northampton, [Massachusetts]. iv, [1], 6–34 pp.

2. [Anonymous]. 1805. *Tommy Trip's History of Beasts*. John Adams, Philadelphia. [Reference not seen].

3. [_____]. [ca. 1809]. *A Natural History of Foreign Birds*. W. Davison, Alnwick. 36 pp. (Quote, p. 4).

4. [_____]. [ca. 1809]. *A Natural History of Water Birds*. W. Davison, Alnwick. 36 pp. (Quote, p. 27).

5. [_____]. [ca. 1810]. *History of Birds*. P. Merrifield, Windsor, V[ermon]t. 10 pp. (Quote, p. 9).

6. [_____]. 1812. *Garden Amusements for Improving the Minds of Little Children*. Thomas B. Wait and Co. and Charles Williams, Boston. 34 pp. (Quote, p. [5]).

7. [_____]. [ca. 1830]. *About Plants*. Jonathan Grout, Jr., Worcester, [Massachusetts]. 24 pp.

8. [_____]. [ca. 1840]. *Book of Fruits*. Mahlon Day & Co., and Baker, Crane & Co., New-York. 30 pp. (Quote, p. [4]).

9. [_____]. [ca. 1840]. *My Flower-pot*. Rufus Merrill, Concord, N[ew] H[ampshire]. 16 pp. (Quote, p. 10).

10. [_____]. [ca. 1850]. *Child's Own Book*, No. 3. S. Hastings, Boston. [16] pp. (Quote, p. 16).

11. [_____]. [ca. 1850]. *The Natural History of Beasts*. Rufus Merrill, Concord, N[ew] H[ampshire]. 24 pp. (Quote, p. 16).

12. [Cobb, Lyman]. 1835. *Cobb's Toys, Third series, No. 5. Stories about the Vulture, Dodo, Spoonbill, Pelican, Cormorant, and Goldfinch, in Words of One, Two, and Three Syllables*. B. and S. Collins, New York. 16 pp. (Quote, p. 2).

13. [_____]. 1836. *Cobb's Toys, Fourth series, No. 2. Stories about the Rhinoceros, Hippopotamus, Musk Ox, Bison, Buffalo, Urus, Raindeer, Camelopard, Hyena, Catamount, Wild Cat, Jackal, Antelope, Chamois, and Armadillo*. Hickok & Blood, Chambersburgh, [Pennsylvania]. 24 pp. (Quote, p. [2]).

14. Darton, F. J. Harvey. 1958. Chapter 5. The pedlar's pack: The running stationers, pp. [70]–84. *In* F. J. Harvey Darton, *Children's Books in England. Five Centuries of Social Life*. 2nd ed. University Press, Cambridge, [England]. (Quote, p. 73).

15. Harvard University. Library. 1905. *Catalogue of English and American Chap-books and Broadside Ballads in Harvard College Library. Bibliographical Contributions*, Vol. 4, No. 56. Library of Harvard University, Cambridge, Massachusetts. xi, 171 pp.

16. Irving, C. 1829. *A Catechism of Botany; Containing a Description of the Most Familiar and Interesting Plants, Arranged According to the Linnaean System*. With an Appendix on the Formation of an Herbarium. Adapted to the Use of Schools in the United States. 3rd American ed., improved and enlarged. Collins and Hannay, New-York. 82 pp.

17. Kuhn, Thomas S. 1970. *The Structure of Scientific Revolutions*. 2nd ed. *International Encyclopedia of Unified Science, Foundations of the Unity of Science*, Vol. 2, No. 2. University of Chicago Press, Chicago. 210 pp.

18. Mavor, William. 1820. *Catechism of Botany; or, An Easy Introduction to the Vegetable Kingdom: For the Use of Schools and Families*. Samuel Wood & Sons, New York; Samuel S. Wood & Co., Baltimore. iv, [5]–70 pp.

19. _____. 1821. *Catechism of Animated Nature; or An Easy Introduction to the Animal Kingdom*. For the Use of Schools and Families. Samuel Wood & Sons, New-York; Samuel S. Wood & Co., Baltimore. 70 pp.

20. Pinnock, William. 1820. *A Catechism of Ichthyology; Being a Familiar Introduction to the Natural History of Fishes; Scientifically Arranged*. Davidson, for Pinnnock and Maunder, Mentorian Press, London. 66, [1] pp.

21. _____. 1824. *A Catechism of Conchology, Containing a Pleasing and Familiar Description of the Construction and Classification of Shells, According to the Linnean System; With Examples of Each Class*. G. & W. B. Whittaker, London. 71, [1] pp.

22. [_____]. 1831. *Pinnock's Catechism of the Elements of Botany*. A new & improved ed. Whittaker, Treacher & Co., London. 72 pp.

23. Weiss, Harry B. 1928. A chapbook on natural history. *The American Collector* 5: 212–214.

Publication Notes

Modified and edited from a manuscript[a] written in 1972 and submitted for publication to a journal on children's literature. The manuscript was never published.

[a] 1972. "Natural History in Nineteenth Century Chapbooks for Children." Typewritten, 9 pp. [Manuscript].

[b] 1972. Natural history chapbooks: Throwbacks to Pliny and the *Physiologus* as well as windows to 18th century biology, p. 3. *In* Abstracts of Papers Presented at [a Combined Meeting of] The Ohio Academy of Science History of Science Meeting [and the Annual Meeting of] the Ohio Academy of Medical History, 8 April 1972, The Ohio Historical Center, Columbus. Mimeographed. [Abstract].

Plant Personification

Carl Linnaeus (1707–1778)

Paper No. 10 (1982, 1983)

Plant Personification from Linnaeus to the Present

Personification of animals is common in human history. It is understandable because animals and their behavior lend themselves readily to this kind of interpretation. The personification of plants, however, is rare, and one wonders why it ever happened. Analysis of selected examples of plant personification, omitting the older tales about mandrake plants and other similar kinds of plants, will show that they can be classified by their purposes into those written (1) to instruct about plants, (2) to provide moral interpretations, and (3) to entertain.

Carl Linnaeus (1707–1778), the great eighteenth century naturalist, lived at a time when philosophers were much concerned with the similarities between plants and animals. As a young university student, Linnaeus wrote a New Year's essay to honor his benefactor, Olaf Celsius (1670–1756), on *The Nuptials and Sex of Plants, in which the Judgements and Observations of Recent Botanists are Reviewed* (1729).[10] Linnaeus wrote:

> Yes, Love comes even to the plants. Males and females, even the hermaphrodites, hold their nuptials (which is the subject that I now propose to discuss) . . . . The actual petals of a flower contribute nothing to generation, serving only as the bridal bed which the great Creator has so gloriously prepared, adorned with such precious bed-curtains, and perfumed with so many sweet scents in order that the bridegroom and bride may therein celebrate their nuptials with the greater solemnity. When the bed has thus been made ready, then is the time for the bridegroom to embrace his beloved bride and surrender himself to her . . . .

The major part of the essay is not so vivid, but it did convince the aging Olof Rudbeck the Younger (1660–1740) to appoint Linnaeus as botanical lecturer for the university students. At the end of the eighteenth century, Erasmus Darwin (1731–1802) in his popular poetic work *The Botanic Garden* (1790–1791, reprint 1978),[5] also used personification to instruct about plant structure and classification. In the second volume of this book Darwin gave an example about *Drosera*, which serves to illustrate his use of personification.

> Queen of the marsh, imperial *Drosera* treads
> Rush-fringed banks, and moss-embroider'd beds;
> Redundant folds of glossy silk surround

Frontispiece Credits: Plant Personification. Carl Linnaeus (1707–1778). (From an illustration printed by Harry Lubrecht; in the Rudolph photograph collection)

> Her slender waist, and trail upon the ground;
> Five sister-nymphs collect with graceful ease,
> Or spread the floating purple to the breeze;
> And five fair youths with duteous lore comply
> With each soft mandate of her moving eye . . .

Darwin's[5] philosophical note is appended in part:

> *Drosera* . . . Sun-dew. Five males, five females. The leaves of this marsh-plant are purple, and have a fringe very unlike other vegetable productions. And, which is curious, at the point of every thread of the erect fringe stands a pellucid drop of mucilage, resembling a ducal coronet.

Darwin's books were widely read into the first quarter of the nineteenth century, serving well their instructional purpose which seems next to appear in a number of children's books by female authors toward the end of that century. Examples are *Plants and Their Children* by Mrs. William Starr Dana (1896),[4] The *Plant Baby and Its Friends* by Kate Louise Brown (1897),[2] and *Seed-babies* by Margaret Warner Morley (1897).[11] Their approach is to tell an instructive story using plant characters. In the preface of her book (Fig. 10.1), Kate Brown (1897) wrote:

> As a rule, technical terms have been avoided. Experience has shown, however, that little people of six and seven can learn to use such terms as *cotyledons* and *corolla* as readily and understandingly as many other words which they appropriate without instruction. *Cotyledon*, meaning the half of the plant baby's cradle, is no more difficult for them to master than Angelina, their little mate's name.

An example of her method is:

> The baby presses his feet down and lifts his hands up when he awakes. So does the plantlet. We call its foot the *radicle* . . . . We call its hand the *plumule* . . . .

Margaret Morley (1897)[11] wrote about a plant:

> I get my food from the air and the earth. I am fond of gas, and when I turn green I can digest it. You know the air is nothing but gas. Well, I can eat air.

These children's books continue into the twentieth century. The book *Blossom Babies; How to Tell the Life Story to Little Children* by M. Louise Chadwick (1913),[3] provides the following instructional example:

> The Pistil Mothers and the Stamen Fathers smiled into each other's faces . . . . Some of the Stamen Fathers, so happy were they to share the seed babies with the Pistil Mothers, twined lovingly around the Pistils, pressing their dusty heads close against them, as if to say, 'We are much happier, little Pistil Mothers, now that we share with you in giving life to the seed babies.'

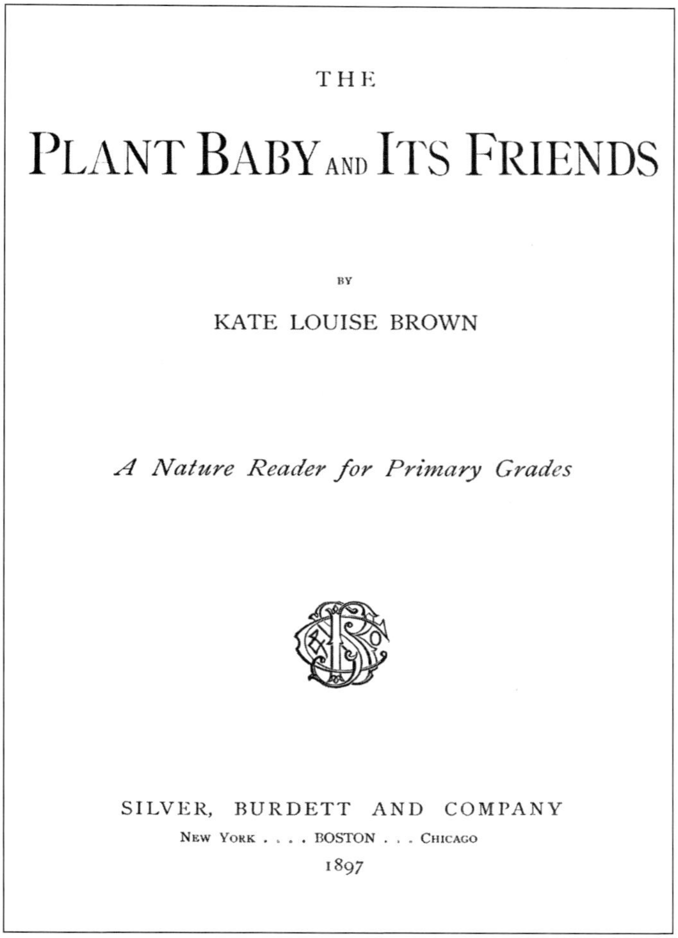
THE

PLANT BABY AND ITS FRIENDS

BY

KATE LOUISE BROWN

A Nature Reader for Primary Grades

SILVER, BURDETT AND COMPANY

NEW YORK . . . . BOSTON . . . CHICAGO

1897

Fig. 10.1. Title page from *The Plant Baby and Its Friends*, by Kate L. Brown (1897).

(In the Rudolph photograph collection)

A more recent unusual example of a book with instructional intent for young adults or adults is *Just Like the Flowers Dear: A Botany for Parents* by Robert B. Donnelly (1932).[6] His more breezy approach is exemplified by this quote:

> The petals next to the magnetic organs are much more attractive, which is human and as it should be . . . . In flowers, these garments (which correspond to chemise, panties, or B.V.D.'s, as the sex may be) are fashioned of the most attractive soft materials and vary widely in color and cut. They are justified as protection against weather and marauders, but their real purpose is to attract attention to pollinating charms.

A second purpose, not very common, is to use plant personification to illustrate moralistic lessons. A late eighteenth century example is *The Fables of Flora* by John Langhorne (1771, reprint 1810)[8] in which flower people exemplify human frailties. In the tulip and the myrtle he wrote:

> 'Twas on the border of a stream,
> A gaily painted Tulip stood,
> And, gilded by the morning beam,
> Survey'd her beauties in the flood.
> . . . 'Deluded flower!' the Myrtle cries,
> Shall we thy moment's bloom adore?
> 'The mean'st shrub that you despise,
> The Meanest flower has merit more.

A more heavy-handed approach is taken by John E. Taylor in his *The Sagacity & Morality of Plants* (1891).[12] "There is hardly a single department of the life of most kinds of plants where we do not find the laws of political and social economy in operation."

One of his moralistic examples is:

> It is worth our while to observe the unequal duration of the lives of plants collected into the same genus . . . . Perhaps in a general way it can be best accounted for by some plants being more thrifty than others—this is, they have learned to accumulate more than they expend; and their excess is laid up underground in a thickened stem, like *Ranunculus bulbosus*, or in root-stalks, rhizomes, and the like.

The third purpose for plant personification, the purpose of entertaining, has the most examples for both adults and children. Both fantasy and whimsy are used, often with imaginative pictures drawn by outstanding illustrators. The works by J. J. Grandville (1847)[7] and by Edward Lear (1904)[9] provided a major impetus for the entertaining aspect of plant personification. The Frenchman, J. J. Grandville (1803–1847), whose real name was Jean-Ignace-Isadore Gerard, was a well known illustrator who prepared in the last years of his life, a work whose English title is *The Flowers Personified: A Series of Fifty-two Female Figures, Adorned with Flower Costume,* (1847).[7] Using elegantly elaborate flower figures, together with a fanciful text, he poked fun at human behavior. For example, he had the cactus say:

> The entire history of my life on earth may be summed up in four words, 'I have been cold.' Had I remained in the torrid zone, I should have had no reason to complain. But I was so foolish as to follow a botanist to Europe, and the consequence was that my limbs become stiff with rheumatism. Even in a greenhouse one must always be the victim of some treacherous draft of air. Then this artificial heat gave me the headache and caused an insupportable heaviness of feelings.

Followers of Grandville include many children's book writers and illustrators. Perhaps the most famous is the English illustrator Walter Crane (1845–1915) who at the turn of the century elaborately illustrated very popular story books. In all cases the pictures add to the amusement of these fantasy story books. The second seminal book is by the English artist-illustrator Edward Lear (1888, reprint 1904)[9] whose whimsical books include sections of nonsense botany in which personified plants are given pseudo-scientific names—for example, *Queeriflora Babyöides*, *Manypeeplia Upsidownia*, and *Phattfacia Stupenda*. Fewer writer-illustrators imitate Lear.

Are there any lessons to be learned from this brief survey of plant personification? First, instructional use of this technique was made by naturalists such as Carolus Linnaeus and Erasmus Darwin because it was a vivid pedagogical device. Later, it was adapted to a story format by non-scientists who mostly wrote for children. It is of note that almost all of these instructional examples seem to be concerned with plant reproduction. Moralistic use of plant personification is quaint and now of interest only as a curiosity. It illustrates the thinking of another age. In recent times, scientists and popularizers have not seriously personified plants. The use of plant personification for entertainment purposes was given a major impetus by illustrators in the middle of the nineteenth century; and its continued use to the present-day by popularizers and illustrators, particularly for the gardening enthusiast, shows its staying power. People of all ages and backgrounds seem to enjoy having their imaginations tickled.

References Cited

1. Blunt, Wilfrid, with the assistance of William T. Stearn. 1971. *The Compleat Naturalist: A Life of Linnaeus.* A Studio Book. The Viking Press, New York. 256 pp. (Quote, p. 34).

2. Brown, Kate Louise. 1897. *The Plant Baby and Its Friends.* A Nature Reader for Primary Grades. Silver, Burdett and Co., New York, Boston, Chicago. 155 pp. (Quotes, pp. 7–8, 15).

3. Chadwick, M. Louise. 1913. *Blossom Babies; How to Tell the Life Story to Little Children.* Eaton & Mains, New York; Jennings & Graham, Cincinnati. iv, 169 pp. (Quote, p. 37).

4. Dana, Mrs. William Starr [Frances Theodora Smith Dana Parsons]. 1896. *Plants and Their Children.* Illustrated by Alice Josephine Smith. American Book Co., New York, Cincinnati, Chicago. 272 pp.

5. Darwin, Erasmus. 1978. *The Botanic Garden.* Photo-facsimile [of the original 1791 ed. of part I and 1790 ed. of part II (2d ed.)]. With an introduction for the Garland edition by Donald H. Reiman. Romantic context: Poetry. Garland Publishing, Inc., New York & London. 2 parts (I: xiv, xii, 214, 126 pp.; II: xi, 202 pp.). (Quote, part II, p. 25).

6. Donnelly, Robert B. 1932. *Just Like the Flowers Dear: A Botany for Parents.* Coventry House, New York. 270 pp. (Quote, p. 109).

7. Grandville, J. J. [Jean-Ignace-Isidore Gerard Grandville]. 1847. *The Flowers Personified: A Series of Fifty-two Female Figures, Adorned with Flower Costume Beautifully Colored from Designs by J. J. Grandville*. Translated by N. Cleaveland. Johnson, Fry and Co., New York. 2 vols. (1: 280 pp.; 2: pp. 281–385 + pp. 3–160). (Quote, vol. 2: pp. 285–286).

8. [Langhorne, John]. 1810. The Fables of Flora, pp. 443–450. *In* The poems of John Langhorne, D.D., pp. [405]–475. *In* Alexander Chalmers, [ed.], *The Works of the English Poets, from Chaucer to Cowper; Including the Series Edited, with Prefaces, Biographical and Critical, by Dr. Samuel Johnson: And the Most Approved Translations*. The Additional Lives by Alexander Chalmers, Vol. XVI. Printed for J. Johnson . . . , London. [The Fables of Flora first appeared in 1771]. (Quote, p. 447).

9. Lear, Edward. 1904. *Nonsense Books: I. A Book of Nonsense. II. Nonsense Songs, Stories, Botany, and Alphabets. III. More Nonsense Pictures, Rhymes, Botany, etc. IV. Laughable Lyrics: A Fresh Book of Nonsense Poems, Songs, Botany, etc.* With all the original illustrations. Little, Brown, and Co., Boston. xxii, 85 pp.; [1]–79, [80]–[161] pp.; ix, 107 pp.; [1]–41, [42]–84 pp. (Illustrations cited, Book II: p. [88], p. [89]; Book IV: p. [54]). [Reprint of the 1888 republication].

10. Linnaeus, Carl. 1729. *Caroli Linnaei . . . Praeludia Sponsaliorum Plantarum, in Quibus Physiologia Earum Explicatur, Sexus Demonstratur, Modus Generationis Detergitur, nec Non Summa Plantarum cum Animalibus Analogia Concluditur*. Upsala. [Manuscript (D. 63) in the Royal University Library, Uppsala]. (Reference not seen; quoted from Blunt,[1] p. 34).

11. Morley, Margaret Warner. 1897. *Seed-babies*. Ginn & Co., Boston and London. 75 pp. (Quote, pp. 9–10).

12. Taylor, John Ellor. 1891. *The Sagacity & Morality of Plants: A Sketch of the Life & Conduct of the Vegetable Kingdom.* A new ed. Chatto & Windus, Piccadilly, London. xi, 311 pp. (Quotes, pp. 181, 215).

Publication Notes

Modified and edited from an unpublished manuscript[a] read 10 August 1982, at the meeting of the Historical Section, Botanical Society of America, Pennsylvania State University, University Park; also read 23 April 1983, at the Annual Meeting of The Ohio Academy of Science, Bowling Green State University, Bowling Green, Ohio.

[a] 1982. "Personifying Plants from Linnaeus to the Present." Typewritten, 7 pp.; handwritten, 16 pp. [Manuscript].

1982. Personifying plants from Linnaeus to the present. *Botanical Society of America, Miscellaneous Series, Publication* 162: 52. [Abstract].

1983. A brief history of plant personification. *The Ohio Journal of Science* 83(2): 9. [Abstract].

APPENDIX 10.1.
EXPLAINING PLANT PERSONIFICATION TO COLLEGE STUDENTS IN THE TWENTIETH CENTURY, BY STUCKEY

For nearly four decades, from the 1920s through the 1950s, at The Ohio State University, Botany Department Chairman Edgar N. Transeau and his associates Homer C. Sampson and Lewis H. Tiffany and their followers developed a distinctive course in General Botany. It featured a demonstration-discussion approach for small class sizes. One of the prevailing thought processes that students were expected to learn was how not to ascribe purpose to plants, as was explained in the chapter "Points of View in the Interpretation of Plant Behavior," in their *Textbook of Botany* (1940).[a]

> As far back as human records go, man has attempted to account for himself and his surroundings. Primitive man lived in most intimate contact with nature. Wishing to survive, he was compelled to give attention to other animals, to light and darkness, to flood and drought, and to trees and herbs. He came to regard all these objects and processes of his environment as separate entities. Before he recognized their relation to one another, he had very definitely related them to himself as either benevolent or malevolent. Lacking the facts and the methods for testing the soundness of his ideas, he attempted to explain *why*, without knowing *how* natural phenomena take place. To him all natural objects appeared to be more or less human-like. Accordingly he attributed to other animals, to plants, to the moon and other inanimate objects the same abilities, motives, and emotions he recognized within himself. In other words, he *personified* them.
>
> . . .
>
> For example, who is not familiar with such explanatory statements as these: that trees in a forest grow tall and straight because they are trying to reach the light; that stems and petioles of plants growing at a window bend toward the window in order to get more light; that potato sprouts and stems of seedlings in a dark room grow longer in search of light; that plants produce seeds in order to propagate their kind, or to store food for future use or for the use of animals; that roots grow deep into the soil or toward a cistern in search of water; finally, that plants do this or do that particular thing in order to adapt themselves to their environment?
>
> Do not all these statements imply that the plant is conscious of its present and future needs and strives by various means to satisfy them? Whatever happens, according to this point of view, the plant is always forward-looking. In this type of explanation of natural phenomena every act is assumed to be purposeful.

[a] E. N. Transeau, H. C. Sampson, and L. H. Tiffany. 1940. *Textbook of Botany*. Harper & Brothers, New York. 812 pp. (Quote, pp. 50–51).

Four Investigators Representing Career Models

Charles Darwin
(1809–1882)

Gregor Mendel
(1822–1884)

Luther Burbank
(1849–1926)

George W. Carver
(1864–1943)

Paper No. 11 (1979)

The Botanist as a Career Model for American Children: A 150-Year History

What affects childhood influences a person's adult life, or as Wordsworth put it more poetically, "the boy is father of the man." Yet, how does a boy or girl decide to become a botanist never having seen one? In an attempt to answer this question, biographies of botanists have been surveyed in American children's books published during the past 150 years. Records of 174 biographies, from books or chapters in books, representing 38 different botanists, were compiled in this survey. The term botanist is interpreted broadly to include some individuals who may not be primarily thought of as botanists, for example Luther Burbank (1849–1926), Charles Darwin (1809–1882), Johann Goethe (1749–1832), Jean Lamarck (1744–29), and Gregor Mendel (1822–1884). These choices were made because otherwise the already very small sample would have been pitifully smaller.

It is important to remember that biographies written for children tend to differ from those for adults. May Hill Arbuthnot (1957),[1] in her book on children's literature, noted that "The chief distinction between good biographies for adults and those for children are that, in the latter, sources are rarely stated, unsavory episodes are usually omitted, and recorded events are more likely to be enlivened with imaginary dialogue." Children's books also tend to emphasize the youthful part of the person's life. In considering the lack of stated sources, the problem also exists of including factual errors. For example, in one book, Charles Darwin is said to have been an Episcopalian, and in another a Unitarian, the latter being used to account for some of his ideas! Putting aside this problem of factual error, the concern here is mainly with what emerges as the characteristics of the field of botany that could interest children, or make them think of becoming professional botanists.

Before looking at these characteristics, it might be of interest to know who were the most common subjects of biographies among the 38 botanists, or near-botanists. Those with more than 10 biographies are Charles Darwin with 41, George Washington Carver (1864–1943) with 30, Luther Burbank with 23, Carl Linnaeus (1707–1778) with 16, and Gregor Mendel with 12, all of which account for about 80% of the individuals. Only one of these, Linnaeus, can be said to be a "pure"

Frontispiece Credits: Four Investigators Representing Career Models. Charles Darwin (1809–1882). (From *Science* 130: 1214. 1959)

Gregor Mendel (1822–1884); Luther Burbank (1849–1926); George W. Carver (1864–1943). (From the Hunt portrait collection; permission to reproduce these photographs courtesy of Hunt Institute for Botanical Documentation, Carnegie Mellon University, Pittsburgh, Pennsylvania)

botanist. Only four individuals were in American biographies published before 1900: Charles Darwin, Johann Goethe, Bernard de Jussieu (1699–1777), and Carl Linnaeus. Most of the biographies, about 60 percent, were published after 1950. The 1960s is the decade with the most biographies, about 35 percent.

In these works, the field of botany is generally very poorly defined. The only clear aspect is that botanists study plants. An exception is a good explanation given in a recent book (Klein and Klein, 1968),[7] where the reader is informed that:

> When you experiment on plants, you are studying life – you are a biologist. The biologist who studies plants is called a botanist. Some botanists collect and name plants; they are plant taxonomists . . . Others study the structure of whole plants, parts of plants, or plant cells; they are called plant morphologists . . . There are botanists who investigate why and how a particular plant grows in a particular place and how plants get along with other living things; they are ecologists . . . Plant scientists who study why and how plants grow are physiologists . . . .

The personal characteristics that are suggested to children in these writings, as those typical of botanists, are curiosity, hard work, interest in learning and reading, love of nature, inventiveness, enthusiasm, and artistry. Surely, these are fine attributes, but not particularly descriptive of botanists as compared with other successful people.

Perhaps the children are given better information about the kind of training botanists need and receive. Again, the picture is disappointing. The two forms of training most commonly mentioned are gardening and plant collecting. In several cases information is given about schooling. For example, in his biography about George Washington Carver (1864–1943), Yost (1941),[14] tells the reader:

> In three years—at the age of thirty—he received his B.S. in agriculture and was placed in charge of the work in systematic botany and of the college greenhouses which latter gave added inspiration for his paintings of flowers.

Usually the information about training is more general, as for example the following about Gregor Mendel from a biography by Webster (1963):[13]

> Father Gregor collected facts. He made this his hobby, somewhat as many persons make stamp collecting a hobby.

Perhaps the authors of biographies do not think that children are interested in learning about formal education and its details; at least they do not describe it.

Job descriptions of botanists are more informative than the personal characteristics. The readers learn that botanists travel and explore, collect plants, try experiments with plants, work in laboratories, teach, and serve on governmental committees.

Some specific quotes are illuminating. In a biography of Linnaeus by Waring (1829),[12] the reader learns that

> . . . he began to keep a regular herbal; a plan, of which he soon discovered the importance. It induced him to be more diligent in collecting flowers, and more careful in examining their internal structure, on which he was hereafter to found his botanical system.

Cane (1959)[3] writes that Darwin

> . . . spent his time in going over the tremendous body of evidence that led to his theory of evolution. A lovable and sweet-tempered invalid, he wandered about his garden, caring for his flowers. He made botanical experiments which he used to test his theory.

Of George Washington Carver, Moore (1971)[9] says that he

> . . . spent two years working in the botany laboratory at Iowa State. In the laboratory he made experiments with plants, trying different ways to make them grow bigger and better. He took long walks in the woods, looking for plants that had a fungus—a kind of plant disease. He took these plants back to the laboratory and studied them under a microscope.

About a modern botanist, Brewer (1965)[2] tells the reader that:

> There is no such thing as a typical work day in Dr. Jaynes' research. He might start off a day by preparing new cultures of *Endothia parasitica*. After that, he might take a tour through the station to see how its several thousand chestnuts are doing. Occasionally, he travels outside Connecticut, visiting plots in other states where hybrid trees have been planted.

In some other biographies, however, the child might get a somewhat negative idea of the viability of a botanical career, as from those statements in Milne and Milne (1952):[8]

> Many of these naturalists are physicians, or bankers, or bricklayers, or clerks. They may find time for natural history only during the hours after work. More fortunate in many ways are those who can afford to live on a small salary, as a museum worker, or a teacher, so that all their time can be related to nature. . . . Some of the active modern naturalists have turned more and more to writing. Donald Culross Peattie [1898–1964], born in Chicago . . . and educated as a botanist at Harvard University, took a position in David Fairchild's Office of Foreign Seed and Plant Introduction in the Department of Agriculture. Then he married a writer, and a few years later gave up his position to enter a literary career.

Creevey (1893)[4] makes similar negative statements:

> While it is not claimed that botany, like Greek or Mathematics, can produce mental brawn, yet it certainly does cultivate close observation, prolonged attention to minutiae, a habit of comparison and deductive reasoning—all mental qualities worth possessing.

The inspiration for being a botanist is sometimes vividly presented in these works. For example, Day (1890)[5] writes:

> The facts of nature are open to the peasant and the mechanic, as well as to the philosopher and scholar; and all are capable, in greater or lesser degree, of making use of those facts to the best of their power, so that even among the lowliest, the true worker may achieve the loftiest results.

Fisher (1891),[6] informs the reader that "to be a botanist is to be like a traveller, going from country to country, and in each finding new wonders and new beauties." Webster (1963)[13] describes Mendel this way:

> Not that he ever succeeded in satisfying all his urges to unlock the secrets of nature. Wherever he looked, he could see a marvel and a mystery—and he was perpetually eager to interpret them in such a way that he would bring glory to God and satisfaction to mankind.

Stoutenburg and Baker (1958)[11] write:

> More enduring than all the written tributes Douglas received after his death, or those engraved in marble, are the acres of beauty he left to the world. There is scarcely a garden anywhere that does not contain at least one of the flowers he introduced.

These and many more inspirational evaluations of botanists might stimulate children toward botany if they could associate with the heroes of the biographies.

The picture derived from this survey of children's biographies of botanists and near-botanists, is not a particularly favorable one for providing a role model for the readers. Only a few biographies of scientists for children have been published in the past 150 years, and only a very few of these are of botanists, or biologists who did some botany. In fact, the most popular botanists in this sample are usually more noted for some other aspect of their achievements; Darwin for evolution theory, Carver for agriculture, chemistry, and for being successful as a black man; Burbank for horticultural accomplishments. Except in a few recent books, little is said about what the subject of botany is and how one is trained to become a botanist. Certain aspects of what the botanist does—such as exotic exploration, breeding plants to better humankind, and making large collections of plants—are emphasized over the more usual ones of engaging in experiments for many hours, reading the literature, attending scientific meetings, and gathering data. Among these 174 biographies, only one is of a woman, Jane Colden (1724–1766), about whom Payne (1968)[10] writes:

> Why was Jane Colden not honored? Was it an oversight? Hardly, with so many letters to Linnaeus mentioning her competence. Was it then because she had departed from the usual activity of women of her time? Possibly. After all, she did pioneer in a field previously thought to be exclusively a man's domain.

This description is hardly good reason for a modern girl to aspire to become a botanist. "The past is prologue." It has not provided children who read with very many or very informative botanist role models. The future may do better, although fewer biographies of botanists were published in the 1970s than in the 1960s.

References Cited

1. Arbuthnot, May Hill. 1957. *Children and Books*. Revised ed. Scott, Foresman & Co., Chicago. xvi, 684 pp. (Quote, p. 512).

2. Brewer, Fred. 1966. *Challengers of the Unknown: On the Job with Today's Scientists-Plus Related Projects*. Four Winds Press, New York. 128 pp. (Quote, p. 83).

3. Cane, Philip. 1959. *Giants of Science*. Illustrated by Samuel Nisenson. Grosset & Dunlap, New York. 159 pp. (Quote, p. 102).

4. Creevey, Caroline A. 1893. *Recreations in Botany*. Harper & Brothers, New York. xiii, 216 pp. (Quote, p. 8).

5. Day, George. [ca. 1890]. *Naturalists and Their Investigations: Linnaeus, Edward, Cuvier, Kingsley*. Fleming H. Revell Co., New York. 160 pp. (Quote, p. 8).

6. Fisher, Robert. 1891. *Flower-land: An Introduction to Botany*. Frederick A. Stokes Co., New York. viii, 240 pp. (Quote, p. 3).

7. Klein, Richard M. and Deana T. Klein. 1968. *Discovering Plants: A Nature and Science Book of Experiments*. Published for the American Museum of Natural History [by] the Natural History Press, Garden City, N.Y. 124 pp. (Quote, p. 7).

8. Milne, Lorus J. and Margery J. Milne. 1952. *Famous Naturalists*. Dodd, Mead, New York. 178 pp. (Quotes, pp. 170, 173).

9. Moore, Eva. 1971. *The Story of George Washington Carver*. Illustrated with prints by Alexander Anderson and with photographs. Scholastic Book Services, New York. 96 pp. (Quote, p. 58).

10. Payne, Alma Smith. 1968. *Partners in Science*. Illustrated by Polly Bolian. World Pub. Co., Cleveland and New York. 224 pp. (Quote, p. 86).

11. Stoutenburg, Adrien and Laura Nelson Baker. 1958. *Wild Treasure: The Story of David Douglas*. Scribners, New York. 216 pp. (Quote, p. 216).

12. [Waring, Miss S.]. [ca. 1829]. *The Young Botanist: or, A Sketch of the Life of Linnaeus Designed for Young Persons*. Munroe and Francis, Boston; Charles S. Francis, New York. 126 pp. (Quote, pp. 36–37).

13. Webster, Gary. [pseudonym for Webb B. Garrison]. 1963. *The Man Who Found Out Why: The Story of Gregor Mendel*. Illustrated by Greg and Tim Hildebrandt. Hawthorn Books, New York. 188 pp. (Quotes, pp. 64, 93).

14. Yost, Edna. 1941. *Modern Americans in Science and Invention*. Drawings by Margaret Ayer. J. B. Lippincott Co., Philadelphia. xiii, 270 pp. (Quote, pp. 151–152).

Additional References Not Cited

[Goodrich, Samuel Griswold]. 1831. *The Child's Botany*. 4th ed. Carter, Hendee and Babcock, Boston. 103 pp.

Hyde, Margaret O. 1960. *Plants Today and Tomorrow. Illustrated by P. A. Hutchinson*. Whittlesey House, New York. 157 pp.

Logasa, Hannah. 1940. *Biography in Collections Suitable for Junior and Senior High Schools*. 3rd. ed., revised & enlarged. H. W. Wilson, New York. 152 pp.

Sparke, Archibald. 1927. Author Wanted. *Notes and Queries* 152(6): 107. [on Miss S. Waring].

Stanius, Ellen J. (comp.). 1971. *Index to Short Biographies: For Elementary and Junior High Grades*. Scarecrow Press, Metuchen, New Jersey. 348 pp. [Abstract].

Publication Notes

Modified and edited from an unpublished manuscript[a] read 14 August 1979, at the meeting of the Historical Section, Botanical Society of America, Oklahoma State University, Stillwater.

[1979]. ["The Botanist as Career Model for American Children: A 150 Year History"]. Handwritten, 10 pp. + 18 pp. of quotations. [Manuscript].

[a] [1979]. "The Botanist as Career Model for American Children. A 150 Year History." Typewritten, 6 pp. [Manuscript].

1979. A 150 year history of the botanist as a career model for American children. *Botanical Society of America, Miscellaneous Series, Publication* 157: 29. [Abstract].

Part Three

BOTANICAL TEACHING

Emanuel D. Rudolph's two contributions on classroom and laboratory teaching are quite different from each other, but are placed together here (**Part III**). The one on classroom teaching (**Paper No. 12**) emerges from his National Science Foundation (NSF) Summer Institute for College Teachers in the Biological Sciences, held during the summers of 1968 and 1969. The other on botanical teaching laboratories (**Paper No. 13**) discusses their development during the nineteenth and early twentieth centuries in the United States.

Introduction No. 12

History of Science in the Classroom (1968–1969)

In the mid-1960s, when Rudolph was in the early phases of his research on the history of botany, he held the view that most college science teachers lacked training to treat adequately topics on biology courses in an historical context, or to integrate even a limited amount of historical perspective into their classroom teaching. In an effort to remove this deficiency, Rudolph[1] proposed to the National Science Foundation (NSF) in the United States that he conduct a summer institute at The Ohio State University to provide primary information on the history of biology and the history of science to college biology teachers. Rudolph believed the results of this effort would assist these teachers in the use of historical information emphasizing primary sources in their biology courses. He invited to the campus well-qualified researchers and teachers in the history of science to lecture and lead discussions with the institute participants. Ten invited guest historians of science and five OSU resident faculty provided a broad coverage of biological research and historical study for the group during the summer of 1968. The institute was funded for a second summer during 1969.[2]

The intellectual talent that Rudolph assembled for the Institute now reads like a "who's who" in the history of the biological sciences in the United States. The guest lecturers were Richard G. Beidleman, Frederick B. Churchill, Charles A. Culotta (1938–1974), Ralph W. Dexter (1912–1991), Joseph Ewan (1909–1999), George H. M. Lawrence (1910–1978), Everett Mendelsohn, Phillip C. Ritterbush, Jerry Stannard, and Conway Zirkle (1895–1972). The resident faculty were Institute Director Emanuel D. Rudolph (1927–1992), in botany; Assistant Institute Director June Z. Fullmer (1920–2000), in history; John C. Burnham, in history; Aurele LaRocque (1909–1990), in geology; and Ronald L. Stuckey, in botany (Figs. 12.1, 12.2). In the summer of 1969, historians John C. Greene, Frederic L. Holmes, and Genevieve Miller were invited participants, in addition to most of the previous year's guest lecturers and the five resident faculty. Invited also were Thomas S.

---

[1] Emanuel D. Rudolph. 1967 [1968]. "A Proposal to the National Science Foundation for Support of a History of Biology Summer Institute for College Teachers in Biological Sciences [at] The Ohio State University." Typewritten, [4], 27 pp. [Grant].

[2] [Emanuel D. Rudolph]. 1968 [1969]. "A Proposal to the National Science Foundation for Support of a History of Biology Summer Institute for College Teachers in Biological Sciences [at] The Ohio State University." Typewritten, [4], 27 pp. [Grant].

Fig. 12.1. Ronald L. Stuckey (l.) and George H. M. Lawrence (r.) in Stuckey's office at The Ohio State University during the NSF Institute.

(Taken July 1968 by William H. Anderson; in the collection of RLS)

Kuhn of Princeton University and Peter J. Vorzimmer of Temple University, both of whom declined with regrets. In addition to a combination of lectures and discussions provided by the guest and resident lecturers, the 30 participants discussed readings on selected topics and each one reported on a mini reading-research project. These achievements were part of Rudolph's intended objectives, stated as follows:[1,2]

1. To indicate how information from the history and philosophy of biology can be meaningfully integrated into the undergraduate teaching of biology.

2. To familiarize the participants with the extent and amount of current and past research in the history of biology.

3. To require that the participants read important primary and secondary source materials in the history of biology and learn how to use them in preparing historical approaches to the teaching of topics in biology at various levels in the college program.

4. To promote discussions among the participants and the guest lecturers with the aim of understanding how research in the history of biology can be used by the college teachers.

5. To provide a working knowledge of the literature in the history of biology and to distribute to the participants selected bibliographies which can be useful in their future studies.

6. To stimulate those participants who have greater interest in the history and philosophy of biology to participate in research on their own by providing them with a background and techniques used in historical research.

7. To provide an understanding of the relationship of the history of biological thinking to the thinking in other fields of the history of science and to the broader field of the history of ideas.

Rudolph prepared and distributed schedules of the 34 titled presentations that were delivered each summer by the guest lecturers and resident faculty.[3,4] At the request of George H. M. Lawrence, Director of the Hunt Botanical Library, Pittsburgh, most of the lectures during the summer of 1968 were recorded with permission of each presenter. Copies of the reel-to-reel tapes were deposited in the archives of the Hunt Library, and Rudolph retained the originals. All 31 tapes, including one

Fig. 12.2. Ronald L. Stuckey (l.) and Joseph Ewan (r.) in Stuckey's office at The Ohio State University during the NSF Institute.

(Taken July 1968 by William H. Anderson; in the collection of RLS)

[3] [Emanuel D. Rudolph]. 1968. "Lecture Schedule, The Ohio State University NSF Supported History of Biology Summer Institute, 1968." Typewritten, 1 p.

[4] [Emanuel D. Rudolph]. 1969. "Lecture Schedule, The Ohio State University NSF Supported History of Biology Summer Institute, 1969." Typewritten, 2 pp.

that did not record properly according to Lawrence's list,[5] are in Rudolph's archives. Lawrence[6] referred to the tapes of the Institute as "a clearly valuable collection, and I am confident that it will be sought and consulted perhaps after all concerned are gone and there is an interest by others to hear their voices. I feel this way particularly about the one[s] by Joe Ewan and Conway Zirkle."

The names of the guest and resident speakers, along with their presentation titles are listed here, taken from the Lecture Schedules.[3,4] Only those titles of botanical interest are selected, and only those lectures given during 1968 are available on the recorded tapes (Appendix 12.1).

Regrettably, Rudolph's lecture on "The Popularization of Botany since 1750," presented both summers, was not taped. Furthermore, no manuscript or notes have been located for those presentations. He gave them as informal talks without preparing a handwritten or typewritten script, as was usually his custom for informal presentations. Rudolph brought books to the lecture to illustrate themes on the popularization of botany. During his lecture of 19 July 1968 on "The Origin of and Reactions to Molecular Biology," Rudolph showed some of the books relevant to the subject, and he read excerpts or made comments about them. This lecture survives on tape in his own archives and at the Hunt Botanical Library. Because of the manner in which the lecture was presented the editors were unable to prepare an appropriate summation of this presentation. Rudolph retained notes from the lectures of all of the 1968 guest speakers, but no notes of his own lectures are known to exist.

Some of the guest lecturers and participants suggested that their contributions be published, but no publications appeared as a project of the Institute. Most of the information sheets prepared by the guest botanists and OSU personnel participating in the Institute have been deposited in the Archives of the Hunt Institute for Botanical Documentation (Hunt Botanical Library) by Ronald L. Stuckey. Rudolph's files in the University Archives at The Ohio State University contain the same items, in addition to his correspondence with the guest lecturers and his handwritten notes made of the presented lectures for both summers.[c] Information from Rudolph's 1968 proposal[a] to the National Science Foundation and from the brochure advertising the Institute gives his reasons for introducing topics in the history of science into the classroom teaching of the biological sciences. That information was used to develop the paper printed here. **(Paper No. 12)**.

---

[5] George H. M. Lawrence. 1969. "List of Tapes of Talks at [The] Ohio State University's Summer Institute in [the] History of Biology – 1968." Typewritten, 2 pp. [sent with letter, 18 April 1969, to Emanuel D. Rudolph, see below].

[6] George H. M. Lawrence. 1969. Letter of 18 April to Emanuel D. Rudolph, Department of Botany, The Ohio State University, Columbus. 1 p.

Participants in the History of Science NSF Institute

Emanuel D. Rudolph

June Z. Fullmer

John C. Burnham

Aurele LaRocque

Biology Institute Set

Ohio State will attempt to bridge the gap between "two cultures" in a history of biology institute for college teachers scheduled for June 17 to July 27.

The two cultures according to Emanuel D. Rudolph, director of the institute, are represented by those students trained in science and those trained as nonscientists.

Rudolph, associate professor in the College of Biological Sciences, said there is concern with the problem of making undergraduate courses in biology meaningful to all students, whether or not they are science majors.

History Gap is Wide

He said there is a wide gap in knowledge between the two groups, but believes the historical approach for certain topics in biology courses can be used profitably both in developing a sounder understanding on the part of the students and in helping to close the so-called culture gap.

Invited to take part in the National Science Foundation-sponsored institute have been college teachers of biological sciences who have not had an opportunity to study the history of science and who want to know more about using a historical approach in their teaching and research.

The program will aim to provide the teachers a background of the history of biology and a broad coverage of research in the methods of historical study.

Leading researchers from throughout the country will participate as guest lecturers and will advise the teachers in individual projects.

Newspaper article from the *OSU Lantern*, 9 April 1968

Ronald L. Stuckey

Joseph Ewan

George Lawrence

Genevieve Miller

Conway Zirkle

Jerry Stannard

Paper No. 12 (1968, 1969)

Incorporating the History of Science into Classroom Teaching of Biology

In recent years, some educators have expressed concern toward making undergraduate courses in biology meaningful to non-major, as well as major students. They claim that a serious gap exists in knowledge between students trained as scientists and those trained in other disciplines. Considerable credibility for this claim appears in *The Two Cultures* by C. P. Snow (1965).[4]

Charles Culotta

Everett Mendelsohn

Philip Ritterbush

Richard Beidleman

Ralph W. Dexter

John C. Greene

Frontispiece Credits: Participants in the History of Science NSF Institute (1968). Emanuel D. Rudolph (1927–1992), John C. Burnham, Joseph Ewan (1909–1999), Conway Zirkle (1895–1972), Ronald L. Stuckey, George H. M. Lawrence (1910–1978), Jerry Stannard, Charles A. Culotta (1938–1974), Everett Mendelsohn, Philip C. Rittenbush, Richard G. Beidleman, and Ralph W. Dexter (1912–1991). (Photographs by William H. Anderson, student assistant in the Institute; in the collection of RLS); June Z. Fullmer, Aurele LaRocque (1909–1990). (From Photo Archives, The Ohio State University); Genevieve Miller. (From the Howard Dittrick Museum of Medical History, Cleveland); John C. Greene. (From John C. Greene, Storrs, Connecticut); Newspaper article about the Summer Institute for the History of Biology. (From *The Lantern*, The Ohio State University, 9 April 1968; permission to reproduce article courtesy of the *OSU Lantern*)

College science teachers who have responsibility for biological courses should have available information about the historical development of biological thinking. By using the historical approach for certain topics in biology courses, at the beginning as well as advanced levels, students may develop a sounder understanding of the subject and partly bridge the gap between the "Two Cultures." College science majors should also learn about the history of their discipline. James B. Conant (1960)[2] has stated it this way:

> It is as a person who must function outside the laboratory that the scientist has need of an understanding of history. And the time to begin the development of this understanding is while he is still in college.

The history of science is a field of vigorously active research and the history of the biological sciences is a natural smaller part of it. In the United States, *Isis*, the official journal of the History of Science Society, contains many papers in the history of biology; and an additional resource, the *Journal of the History of Biology*, edited by Everett Mendelsohn, began publication at Harvard University in the spring of 1968. The Botanical Society of America in 1963 founded an Historical Section that provides a venue for the presentation of papers in the history of botany at the Society's annual meetings. Many historians are engaging in research on the history of biology, and biologists are analyzing their science in an historical context.

By reading original primary sources of published reports and papers students can add excitement in the discovery in biology, and this information can be incorporated into the teaching of biology. This excitement is often lost by the mere presentation of textbook results and conclusions. Discussing and analyzing the original researchers' words and thoughts in the presentation of certain biological topics in courses may help bridge the gap between the sciences and the humanities, and make science more meaningful to non-science students. A wealth of source materials is available in journals, anthologies, and autobiographies. Selected excellent sources for historical information on certain specific biological topics are: *The History of Cell Respiration* by David Keilin (1966),[3] *Phage and the Origins of Molecular Biology*, edited by John Cairns, Gunther S. Steut, and James D. Watson (1966);[1] *A History of Genetics* by A. H. Sturtevant (1965);[6] and *The Origin of Genetics* edited by Curt Stern and Eva R. Sherwood (1966).[5]

No substitute, however, exists for the primary sources which provide exciting accounts of biological developments in the words of the researchers themselves. These sources should be the ones used by the teacher to develop an historical perspective in the student's thinking (Appendix 12.2). As more and more institutions add undergraduate level courses in the history of science to their curricula, the information generated from the study of the history of science should become a valuable data resource for teachers in all disciplines. College teachers of biology should be given the opportunity to learn about the wealth of information now available as a result of the continuing expansion of the literature and research in the history of science.

References Cited

1. Cairns, John, Gunther S. Steut, and James D. Watson. 1966. *Phage and the Origins of Molecular Biology*. Cold Spring Harbor Laboratory of Quantitative Biology. Cold Spring Harbor, Long Island, New York. x, 340 pp.

2. Conant, James B. 1960. History in the education of scientists. *Harvard Library Bulletin* 14: 315–333. (Quote, p. 322).

3. Keilin, David. 1966. *The History of Cell Respiration and Cytochrome*. Prepared for publication by Joan Keilin. At the [Cambridge] University Press, Cambridge, [England]. xix, 416 pp.

4. Snow, C. P. 1965. *The Two Cultures: And A Second Look. An Expanded Version of The Two Cultures and the Scientific Revolution*. At the [Cambridge] University Press, London. 107 pp.

5. Stern, Curt and Eva R. Sherwood (eds.). 1966. *The Origin of Genetics*. A Mendel Source Book. W. H. Freeman and Co., San Francisco and London. xi, 179 pp.

6. Sturtevant, A. H. 1965. *A History of Genetics*. Harper & Row, Publishers, New York. viii, 165 pp.

Publication Notes

Modified and edited from the narrative portion[a] on subject matter coverage and educational objectives of a proposal to the National Science Foundation in the United States for support of "A History of Biology Summer Institute for College Teachers in Biological Sciences," 17 June–27 July 1968 at The Ohio State University, Columbus. An addendum[b] comprises an appendix to this paper (**Paper No. 12**).

[a] 1967 [1968]. "A Proposal to the National Science Foundation for Support of a History of Biology Summer Institute for College Teachers in Biological Sciences [at] The Ohio State University." Typewritten, [4], 27 pp. [Grant].

[b] Maya Gasuk. 1996. Books still inspire great minds at Mann [Library]. *Agriculture and Life Sciences News, Cornell University*. December 1996: 8.

[c] Ronald L. Stuckey. 2000. "NSF-Sponsored Six-Weeks Summer Institute in the History of Biology for College Teachers: An Inventory of the Archives." The Ohio State University, Columbus, Ohio. Typewritten, 6 pp.

APPENDIX 12.1. SELECTED LIST OF EIGHT LECTURERS AND PRESENTATION TITLES WITH REFERENCE TO BOTANY IN THE HISTORY OF BIOLOGY, NSF INSTITUTE, 1968, 1969, BY RUDOLPH

Lecturer	Presentation Title	Summer Given
Richard G. Beidleman, Colorado College	European Scientists in the New World Wilderness	1968
	The Pattern of Biological Endeavor in Frontier America	1968
Joseph Ewan, Tulane University	Traffic in Seeds, Shells, Skins from American to European Gardens and Cabinets to 1860	1968, 1969
	The American Sources of Linnaeus	1968, 1969
	Bibliographic Tools to Answer Biohistorical Questions: Helps and Hints	1968
John C. Greene, University of Connecticut	Cuvier and Lamarck: Polar Types in the History of Science, I, II	1969
George H. M. Lawrence, Hunt Botanical Library (now Hunt Institute for Botanical Documentation)	Preservation and Conservation of Archival Materials Pertinent to Field of Botany	1968, 1969
	Botanical Literature and Its Historical Use	1968, 1969
Emanuel D. Rudolph, The Ohio State University	The Origins of and Reactions to Molecular Biology	1968, 1969
	The Popularization of Botany since 1750 (not taped)	1968, 1969

APPENDIX 12.1. (CONTINUED)

Lecturer	Presentation Title	Summer Given
Jerry Stannard, University of Kansas	Theophrastus and the Beginnings of European Botany	1968
	The Presuppositions of Aristotle's Biology	1968
	Plant Science in the Twelfth Century	1968
	Scholar Naturalists of the Sixteenth Century	1968
	Ancient and Medieval Biology	1969
Ronald L. Stuckey, The Ohio State University	Botanical Explorers of the Old Northwest Territory	1968
	Botanical Exploration of the Eastern United States, 1800–1880, I, II	1969
Conway Zirkle, University of Pennsylvania	The Discovery and Rediscovery of Mendelian Heredity	1968, 1969
	The Development of Plant Genetics and Cytology in the United States	1968, 1969

APPENDIX 12.2.
ADDENDUM ON HISTORY OF SCIENCE TEACHING,[b] BY BURK

One of the most important ideas to emerge from Rudolph's Biology Institute was the use of primary sources of historic significance in the teaching of biology courses. Now, nearly 30 years after the time of Rudolph's Institute, the paragraph below makes reference to an historic botanical book published in 1853. The use of this book is described in the present-day teaching of a botany course on the campus of Cornell University, Ithaca, New York.[b] The book mentioned is the same one Rudolph discussed in **Paper No. 1** of this memorial volume.

Plant biologist Randy Wayne finds the material in Mann Library fuel for his ideas. "I must be in there every day," he says. During one expedition in Mann, Wayne found *The Poetry of the Vegetable World; A Popular Exposition of the Science of Botany and its Relations to Man* by M. J. Schleiden. Published in 1853, Schleiden's book chronicled the thought and emotion surrounding new discoveries. Written for the general public, this book made connections between science and the larger world, and was not merely a documentation of experiments and results. In this text, Wayne saw the potential for inspiring undergraduates. Wayne's course BioPl 444: Plant Cell Biology is now based largely on this philosophy, integrating many disciplines, with an extensive reading list of books that take Schleiden's approach.

A simple laboratory table arranged near the windows of a school room.

Each shelf between the windows holds two bottles, one for water and one for alcohol. These are provided with flexible siphon tubes terminating in a pipette-like point, and closed by a pinch-cock. The siphons hang over a metal tray having bars across the top, on which slides may be placed while staining. A stand contains the necessary reagents. Each set of apparatus can be used by two pupils during each session. It is advisable to support the table by braces running to the wall, in order to escape the vibration of the floor.

Fig. 12.3. Simple laboratory table and student by window of a school room.

(From Liberty Hyde Bailey. 1901. *Botany: An Elementary Text for Schools,* Macmillan, New York)

Introduction No. 13

The Botanical Teaching Laboratory (1986–1991)

Closely allied to Rudolph's study of botanical textbooks was his interest in the manner that these books were used for teaching. As an educator in both the botanical and biological sciences, Rudolph could well appreciate the efforts of scientists and teachers who worked in classrooms and laboratories.

A study of the origin and development of the botanical teaching laboratory was one of these educational approaches (Fig. 12.3). In 1986, Rudolph was invited to give a presentation at an international symposium on "Laboratories: The Place of Experiment," organized by the Royal Institution Centre for the History of Science and Technology, held in London, England. His chosen title, "Laboratory Teaching of Botany: The United States Experience," was delivered on 19 September 1986, to be followed by publication in a symposium volume. Rudolph prepared his contribution[b] for that project, but no publication of these symposium papers resulted. He then modified various aspects of his manuscript[c,e] and gave presentations at the contributed paper sessions on 11 August 1987 at the Historical Section of the Botanical Society of America, The Ohio State University, Columbus, and on 11 July 1991 at the meeting of the International Society for the History, Philosophy, and Social Studies of Biology (ISHPSSB), Northwestern University, in Evanston, Illinois. Listed in the program under the title, "The Origin and Early Development of the Botany Teaching Laboratory in the United States," his presentation at this meeting was the last time he read a formal paper.

The manuscript prepared following the London symposium, referred to as the "long version,"[b] in contrast to the "short version"[e] for the Evanston meeting, was edited by Stuckey and Burk as a special publication[f] in the *American Journal of Botany*.[1] Paper No. 13 is developed from the two unpublished abstracts[a,d] appearing in the printed programs of the two symposium presentations. **(Paper No. 13)**.

---

[1] Emanuel D. Rudolph. 1996. History of the botanical teaching laboratory in the United States. *American Journal of Botany* 83: 661–671. [Edited by Stuckey and Burk].

Nineteenth Century Educators in the Teaching Laboratory

William J. Beal
(1833–1924)

Amos Eaton
(1776–1842)

Asa Gray
(1810–1888)

Louis Agassiz
(1807–1873)

Charles E. Bessey
(1845–1915)

Paper No. 13 (1986–1991)

Laboratory Teaching of Botany: The United States Experience

Initiating students to the investigation and experimentation with plants in a laboratory is an important development in the history of botany. This development of laboratories in teaching botanical sciences in the United States differs significantly from that in Great Britain and Continental Europe because of several unique circumstances.

First, a large number of secondary-level free schools and private academies developed before 1865. These schools taught scientific subjects, including botany. Amos Eaton (1776–1842) is credited as the first major educator in science to establish laboratory and field-laboratory teaching in the United States, early in the nineteenth century at the Rensselaer School, Troy, New York. He was influential in spreading the trend with his wide travels in the States and through his many students.[a]

Second, women were very early involved in the laboratory teaching of botany. In the second decade of the nineteenth century,[a] Amos Eaton encouraged women to attend his school and give public lectures. One of them, Laura Johnson (fl. 1834–1840), taught one of the courses to other women. Living or freshly cut plants were available in the laboratory for the students to analyze on their own and to determine the plants' names. Asa Gray (1810–1888), soon after becoming professor of natural history at Harvard University in 1837, required some practical work of all of his students in botany, somewhat similar to that developed by Eaton.[e]

Third, after 1865, land-grant state universities were founded in many states. Their

---

Frontispiece Credits: Nineteenth Century Educators in the Teaching Laboratory. William J. Beal (1833–1924). (From Ernst A. Bessey, *Botanical Gazette* 79: 103. 1925)

Asa Gray (1810–1888). (From Deane, *Bulletin of the Torrey Botanical Club* 15: opp. p. 59. 1888)

Amos Eaton (1776–1842). (Between pages 256 and 257, from A. Hunter Dupree. 1959. *Asa Gray 1810–1888*. The Belknap Press of Harvard University, Cambridge; photograph in Jane Gray's Collection of Botanists' Autographs and Biographies; permission to reproduce this photograph of A. Eaton courtesy of Archives, Library of the Gray Herbarium, Harvard University, Cambridge, Massachusetts)

Louis Agassiz (1807–1873). (From the Hunt portrait collection; permission to reproduce photograph courtesy of Hunt Institute for Botanical Documentation, Carnegie Mellon University, Pittsburgh, Pennsylvania)

Charles E. Bessey (1845–1915). (From Coulter, *Botanical Gazette* 60: 72. 1915)

explicit aim was to foster the practical aspects of agricultural and mechanical studies. Because these universities frequently used older established non-practical institutions as their models, the land-grant universities had a unique mixture of theoretical and practical approaches that is reflected in their botanical laboratory teaching.[a]

Standardized laboratory exercises for undergraduate students of botany were used in the 1870s at some of the agricultural colleges. The English translation of the grand German textbook of botany by Julius von Sachs (1875)[1] appeared, introducing structure and development of cells and tissues, as well as physiology. Emboldened with the fervor of this Germanic-modeled "New Botany," two of Asa Gray's students, Charles E. Bessey (1845–1915) at the Iowa Agricultural College, and William J. Beal (1833–1924) at the Michigan Agricultural College initiated such laboratories, used them in their courses, and proselytized for their general use (Figs. 13.1, 13.2). The times were ripe for this approach, and it rapidly spread. Before the end of the century, it became unthinkable to teach "modern" botany without a teaching laboratory component, and, in some cases, an accompanying series of field experiences.[d]

Fig. 13.1. Botanical laboratory of Charles E. Bessey with students, at the University of Nebraska, 1914. The two individuals standing at the left are the graduate assistants. Paul B. Sears is at the left; the other student is unknown.

(Permission to reproduce photograph courtesy of University Archives, University of Nebraska, Lincoln)

Fig. 13.2. Botanical laboratory of William J. Beal, at the Michigan State Agricultural College, with him standing in center and students sitting at tables, before 1900.

(Permission to reproduce photograph courtesy of the Archives and Historical Collections, Michigan State University, East Lansing)

The impetus for starting students in biology courses, rather than those in botany and zoology, came from Thomas H. Huxley (1825–1895) and Henry Newell Martin (1848–1896). They introduced laboratory exercises in their elementary biology course at Johns Hopkins University. Their program was to eclipse partially the "New Botany," and to gain the support of those individuals who believed that students should be introduced to organisms with a general biology course and its laboratory, rather than with botany and zoology courses.[e]

Fourth, the German model for teaching botany became important at the end of the century through professors who had done advanced study in Germany. The typical land-grant practical approach, however, tended to modify the European models, and this difference carried into the twentieth century.[a] Even greater emphasis on laboratory teaching came with the development of specialized supply companies, and laboratory manuals enhanced the botanical teaching laboratory during the twentieth century. Beginning botany courses with laboratory components disappeared at this time in favor of a general biology course, except in the land-grant institutions which favored the practical aspects of studying plants. Other developments were Louis Agassiz's influence in the use of field trips and field stations for laboratory teaching. The laboratories were used increasingly to conduct experiments and were supplied with more readily available American made microscopes. Laboratory use even spread to some secondary schools. Taken together, these themes show the American experience to be practical in nature, and egalitarian in intent.[e]

Reference Cited

1. Sachs, Julius. 1875. *Text-book of Botany, Morphological and Physiological*. Translated and annotated by Alfred W. Bennett, assisted by W. T. Thiselton Dyer. Clarendon Press, Oxford, [England]. xii, 858 pp.

Publication Notes

Compiled and edited from abstracts,[a,c,d] manuscripts,[b,e] and notes of papers read at meetings identified in the introduction to this paper and in the citations below.

[a] 1986. Laboratory Teaching of Botany: The United States Experience, 1 p. *In* Symposium. Laboratories: The Place of Experiment. Center for the History of Science and Technology, Royal Institution of Great Britain, London. Typewritten. [Abstract].

[b] [1986]. "Laboratory Teaching of Botany: The United States Experience." Read 19 September 1986 at London. Typewritten, 39 pp. [Manuscript, long version, including several revisions].

[c] 1987. The development of teaching laboratories for botany in nineteenth century America. *American Journal of Botany* 74: 676. [Manuscript, short version, read 11 August 1987 at Columbus, Ohio]. [Abstract].

[d] 1991. The origin and early development of the botany teaching laboratory in the United States, p. 13. *In* Abstracts, International Society for the History, Philosophy & Social Studies of Biology, Summer Meetings, July 11–14, 1991, Northwestern University, Evanston, Illinois. [Abstract].

[e] [1991]. "The Botanical Teaching Laboratory." Typewritten, 11 pp. Read 11 July 1991 at Evanston, Illinois. [Manuscript, short version].

[f] 1996. History of the botanical teaching laboratory in the United States. *American Journal of Botany* 83: 661–671. [Edited by Stuckey and Burk].

Part Four

BOTANICAL EDUCATORS

Botanical educators are both the creators and perpetuators of botanical history. Emanuel D. Rudolph's papers in this group (**Part Four**) discuss seven educators in four papers. Those on Amos Eaton (**Paper No. 14**), Daniel Drake (**Paper No. 15**), and Bruce Fink (**Paper No. 17**) are examples of botanical educators, each with a different approach to their subjects. Rudolph's paper is not about Drake's contribution as a botanical educator, but rather is on educational reform and is the only study in this group on which Rudolph published. The four physicians (**Paper No. 16**) who taught botany are Benjamin S. Barton, John Torrey, Asa Gray, and George Engelmann. All seven of these educators worked entirely in the nineteenth century except Bruce Fink, who carried over into the twentieth century.

Introduction No. 14

Amos Eaton, Admired Teacher (1988)

Because of Rudolph's teaching instincts, he admired the pioneering educational efforts of Amos Eaton (1776–1842). Rudolph considered Eaton a pivotal person in the teaching of American botany. A truly dedicated teacher, Eaton was most interested in acquiring ideas and methods of others which he presented in the clearest and simplest possible way to students in the classroom. As an example of a method for organizing information on plants, he remained loyal and taught throughout his life the Linnaean System of classification in botany (Fig. 14.1).

In his book, *Williamstown and Williams College* (1899), Arthur L. Perry (1830–1905),[1] professor of history and economics at the College, described three personal attributes of Professor Eaton that made him a popular and successful teacher of science: (1) He displayed self-confidence in matters relating to his knowledge of natural history and imparting it to students; (2) He trusted others, and in displaying this virtue gained the respect and support of his students; (3) He possessed a natural, inexhaustible, harmless, and enjoyable humor from which he "could enliven at will the driest scientific presentations." The result was that those points Eaton made were more likely to be remembered by his students, while at the same time making "himself sure to become more esteemed and endured" to them.

Eaton had the best of any teaching situation. He became allied with General Stephen Van Rensselaer (1764–1839), one of the greatest patrons and earliest benefactors of practical science in the United States, who established and endowed the "Rensselaer Institute," in Troy, New York. This school, superintended by Eaton until his death, was of the highest quality. At the opening of the Rensselaer School in 1824, Amos Eaton was the Senior Professor and taught botany until 1838, geology until 1835, and chemistry until 1835. He was followed by John Wright who was professor of botany and zoology (1838–1845). The junior professors were Lewis Caleb Beck who taught natural history (1824–1829), Eaton's son Hezekiah Hulbert Eaton (1829–1830), Paul Eugene Stevenson (1830–1835), and Ebenezer Emmons who taught geology (1835–1839). The adjuncts and assistants to senior professor Eaton (Appendix 14.1) were Fay Edgerton (1828) and John M. Barrows (1829), and those who assisted the junior professors were Eaton's son Timothy Dwight Eaton (1827), Douglass Houghton (1830), Abram Sager (1831), and James Hall (1833). Many of Eaton's students developed careers in the teaching of botany and

[1] Arthur Latham Perry. 1899. Chapter V. [portion pertaining to Amos Eaton], pp. 765–773. *In Williamstown and Williams College*. Published by The Author, Norwood, Massachusetts. viii, 2, 847 pp. (Quote, p. 770).

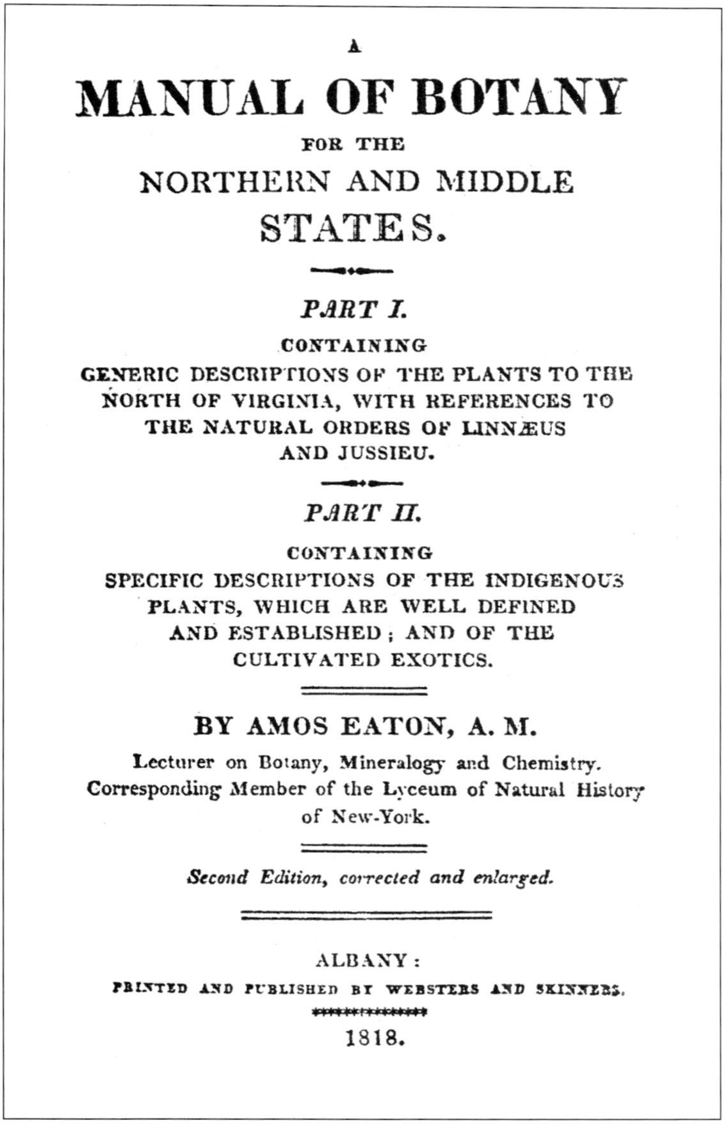

A

MANUAL OF BOTANY

FOR THE

NORTHERN AND MIDDLE

STATES.

PART I.

CONTAINING

GENERIC DESCRIPTIONS OF THE PLANTS TO THE NORTH OF VIRGINIA, WITH REFERENCES TO THE NATURAL ORDERS OF LINNÆUS AND JUSSIEU.

PART II.

CONTAINING

SPECIFIC DESCRIPTIONS OF THE INDIGENOUS PLANTS, WHICH ARE WELL DEFINED AND ESTABLISHED; AND OF THE CULTIVATED EXOTICS.

BY AMOS EATON, A. M.

Lecturer on Botany, Mineralogy and Chemistry.
Corresponding Member of the Lyceum of Natural History of New-York.

Second Edition, corrected and enlarged.

ALBANY:

PRINTED AND PUBLISHED BY WEBSTERS AND SKINNERS.

1818.

Fig. 14.1. Title page of *A Manual of Botany* by Amos Eaton (1818).

(From Rudolph's copy in the Rare Book Room, Herbarium, The Ohio State University)

natural history. Prominent teachers are listed (Appendices 14.1 and 14.2).

Rudolph's manuscript on Amos Eaton was presented on 30 April 1988 at the Annual Meeting of The Ohio Academy of Science at The Ohio State University regional campus, Newark. The manuscript was only partially written, but included notes about Eaton's life arranged chronologically. Rudolph's paper here is developed from his published abstract,[a] his manuscript,[b] and other published sources[c] supplied by the editors. **(Paper No. 14)**.

Amos Eaton, Popularizer of Botany

POPULARIZING BOTANY

Books must follow sciences, and not sciences books.—BACON

AMOS EATON'S interest in the study of botany began while he was studying law in New York City at the opening of the nineteenth century. It was there that he came under the influence of Dr. David Hosack, one of the leading scientists of the day and a member of the medical faculty of Columbia College.[1] Of this early interest in science, Eaton later wrote to his son: "I believe you know, that I was one of the few—(I believe I may say one of the three or four) who endeavored to keep pace with the modern sciences of Europe; when we had neither books nor any other facilities with which the present generation is favored."[2]

This early interest in science was continued by Eaton as an avocation when he established his Botanical School at Catskill in 1810. Hosack's interest in him also continued warm and friendly as evidenced by the very encouraging letter he wrote to Eaton praising the methods of instruction he was adopting in his Catskill school.[3] Of this unique attempt at popularizing botany, Eaton wrote: "In May, 1810, I made the first attempt in this country at a popular course of lectures; with a view to make practical Botanists of young persons of all conditions and pursuits. For this class I compiled a small elementary treatise."[4]

Amos Eaton, the first educator in the United States to popularize botany, from McAllister (1941, p. 212)

Amos Eaton (1776–1842)

Paper No. 14 (1988)

Amos Eaton (1776–1842): A Pivotal Person for American Botany

Amos Eaton (1776–1842), trained as an attorney, became a self-made naturalist, writing and lecturing about natural history and chemistry as a career later in life. He was a popular public lecturer and in that way interested many students in geology, chemistry, and botany. Eaton had a particularly important role in the development of American botany at a time when the country was becoming more interested in public education and in the study of its natural resources. He wrote the first manual of American plants for student use in 1817; helped set the pattern for teaching about plants using laboratory methods in a course he established at the Rensselaer School in Troy, New York; and believed that women should be taught about plants in a manner equal to men, and in doing so, he trained teachers who educated many other women in botanical subjects.[a]

Popularizer of Botany

Eaton, after obtaining an A.B. degree at Williams College in Massachusetts, began to popularize botany. Of these efforts, Eaton[4,6] wrote that ". . . in May, 1810, I made the first attempt in this country at a popular course of lectures; with a view to make practical Botanists of young persons of all conditions and pursuits . . . ." On the completion of a formal set of botanical lectures at Williams College, his students paid for the printing of his *A Manual of Botany for the Northern States* (1817),[2] a slender volume for the identification of plants according to the Linnaean System.[c]

In 1824 Eaton succeeded in inducing Stephen Van Rensselaer (1764–1839), a wealthy land-owner in the Albany area of New York State, to support the founding of a new school to train needed teachers. Eaton's interests led him to emphasize the sciences in the curriculum and establish teaching laboratories for these subjects. Eaton's[7] stated aim was to have:

Frontispiece Credits: Amos Eaton, Popularizer of Botany. Portion of page 212 on the topic "Popularizing Botany." (From Ethel M. McAllister. 1941. *Amos Eaton: Scientist and Educator 1776–1842*; permission to reproduce this item courtesy of University of Pennsylvania Press, Philadelphia)

Amos Eaton (1776–1842). (Between pages 28 and 29, from Ethel M. McAllister. 1941. *Amos Eaton: Scientist and Educator 1776–1842*. University of Pennsylvania Press, Philadelphia; permission to reproduce photograph courtesy of Rensselaer Polytechnic Institute, Troy, New York)

> The pupil in the place of the professor, [where] he necessarily acquires a knowledge of the principles of the science on which he lectures; while the experimental demonstrations of the laboratory render him familiar with the practical application of those principles to agricultural and manufacturing operations.

Thus the pupils were becoming teachers, even in school, and at the same time learning practical knowledge. This teaching technique was Eaton's[8] "pedagogical theory emphasizing 'the application of science to the common purposes of life'; students were to learn by doing, in sharp contrast with the conventional method of learning by rote." Eaton encouraged women to attend his public lectures and some of his school lectures. His student, Laura Johnson (fl. 1834–1840), taught one course to other women. Plant specimens were available in the laboratory for study. Some years earlier, Eaton published a guide for students, the *Botanical Exercises* (1820),[3] which in its introduction gave this advice to teachers:

> At your examination meetings, always have the whole class analyze two or three of the plants described in these exercises. Give each a few flowers of the same kind, and assure them it is contained somewhere in this book. Then let each pursue the search alone, and shew you the supposed name, without making it known to the members of the class . . . . Make practical botanists of your pupils, and they can then read physiological botany in their closets to advantage.[c]

During Eaton's years, 1824–1842, nearly 400 students attended the Rensselaer School, which in 1835 first offered degrees in civil engineering. It would be misleading to think that only engineering students were influenced by him. Eaton's popular lectures and writings, as well as his private instruction, gave him a much wider influence, particularly among women. Three of them, Emma Willard (1787–1870), Almira Hart Lincoln Phelps (1793–1884), and Mary Lyon (1797–1849), were instrumental in initiating an American female seminary movement. Numerous seminaries and academies, both male and female, operated in the early part of the century. Over 1,000 such institutions existed in the United States by 1830, and over 4,000 by mid- century. In many of these institutions, Eaton's manuals of botany and Mrs. Phelps' textbooks of botany, based upon Eatonian thinking, were used in the laboratory. It was Eaton's original aim to train teachers, and that is what he and his disciples did. The number of students influenced by these teachers, male and female, must have been very large indeed. These students were able to identify plants by name and know something about their significance in the vegetation. Thus Eaton and his followers provided the rudiments of systematics and ecology. Field trips also were used by Eaton as a laboratory for instruction of students. These trips involved the study of all aspects of natural history.[c]

Personal Commentaries by Two Students

From among the many students who studied botany under Amos Eaton, the following two selections provide insight into the personality and scientific abilities of the man. In *A History of Williams College* by Durfee (1860),[1] a Professor A. Hopkins is quoted in the account of Amos Eaton as follows:

> Professor Eaton was one of the first to popularize science in the Northern States. For this task he had some special qualifications. He had an easy flow of language, a popular address, and a generous enthusiasm in matters of science, which easily communicated itself to his pupils. I well remember attending a lecture of his in my native town, the first scientific lecture I ever attended, and if I may judge by the sharp outline of it still in my mind, one of the most interesting and impressive. Perhaps the 'leafy month of June,' the subject of the lecture, *Flowers*, and the presence of a large number of interesting young persons, may have added something to the charm; but making all due allowances, I am sure that the lecture itself must have had a good deal of intrinsic merit.
>
> Professor Eaton was at this time (1817, I think) nearly in his prime. His person was quite striking,—a large frame, somewhat portly and dignified, though entirely free from what is commonly called *starch*. His face was highly intellectual,—the forehead high and somewhat retreating, locality strongly marked, and the organs of observation and comparison well developed. His hair at that time was black, and being combed back, rendered his fine physiognomy still more striking.
>
> I well recollect the flowers, which I believe his young pupil [Ebenezer] Emmons (1799–1863) had been employed to collect for the occasion (Fig. 14.2). They were, in the first place, the common lilac, which I had probably seen before; however this may be, the small floret, with its salver-form corolla and long tubular throat, into which the Professor dexterously inserted his penknife, with no murderous intent, but to give us a view of the organs which the great Linnaeus had selected as the basis of his classification,—this little floret, I say, is the first I now recollect to have seen; and seen it was, and still is, with great distinctness. Then followed the pedicularis and some plants more difficult in their analysis. In the analysis of these plants Professor Eaton made use of his manual, descriptive of plants in the vicinity of Williams College, a book which, with some imperfections, was highly valuable as a pioneer work.
>
> Professor Eaton was among the first in this country to study Nature *in the field* with his classes. In pursuance of this idea, he used to make an annual excursion with the Rensselaer School; sometimes leading these expeditions in person, at others deputing some competent teacher to take the lead.

• • •

> In the spring of 1819, while I was a member of Lenox Academy, Mr. Eaton came there, at the request of Hon. H. W. Bishop, and delivered a short course of lectures on Botany. And I retain a vivid recollection of his manly appearance, his constantly flowing conversation, and his instructive lectures.

• • •

The young 17-year-old Asa Fitch (1809–1879) from the small town of Salem, New York, decided to visit Troy in the spring of 1826 and ask Amos Eaton about attending the Rensselaer School that autumn (Fig. 14.3). Fitch[5] in his diary, described Eaton:

> There he sat drawing on a pair of boots . . . that appeared to be made in the fashion of a former age. His remaining dress was full as much out of fashion.

Fitch was impressed with the 50-year-old Professor Eaton, so much so that he agreed that summer to accompany him and nineteen students and former students on a traveling school of science for seven weeks on the Erie Canal in 1826.[b] This was the same field excursion during which the well known botanist, Constantine S. Rafinesque (1783–1840), joined the group. . . .

Fig. 14.2. Ebenezer Emmons (1799–1863).

Fig. 14.3. Asa Fitch (1809–1879).

(From Jeffrey K. Barnes. 1988. Asa Fitch and the Emergence of American Entomology . . . . *New York State Museum Bulletin* No. 461. pp. 30, 40. Permission to reproduce photographs courtesy of New York State Museum and Science Museum, Albany, New York)

References Cited

1. Durfee, Calvin. 1860. [Amos Eaton], pp. 358–371 + portrait. *In A History of Williams College*. A. Williams and Co., Boston. x, 432 pp. (Quote, pp. 359–361).

2. Eaton, A. 1817. *A Manual of Botany for the Northern States, ...* By the Members of the Botanical Class in Williams' College (Mass.), from a Manuscript System Compiled by the Author of Richard's *Botanical Dictionary*. Websters and Skinners, Albany, New York. vi, 164 pp.

3. _____. 1820. *Botanical Exercises, Including Directions, Rules and Descriptions, Calculated to Aid Pupils in the Analysis of Plants; With a Labelling Catalogue, for the Assistance of Teachers*. Websters and Skinners, Albany, New York. 171 pp. (Quote, pp. 5–6).

4. _____ and John Wright. 1840. *North American Botany; Comprising the Native and Common Cultivated Plants, North of Mexico: Genera Arranged According to the Artificial and Natural Methods*. 8th ed. Published by Elias Gates, Troy, N[ew] Y[ork]. vii, 625 pp. (Quote, p. v).

5. Fitch, Asa. 1827. [Notation for 5 April 1826]. Asa Fitch Diary, 23 February 1827, Manuscript Group 215 (Asa Fitch Papers), Manuscripts and Archives, Sterling Memorial Library, Yale University. Cited *In*. Barnes, Jeffrey K. 1988. Asa Fitch and the Emergence of American Entomology . . . *New York State Museum Bulletin* No. 461. Albany, New York. i–viii, 1–20 pp. (Quote, p. 8).

6. McAllister, Ethel M. 1941. *Amos Eaton: Scientist and Educator 1776–1842*. University of Pennsylvania Press, Philadelphia, xiii, 587 pp. (Quote, p. 212).

7. Nason, Henry B. [ed.]. 1887. *Biographical Record of the Officers and Graduates of the Rensselaer Polytechnic Institute, 1824–1886*. [With an Introduction by Benjamin H. Hall]. William H. Young, Troy, New York. vii, 614 pp. (Quote, p. 18).

8. Rezneck, Samuel. 1971. Amos Eaton. *In* Charles Coulston Gillispie [ed.], *Dictionary of Scientific Biography*, Vol. 4, pp. 273–275. Charles Scribner's Sons, New York. (Quote, p. 274).

Publication Notes

Compiled and edited from an abstract[a] and manuscript[b] read 30 April 1988 at the Annual Meeting of The Ohio Academy of Science, The Ohio State University regional campus at Newark, and selected passages from a paper[c] published 1996 in the *American Journal of Botany* written by Rudolph and posthumously edited by Stuckey and Burk.

[a] 1988. Amos Eaton (1776–1842), a pivotal person for American botany. *The Ohio Journal of Science* 88(2): 12. [Abstract].

[b] [1988]. "Amos Eaton." Handwritten, 12 pp. [Manuscript].

[c] 1996. History of the botanical teaching laboratory in the United States. *American Journal of Botany* 83: 661–671.

APPENDIX 14.1.
SELECTED RENSSELAER SCHOOL STUDENTS OF AMOS EATON WHO BECAME PROFESSIONAL EDUCATORS, BY STUCKEY AND BURK

Student	Rensselaer School Degree, Year
Barrows, John Manning 1807–1891	A.B. 1829
Eaton, Hezekiah Hulbert 1809–1832	A.B. 1826
Eaton, Timothy Dwight 1807–1828	A.B. 1826
Edgerton, Fay 1803–1832	A.B. 1828
Fitch, Asa 1809–1878	A.B. 1827
Hall, James 1811–1898	B.N.S. 1832, M.A. 1833
Horsford, Eben Norton 1818–1893	Civil Engineering 1838
Houghton, Douglass 1809–1845	A.B. 1829
Lyon, Mary 1797–1849	Studied with Eaton in the summer of 1825

Degrees from other schools	Position, Year(s), Location
	After teaching in several towns, became professor of natural science (1860–1870), Olivet College, Michigan
	Taught natural science and mathematics (1829–1832), Eclectic Institute, Lexington, Kentucky; botanical companion of Charles W. Short, M.D.
	Adjunct professor in natural history under Lewis C. Beck at the Rensselaer School (1826–1828)
	Professor of chemistry and natural sciences (1828–1832), Utica Gymnasium, Utica, New York
M.D. 1829 Castleton Medical College, Vermont	Lectured in botany (1838–1845) in Stillwater, Granville, and Salem, New York; worked as an entomologist in the State of New York (1845–1853); state entomologist of New York (1854–1870)
	Librarian (1832–1835) and professor of chemistry (1835–1841), Rensselaer School; New York Geological Survey, Albany (1837–1898)
Studied chemistry under Justus Liebig, University of Giessen, Germany (1844–1846)	Professor of mathematics and natural history at Albany Female Academy (1840–1844), New York; lecturer on application of science to useful arts (1847–1863), Lawrence Scientific School, Harvard University, Cambridge, Massachusetts
	Physician (1832–1837); first state geologist of Michigan (1837–1842) and professor of geology, mineralogy, and chemistry (1838–1845), University of Michigan, Ann Arbor; mayor of Detroit (1842–1843)
	Founder and principal of Mount Holyoke Female Seminary (1837–1849), now Mount Holyoke College, South Hadley, Massachusetts

APPENDIX 14.1. (CONTINUED)

Student	Rensselaer School Degree, Year
Phelps, Almira (Hart) Lincoln 1793–1884	Attended Eaton's lectures (1823–1824) and was an outstanding student
Peter, Robert 1805–1894	Non-graduate 1828
Riddell, John Leonard 1807–1865	A.B. 1829
Sager, Abram 1810–1877	A.B. 1831
Stevenson, Paul Eugene, Rev. 1809–1870	A.M. 1839
Wright, John 1811–1846	Non-graduate

Degrees from other schools	Position, Year(s), Location
	Taught botany and other subjects in various female academies in the northeastern United States (1806–1856), including her sister's school, the Troy Female Seminary or Emma Willard School (1824–1831), Troy, New York
M.D. 1834 Transylvania University, Kentucky	Professor of chemistry (1833–1857) and Dean of Medical College (1847–1857), Transylvania University, Lexington, Kentucky; professor of chemistry (1865–1877), University of Kentucky, Lexington; botanical companion of Charles W. Short, M.D.
M.D. 1834 Reformed Medical College of Ohio, Worthington; M.D. 1836, Medical Department, Cincinnati College, Ohio	Lecturer in chemistry and botany (1832–1834), Reformed Medical College of Ohio, Worthington; adjunct professor of chemistry and lecturer on botany (1835–1839), Daniel Drake's Medical Department, Cincinnati College, Ohio; professor of chemistry (1836–1865), Medical College of Louisiana, New Orleans
M.D. 1835 Castleton Medical College, Vermont	Engaged in botanical and zoological studies (1837–1838), Geological Survey of Michigan, under Douglass Houghton; professor of zoology and botany, (1842–1855), University of Michigan, Ann Arbor
A.B. 1833 Union College, Schenectady, New York; graduated 1837 from Theological Seminary, Princeton, New Jersey	Teacher and minister in several towns in Virginia, New York, and Pennsylvania
M.D. 1833 Yale College, New Haven, Connecticut	Engaged in botanical studies (1838), Geological Survey of Michigan, under Douglass Houghton; successor to Amos Eaton in botany and zoology at the Rensselaer School (1838–1845)

Appendix 14.2.
Selected Students of Amos Eaton from Schools Other than Rensselaer Who Became Professional Educators, by Stuckey and Burk

Student	Degree, Year, and School
Burritt, Elijah Hinsdale 1794–1838	Non-graduate, 1816, Williams College
Eaton, Hezekiah Hulbert 1809–1832	Assistant to Amos Eaton, 1822, Castleton Medical Academy
Emmons, Ebenezer 1799–1863	A.B. 1818, Williams College; A.B. 1826, Rensselaer School; M.D. 1830, Berkshire Medical Institute
Ingalls, Thomas Russell 1798–1864	Assistant to Eaton, 1822, Castleton Medical Academy; graduated 1822, United States Military Academy, West Point, New York; M.D. 1826, Vermont Academy of Medicine
Lyon, Mary 1797–1849	Studied under Eaton, 1824, at Amherst, Massachusetts
Pitcher, Zina 1797–1872	Studied with Eaton at Castleton Academy; M.D. 1822, Middlebury College

Position, Year(s), Location
Taught school at Milledgeville, Georgia (1819–1829); conducted boarding and day school (1829–1837), New Britain, Connecticut; organized and directed group of colonists to Texas (1837–1838)
See Appendix 14.1
Lecturer in chemistry (1828–1834), professor of natural history (1833–1859), and professor of geology and mineralogy (1859–1863), Williams College; investigator in the First Geological Survey, New York (1837–1843)
Assistant to Lewis C. Beck and Amos Eaton (probably before 1828), Rensselaer School; professor of chemistry and natural history, Jefferson College, Louisiana (1829–1833); professor of chemistry, Medical College, New Orleans (1834–1836); president, Jefferson College (1836–1841)
See Appendix 14.1
United States Army surgeon in the Michigan Territory and elsewhere (1822–1836); medical practice in and mayor of Detroit (1840, 1841, 1843); member University of Michigan Board of Regents (1837–1852), professor of the institutes of medicine and obstetrics (1851–1872), University of Michigan, Ann Arbor

Introduction No. 15

Daniel Drake, Educational Reformer (1985)

Daniel Drake (1785–1852) known for his achievements in the medical profession and civic activities of Cincinnati, was also a scientist, most interested in the now largely forgotten subject of medical topography. This interest gave him occasion to study geology and botany, to which he was a truly dedicated promoter of the latter (Appendix 15.1).[1]

To celebrate the bicentennial of Drake's birth, a symposium on "The Legacy of Daniel Drake, M.D. (1785–1852)," was organized by Stuckey.[2] Most of the speakers chosen had been invited before Stuckey became ill and was hospitalized from October 1984 to May 1985. The symposium, presented at a joint meeting of the Ohio Academy of Medical History and The Ohio Academy of Science, was held 19 April 1985 under the leadership of Billie K. Broaddus, Director of the History-Health Sciences Library and Museum in the Medical School at the University of Cincinnati. Eight scholars were contributors, including Rudolph, who spoke on "Daniel Drake as a Nineteenth Century Educational Reformer." He later reported on the symposium at a meeting of The Ohio State University Medical History Group, 4 June 1985.

Two anonymous reviewers[3] returned comments with Rudolph's manuscript. One wrote that his contribution "identifies correctly those issues Drake was most concerned about." The other one stated:

> The special virtue of this essay is that it collects Drake's words and identifies Drake's concerns, and expands our vision of his educational work beyond the exclusive focus on medical-scientific education which has dominated the literature on him to date. It is an entirely appropriate contribution to the bicentennial of Drake's birth and an important reminder that "scientists" are also citizens [concerned] for the well-being of their communities.

---

[1] John W. Frederick. 1988. Daniel Drake and the botany of the upper Ohio valley. *Bartonia* 54: 83–97.

[2] Ronald L. Stuckey. 1985. Memorandum to Participants in the Program Commemorating the Bicentennial of Dr. Daniel Drake, 1785–1985. Typewritten, 1 p.

[3] [Anonymous]. [1985]. [Two reviews of manuscript on "Daniel Drake as a Nineteenth Century Educational Reformer," by Emanuel D. Rudolph for *The Ohio Journal of Science*]. 1 p.; 1 p.

A souvenir booklet titled, *Daniel Drake, M.D. Frontiersman of the Mind* (1985),[4] consisting of three of Drake's previously unpublished convocation or valedictory addresses, was published and distributed at the anniversary event. The subjects of the addresses were: (1) *Doctorate Address*, delivered at the Medical Commencement of Transylvania University, March 16, 1827; (2) *Formation of Professional Character*, delivered at the First Session of the Medical Department of the Cincinnati College, November 2, 1835; and (3) *Valedictory Address*, delivered to the Graduates of the Louisville Medical Institute, Third Session, March 10, 1840. The book cover (Fig. 15.1) displayed the dogwood flower with its fully opened associated bracts, which Drake used as the emblem for his journal, *The Western Journal of Medicine and the Associate Sciences* (1828–1838). Drake's biographers believed that his early experiences in the forest and his love of nature contributed to his selection of this beautiful blossom.

Although some of the presentations were published, no symposium volume was written about the event. From Rudolph's paper,[c] published in *The Ohio Journal of Science,* two selected paragraphs are combined with his published abstract[b] from the meeting program to form a resume of his published contribution printed here. **(Paper No. 15)**.

Fig. 15.1. Book cover page of *Daniel Drake, M.D. Frontiersman of the Mind* (1985).

(From the copy owned by Ronald L. Stuckey; permission to reproduce illustration courtesy of the Cincinnati Medical Heritage Center, University of Cincinnati Medical Center)

[4] Charles D. Aring and Cory Oysler. 1985. *Daniel Drake, M.D. Frontiersman of the Mind: Three Unpublished Addresses of Daniel Drake, M.D., to Medical Students ....* Crossroads Books, Cincinnati. [xxviii], 60 pp.

Daniel Drake, Educational Reformer

Daniel Drake (1785–1852)

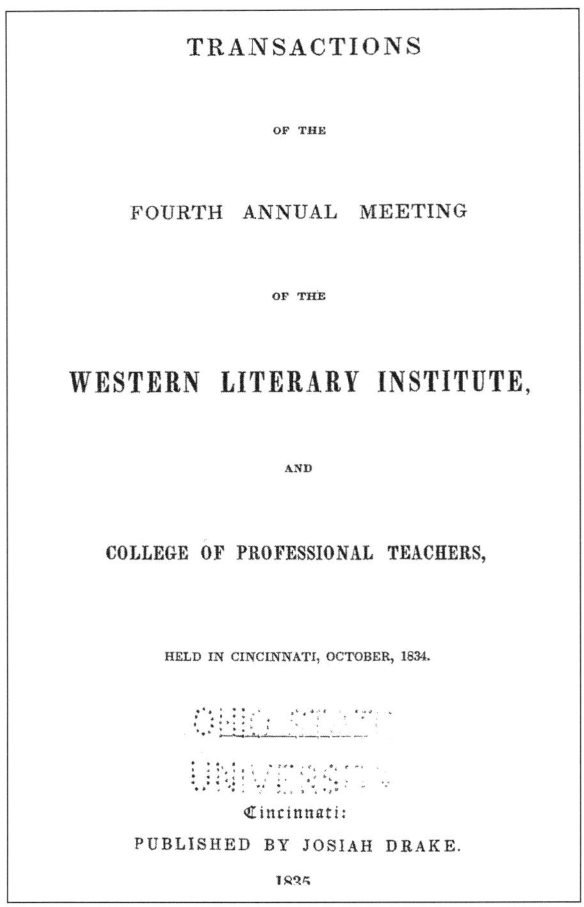

TRANSACTIONS

OF THE

FOURTH ANNUAL MEETING

OF THE

WESTERN LITERARY INSTITUTE,

AND

COLLEGE OF PROFESSIONAL TEACHERS,

HELD IN CINCINNATI, OCTOBER, 1834.

Cincinnati:

PUBLISHED BY JOSIAH DRAKE.

1835

Title page from an educational journal, *Transactions of the Fourth Annual Meeting of the Western Literary Institute and College of Professional Teachers*, published by Josiah Drake (1835)

Paper No. 15 (1985)

Daniel Drake (1785–1852): A Nineteenth Century Educational Reformer

Daniel Drake (1785–1852) was called the "Franklin of the West," because like Benjamin Franklin, he was a versatile person who initiated and became involved with numerous scientific and social projects. On this occasion of the bicentennial of his birth (1985), Drake's activities related to the education of youth are considered, particularly his connection with the Western Literary Institute and College of Professional Teachers from 1831 to 1839. These activities are discussed with reference to Drake's philosophy and to educational theories of those times.[a,c]

If Daniel Drake is remembered as an educator, it is usually for his pioneering efforts in establishing medical schools and related institutions in Ohio, and for his attempts to improve the quality of medical instruction. Drake's entire professional and public life, however, can be interpreted as a continuous educational endeavor. Although Drake's own education was limited, he early learned the value of reading for self-instruction and preached the value of that pursuit for the remainder of his life. After his experiences in Philadelphia, Drake tried to emulate its educational and professional establishments in Cincinnati. As time went on, he developed a more independent approach to education which championed a new "Western Country" attitude. He was active in The Western Literary Institute and College of Professional Teachers, an organization with a broad view of the term "teacher." These meetings, held annually in the 1830s, were Drake's most active involvement with the reforming of non-medical teaching.[b,c]

Issues of public education that were ever fresh were the major concerns of Daniel Drake. Among these were: universal free education, the recognition and rewarding of merit, the relationship of religion to public education; the training for professions in public schools; improving the education of females; the importance of understanding the human body and of exercising it properly. Even if Drake's language is flowery and somewhat archaic, his ideas are still simple and sensible. Some of his thoughts about education and the importance of the West received wider circulation than did the College of Teachers because a few of his writings were excerpted in the McGuffey Readers in the 1840s and early 1850s. Drake's

---

Frontispiece Credits: Daniel Drake, Educational Reformer. Daniel Drake (1785–1852). (Between pages 256 and 257, from Emmet F. Horine. 1961. *Daniel Drake (1785–1852): Pioneer Physician of the Midwest*. University of Pennsylvania Press, Philadelphia; permission to reproduce photograph courtesy of portrait owner)

Title page of *Transactions of the Fourth Annual Meeting of the Western Library Institute*. (From the copy in the library of The Ohio State University)

role was an important one that has had lasting value, even if he was a reformer well ahead of his time in ideas.[a,c]

Publication Notes

Compiled and edited from a manuscript,[a] an abstract,[b] and a published paper.[c]

[a] [1985]. "Daniel Drake as a Nineteenth Century Educational Reformer." Typewritten, 15 pp.; handwritten, 15 pp. [Manuscript].

[b] 1985. Daniel Drake as an educator. *The Ohio Journal of Science* 85(2): 3. [Abstract].

[c] 1985. Daniel Drake as a nineteenth century educational reformer. *The Ohio Journal of Science* 85: 146–151. [See pp. 146, 150–151].

APPENDIX 15.1.
DANIEL DRAKE AS A PROMOTER OF FLORISTIC BOTANY, BY STUCKEY[1]

As a scientist, Daniel Drake was most interested in the natural sciences, especially botany in which he was the first resident of the West to publish a list of trees (*Picture of Cincinnati*, 1815), as well as a phenological record and medical properties of indigenous plants (*Notices*, 1810). He was primarily a promoter of botany, being the first individual to give a course of *botanical* lectures, as advertised in the Cincinnati newspaper, *The Western Spy*, 4 April 1818. More than forty subscribers enrolled when Drake delivered the introductory lecture on Friday evening, 8 May 1818, "at candle light in the First Presbyterian Church." Newspaper comment indicated that the lecture was of much interest, and "it was truly gratifying to see so numerous and respectable an audience, at the first attempt of the kind in the Western Country. ..." (Horine, 1961).[2] The introductory lecture was printed in *The Western Spy*, 17 October 1818.

Other botanical firsts were his unpublished valedictory lecture urging medical students at Transylvania University to study botany (1818), and his guidelines for the collection of plants and preparation of herbarium specimens by medical students, published in the *Western Journal of Medicine and the Associate Sciences* (2: 207–207. 1828). Drake further promoted botany by reviewing publications on the flora of eastern North America, urging the writing of local and regional floras of the Mississippi valley, pointing out the necessity for young physicians to study the specialized sciences, especially botany, and arguing for more professorships in natural history and botany in medical schools.

---

[1] Ronald L. Stuckey. 1985. Daniel Drake as a promoter of floristic botany and his circle of botanical students. *The Ohio Journal of Science* 85(2): 2. (Abstract).

[2] Emmet Field Horine. 1961. *Daniel Drake (1785–1852): Pioneer Physician of the Midwest*. University of Pennsylvania Press, Philadelphia. 425 pp. (Quote, p. 153).

Introduction No. 16

Four Medical Botanists (1980)

Among the earliest teachers of botany in the United States were medical doctors who held professorships in colleges and universities. While they usually taught courses in medical subjects or medical botany, they also presented instruction on the basic knowledge about the function, structure, and identification of plants. Rudolph chose four individuals who had medical degrees, of whom two were engaged in medical practice. These physicians—Benjamin Smith Barton, John Torrey, Asa Gray, and George Engelmann—all taught botany, but were better known for their botanical investigations and publications. Barton is the only one known of these four to have published on medical botany (Fig. 16.1).

Rudolph's study on these physician-botanists was presented 21 April 1980 at the annual meeting of the Ohio Academy of Medical History in Granville, Ohio. To create the paper printed here, Rudolph's manuscript[b] read at this meeting is combined with a four-paragraph introduction taken from a 1972 two-page typewritten manuscript,[a] titled "American Medical Botany." **(Paper No. 16)**.

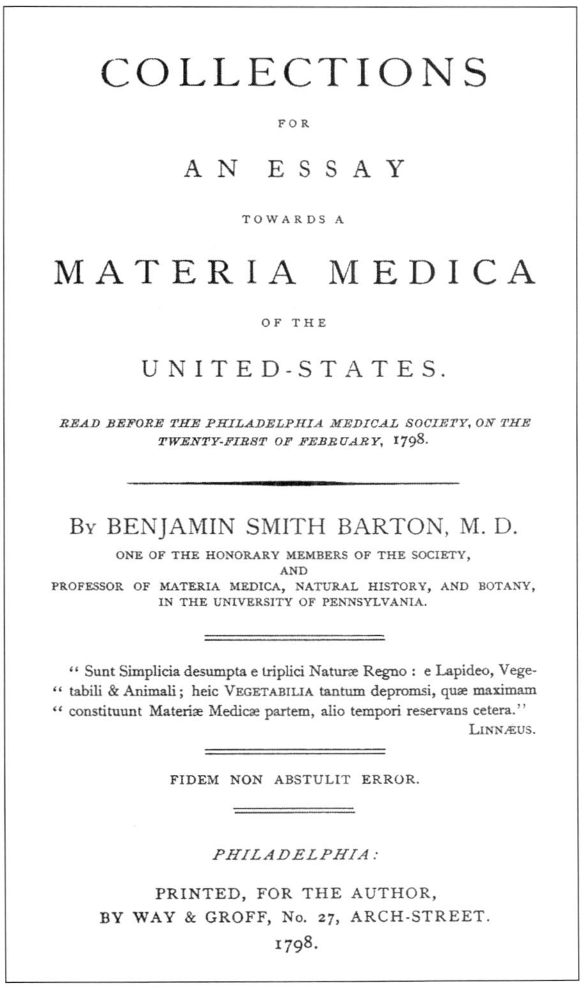
COLLECTIONS

FOR

AN ESSAY

TOWARDS A

MATERIA MEDICA

OF THE

UNITED-STATES.

READ BEFORE THE PHILADELPHIA MEDICAL SOCIETY, ON THE TWENTY-FIRST OF FEBRUARY, 1798.

BY BENJAMIN SMITH BARTON, M. D.

ONE OF THE HONORARY MEMBERS OF THE SOCIETY,
AND
PROFESSOR OF MATERIA MEDICA, NATURAL HISTORY, AND BOTANY,
IN THE UNIVERSITY OF PENNSYLVANIA.

" Sunt Simplicia desumpta e triplici Naturæ Regno : e Lapideo, Vege-
" tabili & Animali ; heic VEGETABILIA tantum depromsi, quæ maximam
" constituunt Materiæ Medicæ partem, alio tempori reservans cetera."

LINNÆUS.

FIDEM NON ABSTULIT ERROR.

PHILADELPHIA:

PRINTED, FOR THE AUTHOR,
BY WAY & GROFF, No. 27, ARCH-STREET.
1798.

Fig. 16.1. Title page of *Collections for an Essay Towards a Materia Medica. . .* by Benjamin Smith Barton (1798).

(From Barton, *Bulletin of the Lloyd Library . . .*, No. 1, *Reproduction Series*, No. 1. 1900; permission to reproduce title page courtesy of the Lloyd Library, Cincinnati, Ohio)

Four Physicians Who Taught Botany

Benjamin Smith Barton (1766–1815)

John Torrey (1796–1873)

Asa Gray (1810–1888)

George Engelmann (1809–1884)

Paper No. 16 (1980)

American Medical Botany: Four Physicians Who Taught Botany to Nineteenth Century American Students

The early history of botany was intimately connected with medicine. Until the professionalization of botany in the late nineteenth century, students interested in plant life realized that medicine was the major professional opportunity available to them that had considerable botanical content. One has only to realize that Linnaeus, the great botanist of the eighteenth century, was a medical doctor by profession. As late as 1875, a British reviewer[39] of a new German botany textbook wrote:

> The circumstances under which most of our text-books have been produced—that is, to supply pabulum suited to the wants of the medical schools of London—are responsible to some extent for this unsatisfactory state of things; . . . .

In the United States, and earlier in the American colonies, medical doctors performed a major role in the development of botany. Botany, as well as other leisure activities, began to be studied on the East Coast when it was the "frontier" of Europe. The pioneer students of plants were European-trained, or at least looked to Europe for their guidance in botanical matters. England and France were the supporters of botanical exploration, and their motivation was often the hope for new medicinal herbs. Centers of scientific activities grew in various places along the coast, and here we will learn about those in New England and Pennsylvania. Equally active were other centers such as those in New York, Virginia, and the Carolinas. As the population moved west, botanical interest increased in new western plants.

Frontispiece Credits: Four Physicians Who Taught Botany. Benjamin Smith Barton (1766–1815). (From Ewan, *Proceedings of the American Philosophical Society* 127: 125–134. 1983. The original portrait is an aquatint by Bass Otis (1784–1861) of Bridgeport, Massachusetts, who came to Philadelphia in 1812; also reproduced in Ewan, *Bartonia* 54: 28. 1988)

John Torrey (1796–1873). (Gift to RLS from Andrew D. Rodgers III, photograph in the collection of RLS. The original is at the New York Botanical Garden)

Asa Gray (1810–1888). (Frontispiece opposite title page from A. Hunter Dupree. 1959. *Asa Gray 1810–1888*, The Belknap Press of Harvard University Press, Cambridge. The original is a crayon portrait by Sir David McNee, P. R. S. A., in the possession of the Royal Botanic Gardens; permission to reproduce photograph of A. Gray courtesy of Royal Botanic Gardens, Kew, England)

George Engelmann (1809–1884). (From Spaulding *Popular Science Monthly* 74: 126. 1909)

Medical botany became a pursuit of some of the residents of the new western frontier. These students, mostly medical doctors, were different from the Easterners because instead of looking to Europe for their inspiration and guidance, they looked toward the established botanical centers in the East.[11,32] Further, the new frontier fostered the development of various irregular medical sects that often had strong botanical interests. These sects had their own medical schools and courses in botanical subjects. With the help of these new Western medical botanists, a truly native American botany was developed with its own characteristics.[33]

The development of the new American botany has a history that mostly remains to be investigated. By the later part of the nineteenth century, botany had become a recognized profession and, in the United States, courses in botany were common in schools, academies, and colleges. The strength of the late nineteenth century American botany and its twentieth century successor was dependent upon the great medical botanists whose story is beginning to be revealed.

Botany, as a scientific discipline, developed as a profession in the United States during the mid-nineteenth century. Its growth and development depended upon the efforts of many individuals, but primarily medical doctors who were trained, as part of their education, in *materia medica* and natural history. Howard A. Kelly, M.D., in his book *Some American Medical Botanists Commemorated in Our Botanical Nomenclature* (1914),[20] provided biographies of 30 physicians all of whom were in one way or another important in the history of American botany in the nineteenth century. Other botanical physicians were not included by Kelly because no plants were named for them. These men were the founders of the profession of botany in this country because they together with other non-medical people, including many women, became the botanical corps, both amateur and professional.

Four individuals among this distinguished group of physicians, who were especially important in the botanical education of students in the nineteenth century, are discussed here. They are Benjamin Smith Barton (1766–1815) of Philadelphia; John Torrey (1796–1873) of New York City; Asa Gray (1810–1888) of Cambridge, Massachusetts; and George Engelmann (1809–1884) of St. Louis. Each in his own way is especially known for substantial contributions made to systematic botany. Only their roles as educators or initiators of the young into botany are discussed here. Of the four, only two, Benjamin Smith Barton and George Engelmann, were actively practicing physicians. Barton's practice was small because of his teaching activities. Each physician is discussed separately, followed by an extensive general summation of their total importance for the embryonic field of botany in this country.

Benjamin Smith Barton (1766–1815)

Benjamin Smith Barton, the son of an Episcopal minister in Lancaster, Pennsylvania, and nephew of the astronomer David Rittenhouse (1732–1796) of Philadelphia, carried on medical studies under the guidance of Doctor William Shippen (1736?–1808) of Philadelphia. Having spent some time during 1785 with his uncle conducting a survey of the western border of Pennsylvania, he left for Scotland to study medicine at the University of Edinburgh. Barton took course work there for two years, leaving without his medical degree, which he is said to have obtained then at the University of Göttingen.[5] Upon returning to the United States, he became the first American teacher of natural history and botany by being appointed to the chair of natural history at The College of Philadelphia in 1789. Later, in 1791, that college became the University of Pennsylvania, and he continued in his chairmanship there. In 1798, he became physician to the Pennsylvania Hospital and in 1813 professor of the theory and practice of medicine, succeeding Dr. Benjamin Rush (1746–1813). He was the founder and editor of the *Philadelphia Medical and Physical Journal* (1804–1808). Always in poor health, he died in 1815 at the age of 49 years.

Besides being the first American to teach botany at a college, Barton was the first American author of a botanical textbook, *Elements of Botany* (1803).[3] The book is a landmark in American botany, and its usefulness merited a second edition (1812–1814)[4] during his lifetime. One of his students, Dr. Charles Caldwell (1772–1853)[22] said he possessed

> . . . at least one very high and important quality [as a teacher of Botany]—an earnest and exciting enthusiasm, by which he induced his pupils to engage in the study of the science with a corresponding earnestness, accompanied by a resolution to teach themselves.

Barton is considered the founder of the study of American plant *materia medica*, about which he wrote and taught. One biographer, however,[24] wrote:

> But, if he is the 'Father of American Materia Medica,' he should be hailed as the 'Father of American Botanical and Zoological Instruction,' and I suspect, as the 'Father of the Scientific Study of the American Indian' as well.

Barton's textbook of 1803 is a masterful summary of the basic descriptive botany of the period. It is detailed in citing previous authors, including many classical writers, the most famous being the Swedish physician-botanist, Carl Linnaeus (1707–1778). It is the Linnaean Sexual System of plant classification that is the capstone of Barton's *Botany* as it was very typical of the botanical information of the second half of the eighteenth century. Linnaeus was viewed by the British and Germans as the great teacher and organizer of botany. Barton thus brought this British attitude to the American students who studied with him or used his book for instruction.[12,23]

John Torrey (1796–1873)

John Torrey lived in Greenwich, New York, where his father was a fiscal agent of the state prison. He studied medicine at The College of Physicians and Surgeons of New York. His teachers were Drs. Wright Post (1766–1828), Samuel Latham Mitchill (1764–1831), and David Hosack (1769–1835). He took his medical degree in 1818 at the age of 22.[28,29] Already at the age of 15, Torrey had become interested in botany by meeting in his father's jail the enthusiastic Amos Eaton, with whom he studied there. Eaton, with his students, later wrote the first *Manual of Botany for the Northern States* (1817).[7] Dr. David Hosack had tried to foster botanical study in New York City by establishing in 1801 The Elgin Botanical Garden which existed for 10 years, but then the property was given to the state and fell into disrepair.[26,27] Hosack's botanical interest reinforced that of John Torrey's.

In 1822 Torrey began to teach botany, but in order to earn a living he taught chemistry at Columbia College. While continuing his botanical teaching, he showed his progressive view of botany by breaking away from the Linnaean Sexual System of classification as early as 1826, when he presented a paper at the Lyceum of Natural History of New York on the plants collected by Dr. Edwin James in 1820 on Major Stephen Long's Expedition to the Rocky Mountains. The 481 plants were arranged according to the natural system of classification of the French botanist A. L. de Jussieu (1748–1836). Torrey's resulting publication[36] was the first paper in which American plants were arranged according to the natural system. It also was the first paper that promoted the use of the natural system for the classification of plants for a geographical region. In addition, Torrey edited for American students John Lindley's London edition of *An Introduction to the Natural System of Botany* (1831),[21] to which he appended a catalogue of North American genera of plants arranged in the system. Through Torrey's personal encouragement many young people studied plants. At the celebration of the 50th anniversary of the completion of Torrey's *A Catalogue of Plants, Growing Spontaneously within Thirty Miles of the City of New-York* (1819),[35] George Thurber (1869)[34] wrote:

> What lover of plants, however young or unskilled, ever failed to receive his patient attention and kind word of encouragement? . . . Every one who has been brought in frequent communication with him knows that he has forgotten the philosopher in the friend, and that he has been made not only a better botanist, but a better man.

Torrey published many technical papers and books on plants, including *A Flora of the State of New-York* (1843).[37] As one of Torrey's friends[1] wrote when he died in 1873:

> His influence upon the development of the science of Botany in this country has been from that time to the present so large and so constant, and his intercourse with its students so extensive, so instructive and so genial, that probably every botanist in America feels that in him we have lost a scientific father and a personal friend.

Asa Gray (1810–1888)

Asa Gray, one of these early medical students, became the most important botanical educator in mid-nineteenth century America. Born in Sauquoit, Oneida County, in upstate New York, the son of a tanner, Gray studied medicine at the Fairfield Medical College where he received a degree in 1831. Four years earlier he had become interested in identifying plants using Eaton's *Manual* (1824).[7] After teaching for a short time, Gray became John Torrey's assistant in 1833 and continued part-time in that capacity until 1838, when he was appointed Professor of Botany at the newly-founded University of Michigan. Two years earlier Gray had written his first book, *Elements of Botany* (1836)[13] which is a clear, well-written introduction to botany and a model for the many later textbooks Gray was to write. Because the University of Michigan did not yet have any buildings in Ann Arbor, Gray was sent to Europe to purchase books for the establishment of their library. This opportunity gave him time to meet European botanists and to learn first-hand more about the subject of botany. Gray never did reside in Ann Arbor because he accepted a position at Harvard University as the Fisher Professor of Natural History in 1841. There he stayed the remainder of his professional career.[2]

Gray's accomplishments in systematic botany were numerous, important, and well known. With reference to his educational contributions, Gray wrote textbooks of botany for every level of student from the elementary school through the university. His books appeared in numerous editions and revisions. They were the standard introductory books in botany from the 1840s through the 1880's. Several generations of students learned their botany from Asa Gray's books. The emphasis on the natural system of classification was a break with some of the other available texts which still used the Linnaean classification. Gray wrote a *Manual of the Botany of the Northern United States* (1848)[15] to accompany his textbooks. It was, unlike Eaton's *Manual*, arranged in the natural system. Its descendant, the eighth edition by Merritt L. Fernald (1950),[10] is available in reprint and used by students of the flora today.

Although Gray never practiced medicine, he believed it to be the best training for a professional botanist. He told students, "Study medicine, and if you then still want to be a botanist, go ahead. Your medicine will keep your botany from starving."[9] One of Gray's students, William Gilson Farlow (1844–1919) became the first professional cryptogamic botanist in the country. Upon Gray's death, Farlow (1890)[9] recalled:

> We learned something about botanists as well as about botany, and often wondered whether Robert Brown were really as great as he was represented; and, on the rare occasions of a visit from a man like Dr. Torrey or Dr. Engelmann, we asked ourselves whether there was any chance that the younger generation of botanists would bear any comparison with the older.

No fear was expected from that happening because the students had been trained by Asa Gray.[6, 8, 19]

George Engelmann (1809–1884)

George Engelmann, born in Frankfurt-am-Main, Germany, was the son of a businessman. He studied at Heidelberg, Berlin, and finally at Würzburg, where he submitted a botanical thesis and obtained his medical degree in 1831. The next year Engelmann visited the American midwest looking at potential property investments for his uncles. He found German settlements in many places, and decided to emigrate to St. Louis, Missouri, where he established his medical practice. In time Engelmann became successful because of his medical ability and progressive ideas. He was the first doctor in St. Louis to use obstetrical forceps for the birth of a child and quinine for the treatment of malaria. Engelmann's importance in the development of professional botany in this country is related to his support of western botanical exploration and his establishment of a center for botanical studies in St. Louis. Because of his careful and useful treatments of various difficult plant groups, such as the cacti, conifers, and grapes, he established a reputation as a very able botanist. All of these accomplishments occurred while carrying on a full and active medical practice. Engelmann was always more than willing to aid young botanists going west on collecting expeditions, and all of them availed themselves of his help. He was responsible for inducing Mr. Henry Shaw (1800–1889), a St. Louis businessman, to establish a botanical garden, which was to become the Missouri Botanical Garden,[31] and to purchase a library and a very important European collection of pressed plants to add to the Garden's herbarium. A founder of the St. Louis Academy of Science and the mentor of St. Louis botany, George Engelmann helped young botanists who passed through that city, and after his death in 1884, the botanical garden that he helped establish continued to aid the development of botany.[30,38,40]

Summary

The development of professional botany in the United States between 1800 and 1888 is inextricably connected with physicians. Four of them, Benjamin Smith Barton, John Torrey, Asa Gray, and George Engelmann, were the giants who, each in his own way, greatly aided the young potential botanists in their quest for information about plants. Barton and Gray wrote textbooks, the former introducing the Linnaean System of classification based upon flower structure alone, and the latter introducing the Natural System based upon many plant characteristics. Gray wrote textbooks for various levels of students, including *The Botanical Text-book* (1842),[14] *Gray's Lessons in Botany and Vegetable Physiology* (1878)[18] for the high school students, and *How Plants Grow* (1858)[16] and *How Plants Behave* (1872)[17] for the elementary school students. To most students in the mid-nineteenth century Asa Gray was "Mr. Botany." He set the tone for botany for the next quarter century, and almost all the professional teaching botanists in colleges and universities at the beginning of this century had been his students.

John Torrey and George Engelmann served young botanists in a different way. They were advisors and encouragers of the plant collecting activities of others.

Torrey, by his gracious willingness to identify or verify identifications of plants for the young individuals, set them onto the path of continued interest in studying plants. He was universally liked and respected. The local botanical club, formed in New York City during his lifetime, first held meetings in 1870 and became incorporated in 1872.[25] Named the Torrey Botanical Club, the organization began the first botanical journal publication in the United States. Still extant, *The Bulletin of the Torrey Botanical Club*, now in over 126 volumes, was renamed *The Journal of the Torrey Botanical Society*. His encouragement of the study of local flora, as well as the western and exotic floras, was a major contribution to the development of American botany. Engelmann, as much as anyone in this country, aided those young collectors who went west to study the western plants. He was in the right place to accomplish that effort, and he knew much about the western flora. His careful studies of plant groups were models, based upon European traditions, for the younger American botanists beginning their careers. Those who look toward the future do much to ensure its success. Barton, Torrey, Gray, and Engelmann, trained as physicians, but with a view of what botany should be, inspired the young and insured the healthy development of American professional botany.

References Cited

1. [Anonymous]. 1873. Death of Dr. Torrey. *Bulletin of the Torrey Botanical Club* 4: [9]–10 + port. opp. p. [9]. (Quote, p. [9]).

2. Bartlett, Harley H. 1941. Asa Gray's nonresident professorship. The beginning of the University in Ann Arbor. *Michigan Alumnus Quarterly Review* 47: 215–226.

3. Barton, Benjamin Smith. 1803. *Elements of Botany: or Outlines of the Natural History of Vegetables.* Illustrated by thirty plates. Printed for the Author, Philadelphia. 2 vols. (1: xii, 302 pp. + 1 pl.; 2: 168, 38 pp. + 29 pls.).

4. _____. 1812–1814. *Elements of Botany: or Outlines of the Natural History of Vegetables.* Illustrated by forty plates. 2nd ed., corrected and greatly enlarged. Printed for the Author, Philadelphia. 2 vols. (1: xviii, 324 pp. + pls. 1–6; 2: iv, 180, 44 pp. + pls. 7–40).

5. Bell, Whitfield J., Jr. 1971. Benjamin Smith Barton M.D. (Kiel). *Journal of the History of Medicine and Allied Sciences* 26: 197–203.

6. Dupree, A. Hunter. 1959. *Asa Gray 1810–1888.* The Belknap Press of Harvard University Press, Cambridge, Massachusetts. x, 505 pp.

7. Eaton, Amos. 1824. *A Manual of Botany, for the Northern and Middle States of America . . .* 4th ed., revised and corrected. Printed and Published by Websters and Skinners, Albany. [x], 11–539 pp. [Earlier editions were issued in 1817, 1818, and 1822].

8. Ewan, Joseph. 1969. Introduction, pp. i–xviii. *In* John Torrey and Asa Gray, *A Flora of North America: . . .* [Facsimile of the 1838–1843 ed.]. *Classica Botanica Americana*, Vol. 4. Hafner Publishing Co., New York. 2 vols. (1: xviii, 711 pp.; 2: xviii, 504 pp.).

9. Farlow, William G. 1890. Memoir of Asa Gray. *Annual Report of the Board of Regents of the Smithsonian Institution [for]* 1888: 763–783. (Quotes, pp. 771–772).

10. Fernald, Merritt Lyndon. 1950. *Gray's Manual of Botany.* 8th (Centennial) ed.—illustrated. Largely rewritten and expanded by Merritt Lyndon Fernald, with assistance of specialists in some groups. American Book Co., New York. lxiv, 1632 pp. [Reprinted in 1987, by Dioscorides Press, Portland, Oregon; currently in its 4th (1993) printing].

11. Frederick, John W. 1988. Daniel Drake and the botany of the upper Ohio valley. *Bartonia* 54: 83–97.

12. Graustein, Jeannette E. 1961. The eminent Benjamin Smith Barton. *The Pennsylvania Magazine of History and Biography* 85: 423–438.

13. Gray, Asa. 1836. *Elements of Botany.* G. & C. Carvill & Co., New-York. xiv, 428 pp.

14. _____. 1842. *The Botanical Text-book for Colleges, Schools, and Private Students: Comprising Part I. An Introduction to Structural and Physiological Botany. Part II. The Principles of Systematic Botany . . .* Wiley and Putnam, New-York; Little and Brown, Boston. xii, [13]–413 pp.

15. _____. 1848. *Manual of the Botany of the Northern United States, from New England to Wisconsin and South to Ohio and Pennsylvania Inclusive, (The Mosses and Liverworts by Wm. S. Sullivant,) Arranged according to the Natural System; . . .* James Munroe and Co., Boston & Cambridge, [Massachusetts]; John Chapman, London. lxxii, 710 pp.

16. _____. 1858. *How Plants Grow, a Simple Introduction to Structural Botany. With a Popular Flora, or an Arrangement and Description of Common Plants, Both Wild and Cultivated. Botany for Young People and Common Schools.* Ivison and Phinney, New York; S. C. Griggs & Co., Chicago. 233 pp.

17. _____. 1872. *How Plants Behave: How They Move, Climb, Employ Insects to Work for Them, &c. Botany for Young People.* Part II. Ivison, Blakeman, Taylor, and Co., New York and Chicago. viii, [9]–46 pp.

18. _____. 1878. *Gray's Lessons in Botany and Vegetable Physiology . . . to Which Is Added a Copious Glossary, or Dictionary of Botanical Terms.* Ivison, Blakeman, Taylor & Co., New York and Chicago. xii, 236 pp.

19. Gray, Jane Loring. 1893. *Letters of Asa Gray.* 2nd ed. Houghton, Mifflin and Co., Boston and New York. 2 vols. (1: 368 pp.; 2: pp. [369]–838).

20. Kelly, Howard A. 1914. *Some American Medical Botanists Commemorated in Our Botanical Nomenclature.* The Southworth Co., Publishers, Troy, New York. 215 pp.

21. Lindley, John. 1831. *An Introduction to the Natural System of Botany: or, A Systematic View of the Organization, Natural Affinities, and Geographical Distribution of the Whole Vegetable Kingdom*... 1st American ed., with an appendix by John Torrey. G. & C. & H. Carvill, New York. lxxix, 393 pp.

22. Middleton, William S. 1936. Benjamin Smith Barton. *Annals of Medical History, new ser.* 8: 477–491. (Quote, p. 480).

23. Pennell, Francis W. 1926. The elder Barton—his plant-collection and the mystery of his floras. *Bartonia* 9: 17–34.

24. _____. 1943. Benjamin Smith Barton as naturalist. *Proceedings of the American Philosophical Society* 86: 108–122. (Quote, p. 121).

25. Rickett, Harold William. 1967. The Torrey Botanical Club: A retrospect. *The Garden Journal* 17: 144–150.

26. Robbins, Christine Chapman. 1960. David Hosack's herbarium and its Linnaean specimens. *Proceedings of the American Philosophical Society* 104: 293–313.

27. _____. 1964. David Hosack. Citizen of New York. *Memoirs of the American Philosophical Society* 62: v–vii, 1–246 pp.

28. _____. 1968. John Torrey (1796–1873). His life and times. *Bulletin of the Torrey Botanical Club* 95: 515–645.

29. Rodgers, Andrew Denny, III. 1942. *John Torrey, A Story of North American Botany*. Princeton University Press, [Princeton, New Jersey]; Humphrey Milford, London; Oxford University Press, [Oxford, England]. 352 pp.

30. _____. 1944. *American Botany 1873–1892. Decades of Transition*. Princeton University Press, Princeton, [New Jersey]; Humphrey Milford, London; Oxford University Press, Oxford, [England]. 340 pp.

31. Rudolph, Emanuel D. 1991. One hundred years of the Missouri Botanical Garden. *Annals of the Missouri Botanical Garden* 78: 1–18.

32. Shapiro, Henry D. 1988. Daniel Drake: The new western naturalist. *Bartonia* 54: 39–47.

33. Stuckey, Ronald L. 1978. Medical botany in the Ohio valley (1800–1850). *Transactions & Studies of the College of Physicians of Philadelphia* 45: 262–279.

34. [Thurber, George]. 1869. Address, pp. 41–45. *In* [Anonymous], The Torrey Festival. *The American Naturalist* 2: 41–47. (Quote, pp. 44–45).

35. Torrey, John. 1819. *A Catalogue of Plants, Growing Spontaneously within Thirty Miles of the City of New-York*. Published by the Lyceum of Natural History of New-York; Printed by Websters and Skinners, Albany. vi, 100 pp.

36. _____. 1828. Some account of a collection of plants made during a journey to and from the Rocky Mountains in the summer of 1820, by Edwin P. James, M.D. Assistant Surgeon, U.S. Army. *Annals of the Lyceum of Natural History of New-York* 2: 161–254.

37. _____. 1843. *A Flora of the State of New-York, Comprising Full Descriptions of All the Indigenous and Naturalized Plants Hitherto Discovered in the State; with Remarks on Their Economical and Medicinal Properties*. Carroll and Cook, Printers to the Assembly, Albany. 2 vols. (1: xii, 484 pp. + pls.; 2: 572 pp. + pls.).

38. Trelease, William and Asa Gray. (eds.). 1887. *The Botanical Works of the Late George Engelmann Collected for Henry Shaw, Esq*. John Wilson and Son, University Press, Cambridge, Mass. ix, 548 pp. + 75, 24, 3 pls.

39. T[rimen], H[enry]. 1875. [Notice of Books]. *Text-book of Botany, Morphological and Physiological*. By Julius Sachs. Translated and annotated by Alfred W. Bennett; assisted by W. T. Thiselton Dyer. At the Clarendon Press, Oxford. 1875. *The Journal of Botany, British and Foreign, new ser.* 4: 243–246. (Quote, pp. 243–244).

40. White, Charles A. 1902. Memoir of George Engelmann. 1809–1884. *Biographical Memoirs, National Academy of Sciences*, 4: [1]–21.

Publication Notes

Compiled and edited from a manuscript,[a] written in 1972 and submitted to the *Plant Science Bulletin*, and never published; and also from a manuscript[b] read 26 April 1980 at the annual meeting of the Ohio Academy of Medical History, Granville, Ohio.

[a] 1972. "American Medical Botany." Typewritten, 2 pp. [Manuscript].

[b] [1980]. "Some Medical Doctors Who Taught Nineteenth Century American Students Botany." Typewritten, 7 pp. [Manuscript].

Introduction No. 17

Bruce Fink, Graduate Educator (1986)

Rudolph, a lichenologist and educator (Appendix 17.1, Fig. 17.1, Fig. 17. 2), had a common interest with Ohio botanist, Bruce Fink (1861–1927), also a lichenologist and teacher. Fink wrote *The Lichen Flora of the United States* (1935), a credible book which was completed posthumously for publication by his student, Joyce (Hedrick) Jones (1897–1980). Fink trained many students at the graduate level in botany, and many of these were identified from Rudolph's notes. Some of the names were obtained from Stuckey's files on Ohio botanists.

Rudolph's presentation on Professor Fink was given on 26 April 1986 at the Annual Meeting of The Ohio Academy of Science at the University of Toledo. His manuscript[b] was only partially written, but many notes and copies of publications about Fink were available from Rudolph's research files. The short paper published here, apart from using Rudolph's published abstract and a paragraph from his manuscript,[b] was written by Stuckey. **(Paper No. 17)**.

Fig. 17.1. Emanuel D. Rudolph (r.) assisting an undergraduate student with an experimental study of a lichen.

(Taken 1979; Photo Archives, The Ohio State University)

Bruce Fink, Lichenologist

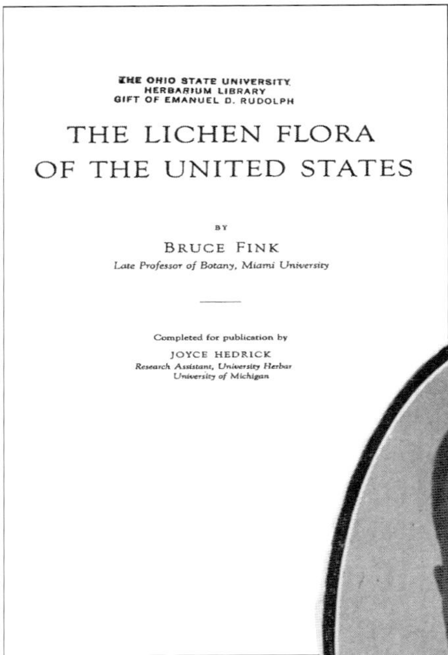
THE OHIO STATE UNIVERSITY
HERBARIUM LIBRARY
GIFT OF EMANUEL D. RUDOLPH

THE LICHEN FLORA
OF THE UNITED STATES

BY

BRUCE FINK
Late Professor of Botany, Miami University

Completed for publication by

JOYCE HEDRICK
Research Assistant, University Herbar
University of Michigan

Title page from *The Lichen Flora of the United States*, by Bruce Fink (1935)

Bruce Fink
(1861–1927)

SMITHSONIAN INSTITUTION
UNITED STATES NATIONAL MUSEUM

CONTRIBUTIONS

FROM THE

UNITED STATES NATIONAL HERBARIUM

VOLUME 14, PART 1

THE LICHENS OF MINNESOTA

By BRUCE FINK

Title page from *The Lichens of Minnesota*, by Bruce Fink (1910)

Paper No. 17 (1986)

Bruce Fink (1861–1927): A Midwestern Botanist and Educator

By using analytical approaches, scientists of the past can be placed upon a continuum, based on lasting contributions to their discipline. This analysis can range from those individuals who initiated and carried through major changes in the ways the natural world was understood by scientists to those who made no lasting impact upon the future direction of their discipline. In this presentation consideration is given to a botanist who can be placed somewhere in the middle of this continuum. Bruce Fink is not someone whose name is recognized now by many botanists, unless they are American lichenologists, of whom only a few are active today in this country. The following paragraph reveals some important biographical information on the man and his life:

> Bruce Fink, for 21 years professor of botany and head of that Department at Miami University in Oxford, Ohio, until his death at age 65, was a major figure in American lichenology. He was an early champion of the originally unpopular proposal of Schwendener that lichens were dual organisms composed of fungi and algae.[16] He devised and published a classification scheme that placed lichens among the fungi, which he believed were parasitic on their algal components. Not only did he study the systematics and floristics of lichens, but he also was concerned with their ecology and physiology. He espoused the anti-tobacco cause and published books on the "tobacco habit."[9,12] Fink's floristic studies are the ones for which he is now remembered, particularly his studies of Minnesota lichens and his lichen flora of the United States, which was posthumously completed by his student, Joyce Hedrick, later Mrs. Volney H. Jones[13] (1897–1980). The lichen flora, although difficult to use, is the only modern one for the United States that considers all lichen groups.[a]

An anonymous author[3] for the *Miami Alumni News Letter* of August 1927 stated that:

> Dr. Fink was the leading American authority on lichens and one of the three most eminent lichenologists in the world. . . . He was an excellent teacher and sent more students to the graduate schools than any other member of

---

Frontispiece Credits: Bruce Fink, Lichenologist. Title page of *The Lichen Flora of the United States*. (From Emanuel D. Rudolph's copy in the Herbarium Library of The Ohio State University)

Bruce Fink (1861–1927). (From a Miami University publication, entitled "At Old Miami," *The Miami Bulletin*, Series D. July 1906, p. 10; from the photograph collection of RLS)

Title page of *The Lichens of Minnesota*. (From the copy in the library of The Ohio State University)

> our faculty. As a teacher he was a man of most generous spirit, of the highest ideals, and held great human interest in his students.

A complete list of Professor Fink's students has not been compiled, but the following were listed by him:[10] Freda M. Bachman, A.B. 1907, A.M. 1908, Ph.D. University of Wisconsin, 1912; Vernon L. Lantis, A.B. 1909, A.M. University of Cincinnati, 1910; Frank M. O'Byrne, A.B. 1907, A.M. University of Florida, 1913; J. Wilber O'Byrne, A.B. 1910, attended Yale Forestry School, 1910–1912; Lee O. Overholts, A.B. 1912, continued as a scholar at the New York Botanical Garden and as a research fellow at Shaw School of Botany, both places in 1912; Wilmer G. Stover, A.B. 1909, A.M. 1910; Audrey C. Richards, A.B. 1912, A.M. 1914, graduate assistant, 1912–1914; Walter D. Smith, A.B. 1912; and Mabel A. Husband, who did "nothing since leaving college, but has remained at home."[11] Stuckey (1992)[17] in his *Women Botanists of Ohio Born Before 1900* lists five Ohio women students who attended or graduated under the guidance of Professor Fink at Miami University. In addition to the two named above, they are Leafy J. Corrington, A.B. 1914; Joyce Hedrick (Mrs. Volney H. Jones), B.S. 1925, A.M. 1929; and Grace M. Kalter, B.S. 1911.

The distinguished botanist Louis H. Pammel (1862–1931) of Iowa State College, Ames, corresponded with Fink for more than a third of a century. Pammel[14] quoted an Oxford, Ohio, news dispatch stating that Fink was "a man of great retiring nature, . . . [and] was yet interested in all about him. His deep interest in the subject and his kindly nature made him a favorite among Miami students." His major contribution in lichenology was his field study of the lichens of Iowa and Minnesota published in the *Proceedings of the Iowa Academy of Science* (1895–1899), and in the *Minnesota Botanical Studies* (1896–1903), the latter of which studies were collected under the title *The Lichens of Minnesota* and published as a *Contribution of the United States National Herbarium* (1910).[5] Fink's collection of Minnesota lichens totalled 5,500.[18] Riddle (1911)[15] noted in a review of this work that for the specialist, "this publication is of great significance, as it represents the culmination of fourteen years of careful work, . . . " Fink was an advocate for incorporating ecological and phytogeographic data into his lichen research, and his studies using these kinds of data were pioneer efforts. Concerning the importance of ecological information, Fink (1902)[4] wrote:

> In the revival of interest in the study of lichens ecologic researches are surely to be a dominant factor. It has not escaped notice that lichens are of extreme interest ecologically. . . . Let me urge those who have or are able to acquire a knowledge of the lichens of a limited area to turn their attention to ecologic studies.

A member of many scientific societies and professional organizations, Bruce Fink was elected president of the Iowa Academy of Science (1902 to 1904), the Sullivant Moss Society (1910), and The Ohio Academy of Science (1912). Speaking on "Two Centuries of North American Lichenology," as the retiring president of the

Iowa Academy of Science, Fink, as quoted in Pammel (1928),[14] stated that:

> All men of science are interested to some extent in the history of the rise and progress of every phase of scientific inquiry, and even for the layman who may favor us with his presence this evening, it is hoped that the record of valuable work will afford something of interest.

As the retiring president of The Ohio Academy of Science, Fink spoke on "Botanical Instruction in College," making a point that every teacher of botany should be an investigator who should at least extend his research to the plant life about him and to the literature that directly or indirectly relates to his teaching.[6] "No science stimulates to investigation and research more than Botany, and the teacher of this science who is not an investigator, even if he be a member of a college faculty, is scarcely worthy of his profession."[7] With his student Walter D. Smith, Fink prepared an evaluation of "Botanical Instruction in the High Schools of Ohio."[8] Among their recommendations were that practical aspects of learning botany are important and essential for the student. Courses in high school botany should use the compound microscope, and engage the student in field trips. A well-lighted equipped laboratory should be available, and an herbarium should be developed under the teacher's direction.

Another one of his students, Robert B. Wylie (1870–1959) of the University of Iowa, described Fink's research as mainly in the field of lichenology. Wylie(1928)[19] wrote that

> . . . his activity as a scientist is attested by over one hundred titles,[1,2] covering published notes, reviews, papers and monographs. Primarily a taxonomist, he contributed also to the biology of lichens with discussions of their [systematic] classification, [ecological] relationships and [geographical] distribution. . . . Numerous papers dealt with fungi and particularly with the Ascomycetes. His botanical interests were broad and he followed with zest the marked development of the plant sciences as a whole which occurred during his scientific lifetime. . . . [Professor Fink is to be remembered] as a man of genial personality and optimistic temperament. These qualities in combination with his modesty and generosity made him a widespread friend. Of simple tastes, love of home and laboratory, he lived a life of marked sincerity and high achievement.

References Cited

1. [Anonymous]. [1912]. "[List of] Botanical Papers by Bruce Fink, to 1912." Typewritten, 4 pp. 68 titles. [Manuscript].

2. [_____]. [1925]. "[List of] Papers and Books by Bruce Fink, to 1925." Typewritten, [5] pp. 103 titles. [Manuscript].

3. [_____]. 1927. Dr. Bruce Fink dies in his laboratory. *Miami Alumni News Letter,* ser. XXV 12: 11. (Quote, p. 11).

4. Fink, Bruce. 1902. Ecologic distribution an incentive to the study of lichens. *The Bryologist* 5: 39–40. (Quote, pp. 39–40).

5. _____. 1910. The lichens of Minnesota. *Contributions from the United States National Herbarium* 14: iii–viii, 1–269, ix–xvii + 51 pls.

6. _____.1911. Concerning botanical investigation in colleges. *Science* 34: 593–595.

7. _____. [1913]. President's address. Botanical instruction in colleges. *Proceedings of The Ohio Academy of Science* 6(2): 72–87. [Presented at the Twenty-second Annual Meeting of The Ohio Academy of Science, 29 November 1912]. (Quote, p. 84).

8. _____ and Walter D. Smith. 1913. Botanical instruction in the high schools of Ohio. *Miami University Bulletin, ser. XI* 12: [1]–39.

9. _____. 1914. *The Tobacco Habit.* Published by Miami University, Oxford, Ohio. 77 pp.

10. _____. 1915. Letter to J. H. Barnhart, New York Botanical Garden, 17 March. 2 pp.

11. _____. 1915. Letter to J. H. Barnhart, New York Botanical Garden, 10 April. 1 p.

12. _____. 1915. *Tobacco.* The Abingdon Press, New York, Cincinnati. 124 pp.

13. _____. 1935. *The Lichen Flora of the United States.* Completed for publication by Joyce Jones. University of Michigan Press, Ann Arbor. x, 426 pp. + 47 pls. [Subsequent printings were issued in 1960, 1961, and 1971].

14. Pammel, Louis H. [1928]. In memoriam: Dr. Bruce Fink. *Proceedings of the Iowa Academy of Science* 35: 34–39 + port., p. [33]. [Also published as *Prominent Men I Have Met. VIII. Dr. Bruce Fink.* [Published by the Author]. 8 pp. + port.]. (Quotes, pp. 34, 38).

15. Riddle, L. W. 1911. An important publication on lichens. *Rhodora* 13: 15. [Review of "Fink, Bruce. [1910]. The Lichens of Minnesota. *Contributions from the United States National Herbarium*, vol. 14, pt. 1, pp. 1–269 . . ."]. (Quote, p. 15).

16. Schwendener, S[imon]. 1867. Über den Bau des Flechtenthallus. *Verhandlungen der Shweizerischen Naturforschenden Gesellschaft in Rheinfelden* 51: 88–90.

17. Stuckey, Ronald L. 1992. *Women Botanists of Ohio Born before 1900: With Reference Calendars from 1776 to 2028.* Ronald L. Stuckey, RLS Creations, Columbus, Ohio. x, 67 pp.

18. Wetmore, Clifford M. 1978. Bruce Fink's lichen collections in Minnesota. *The Michigan Botanist* 17: 103–112.

19. Wylie, Robert B. 1928. Bruce Fink, lichenologist. *Mycologia* 20: 1–2. (Quote, pp. 1–2).

Publication Notes

Compiled and edited from an abstract[a] and an incomplete manuscript[b] and notes read 26 April 1986 at the Annual Meeting of The Ohio Academy of Science.

[a] 1986. Bruce Fink, midwestern botanist and educator, 1861–1927. *The Ohio Journal of Science* 86(2): 3. [Abstract].

[b] 1986. "Bruce Fink." Handwritten, 22 pp. + 4 pp. of handwritten notes by and about Bruce Fink for Rudolph's paper read 26 April 1986 at Toledo, Ohio. [Manuscript].

APPENDIX 17.1.
LICHENS OF CEDAR BOG: PAST AND PRESENT, BY RUDOLPH[1]

Lichens are sensitive indicators of environmental change, and thus, they can be used to provide information about environmental modifications as is shown in the case of Cedar Bog, Champaign County, Ohio. Lichen collections examined from Cedar Bog date from as early as the 1870s when Miss Hannah J. Biddlecome and Mrs. E. Jane Spence gathered specimens there. Some other specimens dating from the nineteenth century are present also in The Ohio State University Lichen Herbarium. In the past ten years, new collections have been made by me and a number of students of The Ohio State University, as well as other interested persons. In having the opportunity of comparing the early collections with the recent ones, it is possible to make comments about the changes in the lichen flora and to infer changes in environmental conditions that these differences indicate.

The taxa listed in the publication[1] and the statistics derived from it indicate trends that are independent of historical information about the bog. For example, a large proportion of the lichen flora (86%) has changed in about 70 years, only 14% has remained unchanged, while the total number of species remained the same. This would certainly indicate drastic changes in the bog and its environment. Of the species that were present in the past century and have now disappeared, 33% were characteristic of deep woods and shady places, while only 5% of the replacement species are of this kind. On the other hand, 53% of the early replaced species were characteristic of open woods or sunny places, while 90% of the replacement species are characteristic of these habitats. These data give some indication of the changes in the characteristics of Cedar Bog; it is now more open and not as wet.

[1] Emanuel D. Rudolph. 1974. Lichens of Cedar Bog - Past and present, pp. 38–40. *In* Charles C. King and Clara May Frederick (eds.). *Cedar Bog Symposium Urbana College November 3, 1973. Ohio Biological Survey Informative Circular* No. 4. Columbus, Ohio. viii, 71 pp. (Permission to reproduce Rudolph's article courtesy of the Ohio Biological Survey, Columbus)

APPENDIX 17.1. (CONTINUED)

Summary of Lichen Species List

	Deep Woods	Open Woods	Both	Total
All Species	11 (23%)	33 (67%)	5 (10%)	49 (100%)
19th Century only	7 (33%)	11 (53%)	3 (14%)	21 (43%)
Recent only	1 (5%)	19 (90%)	1 (5%)	21 (43%)
Persistent	3 (43%)	3 (43%)	1 (14%)	7 (14%)

Fig. 17.2. Emanuel D. Rudolph (third from the right next to fence) participating in OSU Department of Botany field excursion to Cedar Bog, Champaign County, Ohio, during the 1960s. Department Chairman Bernard S. Meyer is second from the right.

(In the Rudolph photograph collection)

PART FIVE

BOTANICAL ILLUSTRATION

Emanuel D. Rudolph's earliest known interest in pursuing research in botanical illustration is noted in a reply to a memorandum of 7 June 1967 from the Office of the College of Biological Sciences at The Ohio State University. Asked to summarize his current research, the fourth item in his list was titled, "American botanical illustrators and their role in the rise of scientific botany in America." On this topic Rudolph[1] wrote:

> Little is known about the men and women who illustrated the important botanical works of the nineteenth and early twentieth centuries in the United States. It was during this period that American botany assumed a position equal with that of the European botany of the time. My contention is that the development of trained illustrators was an important factor in this development.

Two of Rudolph's papers on botanical illustration are placed in this group (**Part Five**). They are on the botanical illustrator Isaac Sprague (**Paper No. 18**) and the botanical prints of Joseph Banks (**Paper No. 19**).

---

[1] Emanuel D. Rudolph. [1967]. Research Projects, Emanuel D. Rudolph. [The Ohio State University, Columbus]. Typewritten, 2 pp.

Introduction No. 18

Isaac Sprague, Botanical Illustrator (1967–1990)

Rudolph thoroughly enjoyed his investigation of Massachusetts-born botanical illustrator, Isaac Sprague (1811–1895), who drew plants for Asa Gray's publications. The high quality of Sprague's work contributed toward the development and international recognition of American botany in the mid-nineteenth century. Rudolph presented this paper on 21 April 1967 at the Annual Meeting of The Ohio Academy of Science in Dayton, and on 28 August 1967 at the meeting of the Historical Section of the Botanical Society of America in College Station, Texas.

During 1967, Rudolph was in active correspondence with George H. M. Lawrence (1910–1978), Director of the Hunt Botanical Library, Pittsburgh. Lawrence was one of the ten guest lecturers in July of 1968 and 1969 at Rudolph's NSF History of Biology Institute for College Biology Teachers. (**Paper No. 12**). At Lawrence's request, Rudolph[1] sent him a copy of his notes used for the paper presented in Texas. The intention was that Rudolph would write the paper for the forthcoming fourth volume of *Huntia*, a publication for journal papers on botanical history edited by Lawrence and published by the Hunt Library. Over the next several months Rudolph[2] worked on obtaining more information about Sprague. He was particularly interested in finding Sprague's botanical drawings.

Having located Sprague's grandson, Isaac Sprague, Jr., Rudolph,[3] spent an evening with him where he lived in the Boston area. Rudolph[2] wrote to Lawrence that from the younger Sprague, he "was able to gather some important information for . . . [his] paper on Sprague." As a further result of their meeting, virtually all of the botanical drawings that Rudolph[4] saw in the grandson's possession were placed in the Hunt Botanical Library, with the exception as noted by Rudolph of "a few

[1] Emanuel D. Rudolph. 1967. Letter of 7 September to George H. M. Lawrence, The Hunt Botanical Library, Pittsburgh, Pennsylvania. 1 p.

[2] Emanuel D. Rudolph. 1967. Letter of 8 October to George H. M. Lawrence, The Hunt Botanical Library, Pittsburgh, Pennsylvania. 1 p.

[3] Emanuel D. Rudolph. 1967. Letter of 6 December to George H. M. Lawrence, The Hunt Botanical Library, Pittsburgh, Pennsylvania. 1 p.

[4] Emanuel D. Rudolph. 1968. Letter of 7 March to George H. M. Lawrence, The Hunt Botanical Library, Pittsburgh, Pennsylvania. 1 p.

pencil sketches and tracings to use in connection with my study and in my teaching." Lawrence[5] did much to encourage and persuade Rudolph with his investigations of Sprague. On 11 December 1967, he wrote Rudolph, saying: "I congratulate you on the perseverance with which you are seeking to unravel as many of the mysteries as possible about the botanical illustrating work . . ." of Isaac Sprague. For reasons unknown, Rudolph laid this research aside and no paper was published in *Huntia*.

During the late 1980s, Rudolph returned to his work on Isaac Sprague and on 14 January 1987, he submitted a manuscript to *Isis, the Journal of the History of Science Society*.[6] Over a year later, on 21 March 1988, the editor[7] replied that the manuscript was "very interesting," but "marginal to . . . [the] central concerns" of the journal. Eight days later Rudolph[8] sent the manuscript to Everett Mendelsohn, editor of the *Journal of the History of Biology*. By June 1988, Mendelsohn[9] accepted the paper for the *Journal* with the understanding that Rudolph "undertake whatever revisions . . . were appropriate," in accordance with the thoughtful comments relating to style and thematic issues made by one of the reviewers. The reviewer's report[9] indicated that Rudolph's "terse style" created a "brevity" of thought which, "while often effective, occasionally leaves the reader guessing," and partly results from the hesitancy "of the author to draw out larger themes from Sprague's career that are relevant to 19th century natural history and botany generally." Upon returning the manuscript to editor Mendelsohn on 9 August 1988, Rudolph[10] referred to the reviewer's thoughts on the brevity and terseness of his manuscript. With a sense of humor, he justified his writing style and vindicated himself by saying:

> . . . My training in writing scientific papers has made me try to be concise rather than expansive. I can only quote John Maynard Smith[11] who recently

[5] George H. M. Lawrence. 1967. Letter of 11 December to Emanuel D. Rudolph, Department of Botany, The Ohio State University, Columbus, Ohio. 1 p.

[6] Emanuel D. Rudolph. 1987. Letter of 14 January to Charles E. Rosenberg, Editor, *Isis*, University of Pennsylvania, Philadelphia, Pennsylvania. 1 p.

[7] Charles Rosenberg. 1988. Letter of 21 March to Emanuel D. Rudolph, Department of Botany, The Ohio State University, Columbus, Ohio. 1 p.

[8] Emanuel D. Rudolph. 1988. Letter of 29 March to Everett Mendelsohn, Editor, *Journal of the History of Biology*, Harvard University, Cambridge, Massachusetts. 1 p.

[9] Everett Mendelsohn. 1988. Letter of 9 June to Emanuel D. Rudolph, Department of Botany, The Ohio State University, Columbus, Ohio. 1 p. + Referee Report, 2 pp.

[10] Emanuel D. Rudolph. 1988. Letter of 9 August to Everett Mendelsohn, Editor, *Journal of the History of Biology*, Harvard University, Cambridge, Massachusetts. 1 p.

[11] John Maynard Smith. 1988. *Games, Sex and Evolution*. Harvester Press, Brighton. vi, 264 pp. (Quote, p. 97).

GENERA

FLORÆ AMERICÆ BOREALI-ORIENTALIS

ILLUSTRATA.

THE GENERA

OF THE

PLANTS OF THE UNITED STATES

ILLUSTRATED

BY FIGURES AND ANALYSES FROM NATURE,

BY ISAAC SPRAGUE,

MEMBER OF THE BOSTON NATURAL HISTORY SOCIETY.

SUPERINTENDED, AND WITH DESCRIPTIONS, &c.,

BY ASA GRAY, M. D.,

FISHER PROFESSOR OF NATURAL HISTORY IN HARVARD UNIVERSITY,
CORRESPONDING MEMBER OF THE ROYAL BAVARIAN ACADEMY, MEMBER
OF THE IMPERIAL ACADEMY NATURÆ CURIOSORUM; OF THE
BOTANICAL SOCIETY OF RATISBON, &c., &c.

VOL. I.

PLATES 1--100.

NEW YORK:
GEORGE P. PUTNAM.
LONDON: PUTNAM'S AMERICAN AGENCY.
1849.

Fig. 18.1. Title page from *The Genera of the Plants of the United States Illustrated . . . by Isaac Sprague*, by Asa Gray (1849 reissue).

(In the Rudolph photograph collection)

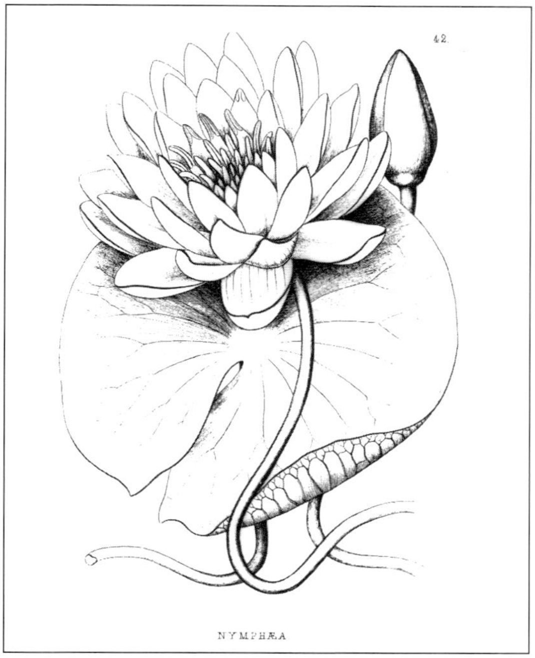

Fig. 18.2. Line illustration of *Nymphaea* drawn by Isaac Sprague.

(From Asa Gray. 1848. [Reissue 1849]. *The Genera of the Plants of the United States Illustrated.* Vol. I. George P. Putnam, New York, pl. 42; in the Rudolph photograph collection)

> wrote, "Everyone knows that the prose written by scientists is constipated, full of jargon and altogether lacking in grace." . . . At least I hope not to have used jargon.

Not until after Rudolph's semi-retirement from classroom teaching in 1989, did his full study of Isaac Sprague appear in the *Journal of the History of Biology* (1990).

Rudolph concluded that Isaac Sprague's importance to American botany was never fully appreciated, and his study intended to make that viewpoint known. Rudolph's publication represents the first comprehensive biographical study of Isaac Sprague, and may also represent one of his best contributions to the history of American botany. His abstract,[a] written for presentation on 28 August 1967 at the meeting in College Station, Texas, and two paragraphs from his paper[b] on Sprague published in 1990, constitute the summary printed here. **(Paper No. 18)**.

Isaac Sprague, Botanical Illustrator

Isaac Sprague (1811–1895)

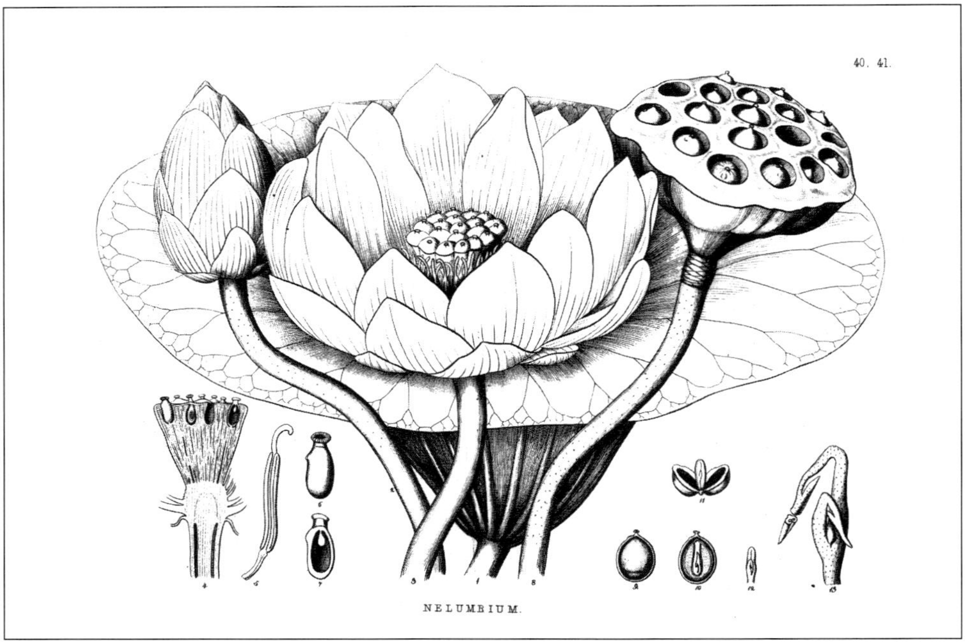

Line illustrations of *Nelumbo lutea* drawn by Isaac Sprague

Paper No. 18 (1967, 1990)

The Role of the Botanical Illustrator Isaac Sprague (1811–1895) in the Ascendency of American Botany during the Nineteenth Century

Mid-nineteenth century American botany achieved international recognition mainly through the activities of John Torrey (1796–1873), Asa Gray (1810–1888), George Engelmann (1809–1884), and William S. Sullivant (1803–1873). In 1845, Gray, who had recently become professor of Natural History at Harvard, employed a Hingham, Massachusetts artist, Isaac Sprague, to make illustrations for his Lowell Lectures and for the revision of his *Botanical Text-book.*[1] Gray was delighted with the abilities of Sprague who, after self-education in art and natural history, had accompanied Audubon up the Missouri River in 1843. Sprague illustrated other works of Gray, among them an elaborate one on the genera of American plants in the style of some European treatments of the time. Only two volumes of the *Genera* appeared (1848, 1849),[2] containing 186 plates. The title pages state, "Illustrated by figures and analyses from nature by Isaac Sprague, member of the Boston Natural History Society" (Fig. 18.1). The illustrations in these and other illustrated works by Sprague drew high praise in the reviews and letters from European and American botanists (Fig. 18.2). European botany had a long tradition of botanical illustrators dating from the Renaissance; thus it was expected that sound botanical taxonomic and morphologic research would be carefully and artistically figured. Sprague was far superior as a botanical illustrator to any previous American and thus played a role, not generally appreciated, in the ascension of American botany.[a]

Although Sprague was self-taught, he had some supervised training during part of a year with Audubon. Unlike Alexander Wilson (1766–1813), Audubon used his artistic skills, together with those of his assistants, to place birds and quadrupeds in naturalistic poses in realistic settings. Audubon's method of using many sketches of the animals, related vegetation, and scenery for his final composite depiction made his illustrations idealistic rather than completely realistic. It seems that Sprague learned from Audubon the technique of composite illustrating; he used various

---

Frontispiece Credits: Isaac Sprague, Botanical Illustrator. Isaac Sprague (1811–1895). (In the Rudolph photograph collection)

Line illustrations of *Nelumbo lutea* drawn by Isaac Sprague. (From Asa Gray. 1848. [Reissue 1849]. *The Genera of the Plants of the United States Illustrated.* Vol. 1. George P. Putnam, New York. pls. 40, 41. In the Rudolph photograph collection)

living and dried specimens to produce sketches that were later combined into drawings of idealized plants. Numerous extant sketches left by Sprague at his death attest to his use of this technique. Sprague was the first of many skillful American botanical illustrators who endeavored to make the written word more lively through the use of complementary accurate pictures. Sprague's aim was to be a delineator of nature; it is only incidental that some of his illustrations were later considered to be of interest for their aesthetic qualities.[b]

Sprague is the prime example of a mid-nineteenth century American illustrator whose drawings were used with increasing frequency in the publications of naturalists. Though expensive, the use of illustrations was rooted solidly in the European tradition, enhancing natural history writings and making them more universally valuable to experienced and beginning naturalists alike. This tradition was much weaker in the United States. Thus, Sprague and other illustrators helped, as silent workers, in the professionalization of nineteenth-century American science, and they deserve to be remembered and credited by historians of science.[b]

References Cited

1. Gray, Asa. 1845. *The Botanical Text-book for Colleges, Schools, and Private Students: Comprising Part I. An Introduction to Structural and Physiological Botany. Part II. The Principles of Systematic Botany*.... 2nd ed. Wiley and Putnam, New-York; Little and Brown, Boston. 509 pp.

2. _____ and Isaac Sprague. 1848–1849. *Genera Florae Americae Boreali-orientalis Illustrata. The Genera of the Plants of the United States Illustrated by Figures and Analyses from Nature by Isaac Sprague.* Superintended, and with Descriptions, &c., by Asa Gray. James Munroe and Co., Boston; John Wiley, New York and London (vol. 1, 1848); George P. Putnam, New York; Putnam's American Agency, London (vol. 2, 1849). 2 vols. (1: 230 pp. + pls. 1–100; 2: 230 pp. + pls. 101–186). [Reissue of vol. 1, 1849 seen].

Publication Notes

Compiled and edited from an abstract[a] written 1967 and selected passages from a paper[b] published 1990 in the *Journal of the History of Biology*.

1966. [Biographical sketch on Isaac Sprague IV], p. 9. *In* C. Nissen, *Die botanische Buchillustration. Ihre Geschichte und Bibliographie*. 2nd ed. Band III: Supplement. Anton Hiersemann. Stuttgart.

1967. Isaac Sprague's role in the ascendency of American botany during the nineteenth century, p. 2. *In* Abstracts of Papers Presented before Special Session on the History of Science, The Ohio Academy of Science Meeting at The University of Dayton, Dayton, Ohio, Friday, April 21, 1967. Mimeographed. [Abstract].

[1967]. "The Role of the Botanical Illustrator Isaac Sprague in the Ascendency of American Botany during the Nineteenth Century." Handwritten, 7 pp.; typewritten, 5 pp. Manuscript in outline format].

[1967]. "Isaac Sprague's Role in the Ascendency of American Botany during the Nineteenth Century." Typewritten, 2 pp. [Abstract].

[a] 1967. The role of the botanical illustrator Isaac Sprague in the ascendency of American botany during the 19th century. *American Journal of Botany* 54: 647. [Abstract].

[1987]. "Isaac Sprague, 'Delineator and Naturalist,' Contributor to the Development of American Botany in the Second Half of the Nineteenth Century." Typewritten, 8 pp. [Manuscript, early draft].

[1988]. "Isaac Sprague, 'Delineator and Naturalist,' a Silent Contributor to the Development of American Botany in the Nineteenth Century." Typewritten, 36 pp. + Appendix, 16 pp.; handwritten note, 1 p. [Manuscript].

[1988]. "Isaac Sprague, 'Delineator and Naturalist.'" Typewritten, 47 pp.; figure captions, 4 pp. [Also copy marked by the printer]. [Manuscript].

[b] 1990. Isaac Sprague, "delineator and naturalist." *Journal of the History of Biology* 23: 91–126. [See pp. 114–115].

Introduction No. 19

Joseph Banks' Botanical Prints (1983)

Rudolph's enthusiasm for botanical illustration blossomed when The Ohio State University Libraries purchased and received the first portion of a set of the unique botanical prints, *Banks' Florilegium*.[1,2] These prints, depicting plant specimens obtained by Sir Joseph Banks (1743–1820) on Captain James Cook's first voyage (1768–1771) around the world, were for the first time made available to botanical and other scholars. The prints, produced from 738 copper plates engraved during the eighteenth century, are from drawings of the plants prepared on the trip. The engravings, completed 13 years after the voyage ended, are now printed by hand in color after 200 years by Alecto Historical Editions, fine printers in London. Restricted to 100 sets, only six were distributed in the United States.

The Ohio State University Libraries and the Friends of the Libraries hosted a reception marking the opening of an exhibit of selections from *Banks' Florilegium* on 4 February 1983 (Fig. 19.1). This event allowed local newspaper reporters[2] an opportunity to explain the importance of these botanical prints to the community. The specimens are of value not only to botanists interested in the detailed morphological features and geography of the plants, but also to students of the history of engraving. As each artist has been identified with each plant illustration, it is possible to study both the craft itself and the particular artists who created the different plates. The botanical prints are regarded as both a teaching and research component of the University and a community resource for scholars and artists who love and appreciate art and plants.[3] Rudolph's paper[a] reviewed the history of

---

[1] Sir Joseph Banks. *Banks' Florilegium*. [1980–1989]. A Publication in Thirty-four Parts of Seven Hundred and Thirty-eight Copperplate Engravings of Plants Collected on Captain James Cook's First Voyage . . . 1768–1771 + Supplementary Part Containing Plates Seven Hundred and Thirty-nine to Seven Hundred and Forty-three. Alecto Historical Editions in association with the British Museum (Natural History), London.

[2] Lucy Clift. 1983. Prints from rare engravings displayed. *The Ohio State Lantern*, 31 January 1983, Monday. p. 2; Tricia and Mat Herban. 1983. Banks botanical study will be available here. *Columbus Citizen-Journal*, 3 January 1983, Monday. p. 10; Earle Holland. 1983. Joseph Bank's [i.e. Banks'] bouquet: A botanist's gift of art. *OSU Quest* 5(3): 12–13; Earle Holland. 1983. Libraries display historic collection. *OSU On Campus* 12(15): 3; Nadine Doan Snyder. 1983. Ohio State buys 18th-century print plates from England. *The Ohio State Lantern*, 16 February 1983. p. 9.

[3] Tricia and Mat Herban. 1983. Banks botanical study will be available here. *Columbus Citizen-Journal*, 3 January 1983. p. 10.

the Banks botanical expedition, the fate of the collected plants and their illustrations, and the revival of interest in them through the publication of the *Banks' Florilegium.* **(Paper No. 19)**.

Fig. 19.1. Robert Tibbetts and Linda Bowers of University Libraries, The Ohio State University, examining one of more than 700 botanical prints from *Banks' Florilegium* of Captain Cook's Voyage (1768–1771).

(Taken by Kevin Fitzsimmons 1983, Photo Archives, The Ohio State University)

Botanical Prints from Cook's Voyage

Libraries display historic collection

By Earle Holland

One of the most comprehensive collections of botanical prints in the world has made its way to University Libraries for the first American public showing early next month.

More than 50 prints from the Banks Florilegium will be on display from until Feb. 11 in the Skylight Exhibit Hall of the Main Library.

Following the exhibit, the collection will be housed with the special collections at the library. The prints will be displayed a second time for public viewing later this year.

The Banks' Florilegium is a limited edition collection of illustrations of plant specimens from South America, New Zealand and Australia gathered during Captain James Cook's first voyage circumnavigating the Pacific Ocean from 1768 to 1771.

A team of illustrators and draftsmen, assembled by Sir Joseph Banks for Cook's voyage, produced the collection of original drawings. Each time Cook's ship dropped anchor, Banks' group would go ashore and gather samples of the local plant life.

Using these samples, the artists would produce finely detailed sketches of each new specimen. It was young artist Sidney Parkinson's task to do most of the initial watercolor drawings as a guide for the later engravers.

Upon his return to England, Banks directed the job of turning the illustrations into copper engraving plates, a 13-year task he supported from his own fortune.

The collection itself consists of 738 prints made from the original copper plates. Banks' intent was to publish this massive collection for scientific study by botanists.

But his fame grew, his fortune dwindled and the Napoleonic Wars loomed in the near future. The project was halted before publication.

Banks held the position of president of the Royal Society of London for 42 years and was well respected as an amateur botanist in his own right. Upon his death, he bequeathed the collection of finished plates to the British Museum.

Only 100 sets of the new publication of Banks' Florilegium will be printed, according to Linda Bowers, director of the Friends of the Library. Less than a half-dozen institutions around the country have received copies of the collection.

It was the Friends group, and more specifically Dixie Miller, a member of the Friends' Board of Directors, who was instrumental in obtaining a set of the prints for University Libraries.

Miller negotiated contributions from the Columbus Foundation, one-third, and from private donors, one-third, to match the University's portion. The total cost of the set is $93,000.

So far, the University Libraries has received 90 of the expected 738 prints. Remaining portions of the collection will be received as they are published over the next several years, Bowers says.

Newspaper article about the botanical prints of *Banks' Florilegium*. Illustration, botanical line engraving of *Banksia serrata*

Paper No. 19 (1983)

Remarkable First Cook Voyage Flower Prints (1771–1784)

Imagine the excitement of seeing a beauty that was put to sleep two hundred years ago suddenly come alive. That is exactly what is happening now with the first color printing of flower illustrations from 738 copper-plates prepared between 1771 and 1784 for Sir Joseph Banks (1743–1820)[4] (Fig. 19.2). These copper-plates have been retained in the British Museum (Natural History) since Banks' death, unprinted except for some black and white early trials and a selected black and white printing of 30 of them some years ago.

Many unusual plants were drawn and collected as dried specimens during the pioneer voyage of Captain James Cook (1728–1779) in the ship Endeavour, on his first voyage of discovery from 1768 to 1771[1,2] (Fig. 19.3). Joseph Banks was in charge of the natural history observations. The primary purpose of the voyage was to observe in Tahiti the transit of the planet Venus between the earth and sun during June of 1769. Banks, a gentleman of ample financial means, had requested permission from the Admiralty to make the voyage with seven assistants, including Daniel Carl Solander (1733–1782), a well known student of the famed Swedish naturalist Carl Linnaeus (1707–1778), and three artists to sketch the fresh plants. It is estimated that Banks spent over 10,000 pounds sterling of his own money to finance his part of the expedition. It was a memorable voyage with numerous discoveries in natural history and ethnology, as well as in geography and other scientific disciplines. So many plants new to science were collected at one locality in Australia by Banks and Solander that Cook named it "Botany Bay," a name it has retained. In six days at that place, they collected about 3,000 plant specimens representing over 200 new species, mostly in new genera, and some in new families.

Once back in England, Banks decided to publish a book about his most interesting botanical discoveries. He hired 18 engravers to prepare copper-plates from the sketches and drawings made by the talented young Sydney C. Parkinson (1745–1771) who had died late in the voyage. His drawings were finished by other artists in England and colored from his notes and from the actual pressed specimens. Publication of the book was continually delayed and finally not accomplished during Banks' life. After all, he was a very busy person, being for over 40 years the President

---

Frontispiece Credits: Botanical Prints from Cook's Voyage. Newspaper article about the plant prints from Captain James Cook's Voyage (1768–1771). (From *OSU on Campus* 12(15): 3. 1983. 17 January)

Botanical line engraving of *Banksia serrata*. (From an Announcement of a Sale of Exhibition Proofs for the Friends and Staff of the Missouri Botanical Garden, © Alecto Historical Editions, London, U.K., 1986; reproduced with permission)

of the Royal Society of London and also being a major scientific and foreign affairs advisor to the King and various businessmen. The potential cost of such a publication after the great expenses of the voyage and engraving may have also been a deterrent during a depressed monetary period. Banks' bequest of his massive collections of natural objects and books formed a major addition to the British Museum upon his death. The copper-plates of flowers of the Cook voyage were a part of that legacy.

Now, after remaining unprinted for about 200 years, the copper-plates are being printed. In the mid 1970s Joseph G. Studholme, one of the owners of Alecto Historical Editions, fine print publishers, approached the British Museum (Natural History) for permission to print the Banks plates. The result is the project of printing 110 sets of the plates in color, known as *Banks' Florilegium.* According to the agreement with the British Museum, 100 sets were sold for about $93,000 each to a quota of various countries. About a half dozen were allocated for the United States.

The printing of the copper-plates is a fascinating story in itself. Edward Egerton-Williams is the master printer supervising the operation. Each plate, after careful cleaning, is coated with a light chrome finish to protect the softer copper. Inks of the proper colors, to match the greens and flower tints of the original drawings and specimens, are dabbed onto and worked into the particular areas of the plates with cloths in a color printing process known as "à la poupée." The printing is done on Somerset hand-made paper using hand cranked rolling presses. It may take as long as a month to obtain a good proof of a particular plate and then additional time to put the 110 copies through the press. Each print is individually marked by the printer and publisher, and has information about the original artist and engraver, as well as the plant name and collection locality. The 738 prints in each set will be gathered into 34 sections, or volumes, of 22 to 24 prints and placed into individual Solander cases. These cases are appropriately named for Daniel

Fig. 19.2. Sir Joseph Banks (1743–1820).

(In the Rudolph photograph collection. The original was engraved by H. Robinson and is in the British Museum. Permission to reproduce photograph courtesy ©The British Museum, London, England)

Fig. 19.3. Captain James Cook (1728–1779).

(From W. Chambers. 1895. *Captain James Cook*, p. 278. In A.R. Spofford, Frank Weitenkampf, and J.P. Lamberton (eds.). *The Library of Historic Characters and Famous Events of All Nations and All Ages*, Vol. 5. William Finley & Co., Philadelphia; from a copy in the main library, The Ohio State University)

Solander who originally designed this type of box for use in Sir Joseph Banks' library. At a pace prescribed by the careful quality controls, the printing of the sets will not end before 1987.

One of the subscribed sets has been obtained by The Ohio State University Libraries. On 4 February 1983 an exhibition of some of the prints in the first volumes, together with related books, was held at Ohio State. It preceded the New York City American Museum of Natural History exhibition of the Smithsonian Institution's prints and related materials held on 25 April 1983. Both of these exhibitions were superseded by the one at the Thirteenth International Botanical Congress in Sydney, Australia on 21–28 August 1981.[3] The *Smithsonian* magazine of March 1983 had a story about the *Florilegium*.[5] Having this remarkable set in central Ohio will provide researchers the opportunity to study the scientific aspects of the plants and to pursue historical interests in exploration, artistic interests about eighteenth century methods and styles, and general interests in fine printing and its methods.

The Ohio State University *Florilegium* is the result of the initial interest and initiative of Mrs. Albert Fullerton Miller of Columbus. She met Joseph G. Studholme, one of the copublishers of the set, some years ago in England. What impressed Mrs. Miller about him was his starting a fine print publishing business while still a Cambridge University student. It was natural that Mr. Studholme should share his excitement about his new venture with Mrs. Miller, and that excitement was easily catching. Mrs. Miller took the lead in trying to obtain funds for a subscription. Much of the cost has already been pledged. The result is that the exciting plant life discovered by Sir Joseph Banks on the first Cook voyage of discovery over two hundred years ago in Australia, Brazil, Java, Madeira, New Zealand, the Society Islands, and Tierra del Fuego has come to dwell in central Ohio.

References Cited

1. Beaglehole, J. C. (ed.). 1962. *The Endeavour Journal of Joseph Banks 1768–1771.* The Trustees of the Public Library of New South Wales in Association with Angus and Robertson, Sydney. 2 vols. (1: xxvii, 476 pp; 2: xvi, 406 pp.).

2. Hooker, Sir Joseph D. (ed.). 1896. *Journal of the Right Hon. Sir Joseph Banks Bart., K.B., P.R.S., during Captain Cook's First Voyage in H.M.S. Endeavour in 1768–71 to Terra del Fuego, Otahite, New Zealand, Australia, the Dutch East Indies, etc.* MacMillan and Co., London. li, 466 pp.

3. The Library Council of New South Wales. 1981. *A Fine Field for Botanizing: An Exhibition Arranged in Association with XIII International Botanical Congress, Sydney, 21–28 August 1981.* The Library Council of New South Wales, Sydney, N.S.W., Australia. 42 pp. (See pp. 8–10).

4. Lyte, Charles. 1980. *Sir Joseph Banks: 18th Century Explorer, Botanist and Entrepreneur.* David and Charles, Newton Abbot. 248 pp.

5. Schiff, Bennett. 1983. It took more than 200 years to get this work printed. *Smithsonian* 13(12): [76]–82, 84–85.

Publication Notes

Modified and edited from an article[a] that Rudolph was invited to write for the Friends of the Libraries of The Ohio State University and for publication in the *Ohio Library Association Bulletin.*

[a] 1983. Remarkable first Cook voyage flower prints, now at The Ohio State University Libraries. *Ohio Library Association Bulletin* 53(3): 30–33. (Permission to reproduce article courtesy of the Ohio Library Association)

Part Six

WOMEN IN AMERICAN BOTANY

Perhaps Emanuel D. Rudolph's greatest love in his studies of botanical history was the role of women botanists in nineteenth century America. By the end of the eighteenth century, as a result of the influence of the French philosophical writer Jean Jacques Rousseau (1712–1778) and his English followers, botany was proclaimed a most suitable study for young ladies. Many American women studied botany at home, in schools and seminaries, and in a few universities during the nineteenth century. They were active with growing plants in gardens (Fig. 20.1) and taking excursions into the fields and forests to collect plants for making dried herbarium specimens (Figs. 20.2, 20.3). According to a Botanical Directory of 1878, 16 percent of the botanists in the United States were women. Rudolph's studies of women botanists followed immediately from his investigations on the botany in children's books. Two of these studies on women botanists are included here (**Part Six**). They feature the role of women in nineteenth century botany (**Paper No. 20**) and the contribution of Emily L. Gregory to plant anatomy (**Paper No. 21**).

Introduction No. 20

Studies on Women Botanists (1978–1990)

From 1978 to 1984, Rudolph presented six contributions on the role of women botanists during the nineteenth century. The titles and meeting information are:

> "The Role of Women in Nineteenth Century American Botany." Presented 27 June 1978 to the Historical Section, Botanical Society of America, Virginia Polytechnic Institute and State University, Blacksburg.
>
> "The Contributions of Women to the Development of 19th Century American Botany." Presented 1 December 1979 to The Ohio Academy of Science, The Ohio Historical Center, Columbus. [no meeting abstract].
>
> "Women as a Constituency in Nineteenth-century American Botany, a Quantitative Analysis." Presented 11 April 1981 to the Midwest Junto for the History of Science, The University of Akron, Akron. [no meeting abstract].
>
> "Women in Nineteenth Century American Botany." Presented 7 November 1981 to the Women's Studies Research Forum, Columbus. [no meeting abstract].
>
> "The Role of Almira Hart Lincoln Phelps in the Spread of Botany in Nineteenth Century America." Presented 9 August 1983 to the Historical Section, Botanical Society of America, The University of North Dakota, Grand Forks.
>
> "The First American Plant Anatomy Book and Its Author, [Emily L. Gregory]." Presented 28 April 1984 to The Ohio Academy of Science, Case Western Reserve University, Cleveland; and 7 August 1984 to the Historical Section, Botanical Society of America, Colorado State University, Fort Collins.

On the subject of women in botany during the nineteenth century, Rudolph published four papers: "How it developed that Botany was the science thought most suitable for Victorian young ladies" (1973), appeared first and was developed from his studies of botany by both children and women. The idea that botany was the science especially suited to the tastes, feelings, and capacities of women in the United

States came into being during the second quarter of the nineteenth century. The thinking was that ladies took delight in examining the minute details of the flowers, pressing and preserving specimens, and illustrating them with lead pencils or water colors; whereas boys were less interested, being careless and harsh in treating specimens. From about 1826 until 1846 in the United States, "botany acquired the reputation of being a girls' science, and that association persisted much longer in the popular mind."[1]

In Rudolph's second published paper, "Women in Nineteenth Century American Botany; A Generally Overlooked Unrecognized Constituency" (1982), he noted that botany was thought to be a suitable study for young women in schools and/or an avocation for amateurs. From studying information about 1,581 women active in botany during the nineteenth century, Rudolph learned that most of them resided in New England, and the Middle Atlantic states, nearly three quarters of them were not married, and only 10 percent had higher educational degrees (Appendices 20.1, 20.2, 20.3). Even though only a few women became professional botanists in the nineteenth century, women as a group constituted an important and overlooked constituency in the developing profession of botany. Many amateur women botanists collected plants to make permanent herbarium specimens, wrote and published conversational notes and sometimes scientific papers and books, and became active officers or workers in natural history and botanical societies.

When submitted to the *American Journal of Botany*, Rudolph's paper on the "Unrecognized Constituency" received very favorable prepublication reviews from two senior botanists, both recommending publication. Harold C. Bold[2] wrote that he

> . . . read with interest the paper [and that it was] clearly a contribution from a member of the Historical Section . . . [who] has done a good deal of work in assembling his excellent summary. It gave me cause to reflect when I realized that I knew or had met several of the 'nineteenth century' lady botanists.

Oswald Tippo[2] stated that it was

> . . . an excellent historical paper which will be used eagerly by most of our members and subscribers. I personally found it fascinating and will want to follow up on some of the publications cited . . . The paper is well written and shows evidence of scholarship of high rank.

---

[1] Emanuel D. Rudolph. 1973. How it developed that botany was the science thought most suitable for Victorian young ladies. *Children's Literature* 2: 92–97. (Quote, p. 95).

[2] Harold C. Bold; Oswald Tippo. [1981]. [Reviews of manuscript on "Women in Nineteenth Century American Botany; A Generally Unrecognized Constituency," by Emanuel D. Rudolph for the *American Journal of Botany*]. 1 p.; 1 p.

Rudolph deserved these complimentary remarks as he did not always get them upon submitting his botanical history papers for journal publications. Perhaps these kinds of statements gave him encouragement to write and publish more on women botanists.

Rudolph's third published paper, "Almira Hart Lincoln Phelps (1793–1884) and the Spread of Botany in Nineteenth Century America" (1984), showed how one person can have a profound influence on the development and spread of a discipline and profession. Mrs. Phelps wrote two innovative botany books especially designed for beginning students and amateurs with the aim of cultivating in the reader a life-long interest in botany and teaching. Through her writings, according to Rudolph,[3] Mrs. Lincoln Phelps "helped provide a favorable climate for the developing profession of botany in America." In addition, she should be given "due credit for being a popularizer of science and women's education, but especially for contributing to the emerging field of American botany by her widely used textbooks."

Fig. 20.1. Botany and horticulture students transplanting plants at the Kansas State Agricultural College, Manhattan, Kansas, 1891.

(From a photograph in the University Archives, Kansas State University, Manhattan; permission to reproduce courtesy of the University Archives)

Rudolph's fourth and final published paper, "Women who Studied Plants in the Pre-twentieth Century United States and Canada" (1990), presented three tables of data for 1,454 women known to have been actively interested in plants before 1901. For each woman, Rudolph noted their birth and death years, marital status,

[3] Emanuel D. Rudolph. 1984. Almira Hart Lincoln Phelps (1793–1884) and the spread of botany in nineteenth century America. *American Journal of Botany* 71: 1161–1167. (Quotes, pp. 1161, 1165–1166).

geographical location by state, being a collector of specimens, plant specialty, and active time period. From this information, Rudolph summarized the activity of the women studying plants at 10-year intervals, the number of them for each state, and the collecting of plants, among other aspects (Appendices 20.4 and 20.5)

During the preparation of the fourth paper, Rudolph received prepublication assistance and review from Ronald L. Stuckey, Tod F. Stuessy, and Joseph Ewan. From Stuckey he received biographical information for those women who lived and worked in Ohio, and assistance in the organization of the paper. Following submission of the long manuscript to *Taxon*, Editor Frans A. Stafleu (1921–1997)[4] in a letter to Rudolph wrote, "I am proposing (in view of the reviews of Ewan and Stuckey) to . . . accept this right-away. . . . I like this!" As the principal author of a 7-volume master-work[5] on the world's taxonomic botanical literature, Stafleu certainly would appreciate receiving a paper containing the details of the lives of hundreds of pre-twentieth century American women botanists.

Fig. 20.2. Botany students of Professor William A. Kellerman departing on a botanical excursion from the OSU Lake Laboratory Building, Cedar Point, Ohio, 1904.

(Taken by William A. Kellerman, summer 1904; in the collection of RLS)

[4] Frans A. Stafleu. 1989. Letter of 17 August to Emanuel D. Rudolph, Department of Botany, The Ohio State University, Columbus, Ohio. 1 p.

[5] Frans A. Stafleu and Richard S. Cowan. 1976–1988. *Taxonomic Literature: A Selective Guide to Botanical Publications and Collections with Dates, Commentaries and Types*. 2nd ed. *Regnum Vegetabile*, Vols. 94, 98, 105, 110, 112, 115, 116. Bohn, Scheltema & Holkema, Utrecht. 7 vols.

Fig. 20.3. Botanist E. Lucy Braun (second from left) with family members and friends on a botanical field collecting excursion on the outcrop at the Natural Bridge, Powell County, Kentucky, during the 1920s.

(From the E. Lucy Braun collection; permission to reproduce photograph courtesy of the Cincinnati Museum Center Image Archives, Cincinnati, Ohio)

Rudolph's last written item on women botanists was the foreword to Stuckey's book on "Women Botanists of Ohio Born Before 1900," published in the spring of 1992. In that short message, with some editorial modification, he[6] wrote:

> Botany, often called the science most appropriate for female study in the past, attracted some of the most able women to its pursuit. The roster of women in Ohio . . . reflects that fact. One might well ask why single out women interested in plants for special recognition? Because they have been so underrated and little recognized in the past, they do warrant our appreciative recognition. Furthermore, they were usually not the recognized professional botanists, who published their findings in professional journals, rather they were popularizers and amateurs who did much to aid the professional and profession. Thus, they have often been ignored by later workers.
>
> [The State of] Ohio, settled later than the eastern part of the country, required time for its cultural and scientific development. Despite this [situation], it is significant that so many women who contributed in very substantial ways to understanding plant life, can be identified. Many of these women belong to the non-professional ranks. In the . . . [nineteenth] century, Ohio and Illinois each had more women known to be actively interested in plants than either Connecticut or New Jersey.

---

[6] Emanuel D. Rudolph. 1992. Foreword, p. iii. *In* Ronald L. Stuckey. *Women Botanists of Ohio Born before 1900 with Reference Calendars from 1776 to 2028*. Ronald L. Stuckey, RLS Creations, Columbus, Ohio. x, 67 pp.

Role of Nineteenth Century American Women Botanists

As a summary of Rudolph's published studies pertaining to the role of women in the development of the popularization and professionalization of botany in nineteenth century America, four items have been selected and merged to create **Paper No. 20**. The first paragraph is the abstract[a] of his presentation on 27 June 1978 at the meeting of the Historical Section of the Botanical Society of America, Blacksburg, Virginia. In that presentation Rudolph credited the French philosopher, Jean Jacques Rousseau (1712–1778), in promoting the study of botany among young ladies (Fig. 20.4).

The second paragraph is a reprinting of the abstract[d] in his published paper, "Women in Nineteenth Century American Botany; A Generally Unrecognized Constituency," published in 1982 in the *American Journal of Botany*. Mrs. Almira Hart Lincoln Phelps serves as an example of an accomplished nineteenth century woman botanist. The abstract[e] written about her for Rudolph's presentation on 9 August 1983 at the meeting of the Historical Section of the Botanical Society of America, Grand Forks, North Dakota, constitutes the third paragraph of this paper. The remaining six paragraphs[h] contain the analysis from his last published paper on women botanists, which appeared in *Taxon*, 1990. **(Paper No. 20)**.

Fig. 20.4. Jean Jacques Rousseau (1712–1778).

(From the Hunt Portrait Collection; permission to reproduce photograph courtesy of Hunt Institute for Botanical Documentation, Carnegie Mellon University, Pittsburgh)

Three Influential Women Botanists

Emma Willard
(1787–1870)

Almira Hart Lincoln Phelps
(1793–1884)

Elizabeth G. (Knight) Britton
(1858–1934)

Paper No. 20 (1978–1990)

The Role of Women in Nineteenth Century American Botany

By the end of the eighteenth century as a result of the writings of the French philosopher Jean Jacques Rousseau (1712–1778) and his English followers, Jane Marcet (1769–1858), Priscilla Wakefield (1751–1832), and Elizabeth Fitton (fl. 1817–1840), botany was proclaimed a most suitable study for young ladies (Fig. 20.1). During the nineteenth century many American women studied botany at home, in schools and seminaries, and a few in universities. Some women conducted botanical studies as amateurs, collecting and exchanging specimens, and publishing papers in journals. Although their numbers were significant, few women entered the professional stream. Those few who entered professions did so as school teachers and writers, for example Almira Hart Lincoln Phelps (1793–1884) and Elizabeth Ann Youmans (1826–1914), or as teachers in institutions of higher education, for example Margaret Ferguson (1863–1951) and Elizabeth Gertrude Knight Britton (1858–1934). Opportunities for and acceptance of professional women botanists during the nineteenth century were very rare in the United States.[a]

Botany was thought to be a suitable study for young women in schools and a popular avocation in the nineteenth century. A surprisingly large number of American women identified themselves as being seriously interested in botany. For example, in the first published directory of American botanists in 1873,[1] women were represented by 13 percent of 599 names and that increased to 16 percent of the 982 names in 1878. In Rudolph's paper,[d] 1,581 women have been identified as a sample of those actively interested in botany during the century. Less than two percent of them were active before 1870, and most of them, 67 percent, resided in New England and the Middle Atlantic states. Almost three-quarters of them were unmarried, and only 10 percent had higher educational degrees although 15 percent had some identifiable profession. Some particular individuals are noted for their contributions

---

Frontispiece Credits: Three Influential Women Botanists. Emma Willard (1787–1870), frontispiece opposite title page. (From John Lord. 1873. *The Life of Emma Willard*. D. Appleton and Co., New York)

Almira Hart Lincoln Phelps (1793–1884), frontispiece opposite title page. (From Almira Lincoln Phelps. 1873. *Reviews and Essays on Art, Literature, and Science*. Claxton, Remsen & Heffelfinger, Philadelphia)

Elizabeth G. Knight Britton (1858–1934). (From the Hunt portrait collection; permission to reproduce photograph courtesy of Hunt Institute for Botanical Documentation, Carnegie Mellon University, Pittsburgh, Pennsylvania)

to science, their activity as plant collectors, and their support of botanical societies. Though few American women became professional botanists in the nineteenth century, they constitute an important overlooked constituency for the developing profession of botany.[d] (Appendices 20.1, 20.2, 20.3).

Almira Hart Lincoln Phelps is an example of a young woman who was instrumental in the spread of botany in nineteenth century America. Mrs. Phelps came under the influence of an older sister, Emma Willard (1787–1870), a well known educational reformer, who helped form her educational philosophy, and of Amos Eaton, who helped form her botanical and scientific understanding. Not having a suitable botanical textbook to introduce students to the subject, particularly female students who were becoming more interested in botany, she first wrote *Familiar Lectures on Botany* (1829),[4] using her first husband's surname, Lincoln. "Mrs. Lincoln's Botany" quickly became popular and continued to be revised and reprinted through 1869. In 1833, a second botany text for lower level students, *Botany for Beginners*,[5] appeared. It too, went through many reprintings until 1891. Mrs. Phelps' other books and writings on science and education also were popular. The botanical texts were educationally innovative in starting with flower structure, using common living examples, and later considering other morphological and physiological aspects of plants. The Linnaean System was used to classify the "most common native and foreign" plants that were described mainly from the various editions of Eaton's *Manuals*.[3] Clear figures, often copied from well known authorities helped to instruct teachers and students. Owing to its wide use, even in later years in competition with the textbooks of Asa Gray and Alphonso Wood, Mrs. Phelps' books influenced many students, especially women, and induced some of them to have a life-long interest in botany[e] and in teaching it to others. Through her writings, she was a person who helped provide a favorable climate for the developing profession of botany in America.[e]

Before the middle of the nineteenth century American young women were being educated in botany in their schools and through readily available textbooks. As one author[2] wrote:

> Formerly, the study of Botany was almost exclusively confined to the medical profession, or to a limited number of philosophic minds, and accordingly the treatises on the subject were elaborate, and in many respects unsuitable to be placed in the hands of youth, and more especially of young females . . . . But this delightful science has become at the present time a favorite with the young, refined, and delicate of our sex, and introductory publications on a plan, easy, practical and comprehensive, are now readily procurable by the ladies of our country who desire to be initiated into this branch of knowledge.[h]

In nineteenth century America, attitudes toward women's education and to their botanical training resulted in many young women having a sound background to study plants independently. Many of them seemed to continue their interest after schooling with the result that the developing professional botanists had a wide audience who understood their activities and supported them. Because of increasing interest in the history of women's roles in American society, analyses have been made of their places in the development of the plant sciences by both historians and botanists. The period after the American Civil War (1861–1865) was one of rapid expansion of women's education and the interest of women in plants. This idea is revealed in the increasing percentages, with finally over 50 percent of them being active in the last decade of the century. Most of the identified women were active in the northeastern part of the country, though California, Illinois, Ohio, and Indiana have significant representation. New York, Massachusetts, and Pennsylvania have the largest numbers because these states had the most active groups of botanists and local clubs, which were founded during the later part of the nineteenth century, and had the first female academies in the country. The southeastern states are less represented probably because the Civil War had a more lasting depressing effect there, and because even before the War, fewer women were actively interested in plants in that part of the country. Those southern women who were interested in plants were important in regional botany. Almost a third of the sample were married (30.9 percent), and marriage especially to another botanist or professional was a stimulus to continue an active interest in botany. About a quarter (27.6 percent) are known to have made plant collections; many of their herbaria are now in institutions; and about half of these women (14.1 percent) are known to have exchanged specimens with other collectors (Appendices 20.4 and 20.5).[h]

Undergraduate and post-graduate degrees were infrequent during the nineteenth century with only 6.4 percent of the women identified as receiving a bachelor's degree. Even less frequent were master's degrees (2.8 percent), or other advanced degrees (2.6 percent), and very few received doctor's degrees (1.1 percent). These figures are further proof that women were generally not seeking admission to the botanical profession. With a few notable exceptions, these women were not original contributors to the botanical literature, nor were they researchers. Some of them were educators who instilled in the youth, by word or book, an appreciation of plants, and perhaps sparked an interest in some of the students to become scientists. Such information is rarely recorded. Then too, these women collected plants, usually in their localities, that were added to herbaria providing documentation for floras. They nurtured botany by joining local natural history and botanical clubs. Fundamentally they were part of the educated public who appreciated and supported the plant sciences then developing in America. It is remarkable that so many women in nineteenth century America were seriously interested in plants. The impact of women botanists is only now being recognized.[h]

References Cited

1. Anonymous. 1873. Botanical directory, for North America and the West Indies. *Bulletin of the Torrey Botanical Club* 4: [49]–59.

2. Coxe, [M.]. 1842. *Claims of the Country on American Females*. Isaac N. Whiting, Columbus, Ohio. 2 vols. (1: iv, 243 pp.; 2: iv, 242 pp.). (Quote, vol. 1, p. 201).

3. Eaton, A. 1817. *A Manual of Botany for the Northern States, . . .* By the Members of the Botanical Class in Williams' College (Mass.), from a Manuscript System Compiled by the Author of Richard's Botanical Dictionary. Websters and Skinners, Albany, NY. vi, 164 pp. [The 8th ed., entitled *North American Botany*, was written with John Wright].

4. [Phelps], A[lmira] H[art] Lincoln. 1829. *Familiar Lectures on Botany*. H. and F. J. Huntington, Hartford; G. & C. & H. Carvill, New York; Richardson & Lord, Boston. x, [11]–335, [1], 4 pp. [At least 39 different printings and two major revisions (1845, 1852) were issued through 1869].

5. _____. 1833. *Botany for Beginners*. F. J. Huntington, Hartford. 256 pp. [At least 27 printings were issued through 1891].

Publication Notes

Compiled and edited from an abstract[a] of a manuscript;[b] selected passages[g] from the resulting paper published 1984 in the *American Journal of Botany*; and selected passages from a paper[h] published in *Taxon*.

1973. How it developed that botany was the science thought most suitable for Victorian young ladies. *Children's Literature* 2: 92–97.

[a] 1978. The role of women in nineteenth century American botany. *Botanical Society of America, Miscellaneous Series, Publication* 156: 38. [Abstract].

[b] 1978. "The Role of Women in Nineteenth Century American Botany." Typewritten, 8 pp.; handwritten, 3 pp. + data sheets, earlier drafts of the manuscript, copies of two reviewers' comments and correspondence. [Manuscript].

[c] 1982. "Women in Nineteenth Century American Botany; A Generally Unrecognized Constituency." Typewritten, 28 pp. [Manuscript].

[d] 1982. Women in nineteenth century American botany; a generally unrecognized constituency. *American Journal of Botany* 69: 1346–1355. (See p. 1346).

[e] 1983. The role of Almira Hart Lincoln Phelps in the spread of botany in nineteenth century America. *American Journal of Botany* 70(5, pt. 2): 65. [Abstract].

[f] 1983. "The Role of Almira Hart Lincoln Phelps in the Spread of Botany in Nineteenth Century America." Typewritten, 10 pp. [Manuscript].

1983. "Women Botanists as Role Models for Young Women Considering a Profession. Proposal to United States Department of Education. Typewritten, 9 pp. [Grant Proposal, not funded].

[g] 1984. Almira Hart Lincoln Phelps (1793–1884) and the spread of botany in nineteenth century America. *American Journal of Botany* 71: 1161–1167. (See p. 1161).

[1989]. "Women Who Studied Plants in the Pre-Twentieth Century United States and Canada." Handwritten, 12 pp. + Summary Sheets for 544 names; handwritten, 17 pp.; typewritten, 11 pp. + Table 1, 39 pp., Table 2, 1 p., and Table 3, 1 p. [Manuscript; duplicate copy sent to the editor of *Taxon*].

[h] 1990. Women who studied plants in the pre-twentieth century United States and Canada. *Taxon* 39: 151–205. (See pp. 151, 203–204).

APPENDIX 20.1. NUMBER AND PERCENTAGES OF WOMEN ACTIVE IN BOTANY BY DATE, BASED ON 1,581 INDIVIDUALS, FROM RUDOLPH (1982, TABLE 1)[d]

Period	Number Active (of 1,581)	Percent Active
1800–1810	0	0.0
1811–1820	3	0.2
1821–1830	5	0.3
1831–1840	14	0.9
1841–1850	13	0.8
1851–1860	12	0.8
1861–1870	21	1.3
1871–1880	272	17.2
1881–1890	353	22.3
1891–1900	888	56.2

Appendix 20.2.
Numbers and Percentages of Women Interested in Botany by Region, Based on 1,160 Individuals, from Rudolph (1982, table 2)[d]

Region	Number (of 1,160)	Percent
New England (Maine, New Hampshire, Vermont, Massachusetts, Rhode Island, Connecticut)	263	22.7
Middle Atlantic (New York, New Jersey, Pennsylvania)	514	44.4
East North Central (Ohio, Indiana, Illinois, Michigan, Wisconsin)	134	11.6
West North Central (Minnesota, Iowa, Missouri, North Dakota, South Dakota, Nebraska, Kansas)	58	5.0
South Atlantic (Delaware, Maryland, District of Columbia, Virginia, West Virginia, North Carolina, South Carolina, Georgia, Florida)	54	4.7
East South Central (Kentucky, Tennessee, Alabama, Mississippi)	17	1.5
West South Central (Arkansas, Louisiana, Oklahoma, Texas)	18	1.6
Mountain (Montana, Idaho, Wyoming, Colorado, New Mexico, Arizona, Utah, Nevada)	31	2.7
Pacific (Washington, Oregon, California)	71	6.1

APPENDIX 20.3.
AMERICAN WOMEN BOTANISTS RECEIVING DOCTORATES (PH.D. OR D.SC.), FROM RUDOLPH (1982, TABLE 3)[d]

Name	Date of Degree	Degree Granting Institution	19th Century Profession
Bunting, Martha	1897	University of Pennsylvania	High school teacher
Byrnes, Esther F.	1898	Bryn Mawr College	High school teacher
Carter, Alice	1888	Syracuse University	High school teacher
Cooley, Grace E.	1896	University of Zürich	Associate Professor (Wellesley College)
Cross, Laura B.	1894	University of Pennsylvania	High school teacher
Gregory, Emily L.	1886	University of Zürich	Lecturer (Barnard College)
Gregory, Emily R.	1899	University of Chicago	(Professor in 20th Century)
Hitchcock, Fanny	1894	University of Pennsylvania	None
Holmes, Mary E.	1888	University of Michigan	Editor
Hooker, Henrietta	1889	Syracuse University	Professor (Holyoke College)
Keller, Ida A.	1890	University of Zürich	High school teacher
Nichols, Mary A.	1896	Cornell University	None
Randolph, Harriet	1892	University of Zürich	Demonstrator and reader (Bryn Mawr College)
Schively, Adeline	1897	University of Pennsylvania	High school teacher
Snow, Julia W.	1893	University of Zürich	Instructor (University of Michigan)
Stoneman, Bertha	1896	Cornell University	Professor (Huguenot College, South Africa)
Wilson, Lucy L.	1897	University of Pennsylvania	High school teacher

APPENDIX 20.4.
PERIOD OF ACTIVITY OF WOMEN WITH PERCENTAGES, BASED ON A TOTAL OF 1,486 INDIVIDUALS FOR EACH TIME PERIOD, AS CORRECTED FROM RUDOLPH (1990, TABLE 2)[h]

Period	Number Active	Percent of Total Active
Before 1850	37	2.5
1850–1859	21	1.4
1860–1869	39	2.6
1870–1879	258	17.4
1880–1889	371	25.0
1890–1899	818	55.1
1900	622	41.9

APPENDIX 20.5.
NUMBERS OF WOMEN ACTIVE IN EACH STATE OR PROVINCE, BASED ON A TOTAL OF 1,486 INDIVIDUALS, AS CORRECTED FROM RUDOLPH (1990, TABLE 3)[h]

AL	5	IN	31	NH	15	TX	7
AK	1	IA	27	NJ	47	UT	2
AZ	2	KS	11	NM	2	VT	33
AR	2	KY	10	NY	379	VA	5
CA	99	LA	0	NC	5	VI	0
CO	15	ME	26	ND	4	WA	5
CT	53	MD	32	OH	55	WV	2
DE	4	MA	190	OK	0	WI	11
DC	25	MI	24	OR	6	WY	1
FL	12	MN	24	PA	164	ALB	1
GA	6	MS	3	PR	0	BC	5
GU	0	MO	22	RI	12	ONT	6
HI	2	MT	16	SC	6	QUE	5
ID	0	NE	0	SD	2		
IL	60	NV	1	TN	3		

Introduction No. 21

Emily L. Gregory and Plant Anatomy (1984)

Rudolph's manuscript[a] on "The First American Plant Anatomy Book and Its Author, Emily Lovira Gregory," was presented twice at professional meetings, first on 28 April 1984 to the Plant Sciences Section of The Ohio Academy of Science, and second on 7 August 1984 to the Historical Section of the Botanical Society of America. Each presentation is represented with a published abstract.[b,c] Gregory's book, published in 1895 in Boston, was the first American book devoted entirely to plant anatomy (Fig. 21.1). She is the first American woman known to have earned a Ph.D. degree in botany from a European school, the University of Zürich, in Switzerland. Well versed in the German language, she closely followed the German writers rather than English or American textbook writers. According to Rudolph, Gregory, a woman and notable botanist, deserves more recognition than she has been given in the past. **(Paper No. 21)**.

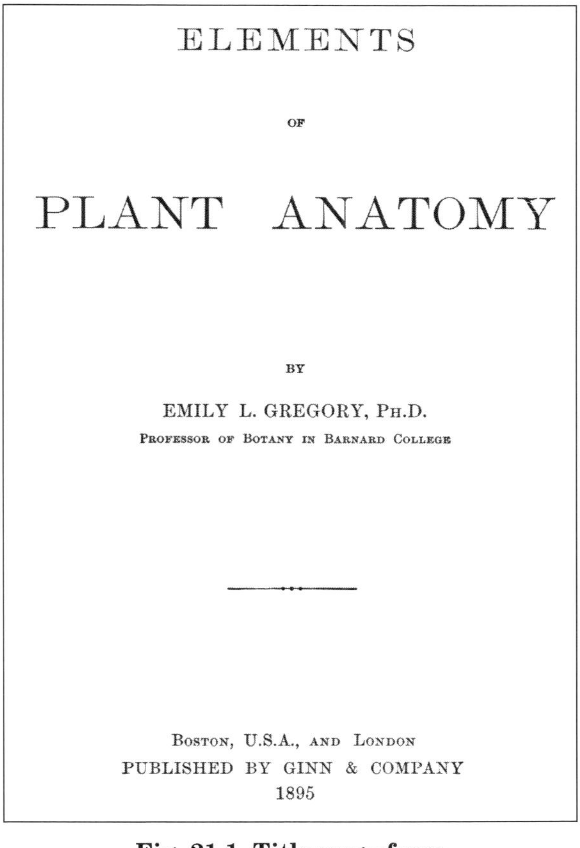
ELEMENTS

OF

PLANT ANATOMY

BY

EMILY L. GREGORY, PH.D.

PROFESSOR OF BOTANY IN BARNARD COLLEGE

BOSTON, U.S.A., AND LONDON

PUBLISHED BY GINN & COMPANY

1895

Fig. 21.1. Title page from *Elements of Plant Anatomy*, by Emily L. Gregory.

(From the copy in the J. N. Couch Biology Library–Botany Section, University of North Carolina, Chapel Hill)

Emily Lovira Gregory and Plant Anatomy

Emily Lovira Gregory (1841–1897)

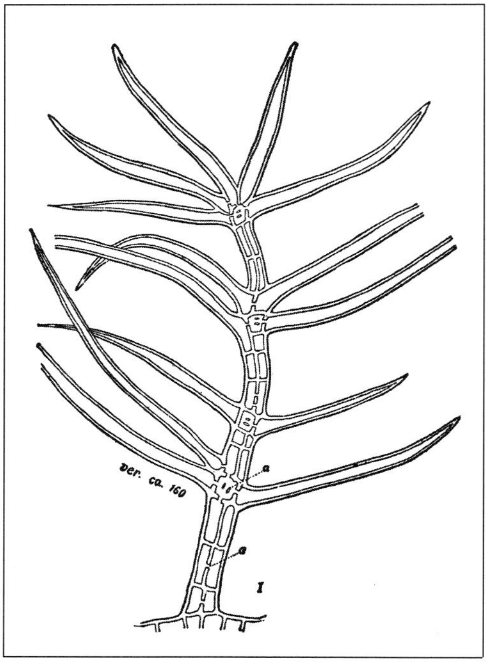

Illustration of hair from under epidermis of *Phlomis russeliana* leaf, from Emily L. Gregory (1895)

Paper No. 21 (1984)

The First American Plant Anatomy Book and its Author Emily Lovira Gregory (1841–1897)

The first American botanical book solely devoted to plant anatomy, *Elements of Plant Anatomy* (1895),[12] published in Boston, was written, for that time period by a remarkable woman, Emily Lovira Gregory (1841–1897). This paper reviews Miss Gregory's life and botanical experiences; analyzes the place of women, particularly hers, in American botany of the 1880s and 1890s; and evaluates the first national plant anatomy textbook in terms of other botanical studies of those times.

Emily Gregory, born 31 December 1841, in Portage, New York, took her early education at Albion Seminary, a school for females in Albion near Lake Erie in upper New York State. She then taught for a period of time at Dunkirk Friendship Seminary for females in Fredonia, New York. In 1876, at age 34, she entered Cornell University as a student with specialization in botany and literature, and five years later earned a bachelor of literature degree from that institution. Women were then entering higher education in large numbers in the state universities and women's colleges. Miss Gregory became a teacher of botany at Smith College, a ten-year-old women's college, in Northampton, Massachusetts, where she taught for two years beginning in 1881. For part of a year, she was in charge of the laboratory work in botany at the Harvard University Annex, while working with Professor George Lincoln Goodale (1839–1923). She spent the next year in Europe, mostly in the laboratory of the elderly Professor Albert Wigand (1821–1886), a former student of Matthias J. Schleiden (1804–1881), noted as a plant anatomist and physiologist of Marburg, Germany. While there, she also visited the laboratories of Professors Johannes Reinke (1849–1931) in Göttingen and Simon Schwendener (1829–1919) in Berlin. After a brief period again at Smith College, she returned to Europe and after a year there, earned her doctor's degree, at the University of Zürich on 23 July 1886. She was the first American woman known to have earned a European Doctor of Philosophy degree in botany.

Upon returning to America, Dr. Gregory became an associate in botany at the new women's college in Bryn Mawr, Pennsylvania, under the supervision of Professor

Frontispiece Credits: Emily Lovira Gregory and Plant Anatomy. Emily Gregory (1841–1897). (From Elizabeth G. Britton, *Bulletin of the Torrey Botanical Club* 24: portrait opp. p. 221. 1897)

Illustration of hair from under epidermis of *Phlomis russeliana* leaf, on page 56. (From Emily L. Gregory. 1895. *Elements of Plant Anatomy*. Ginn and Co., Boston and London)

Edmund B. Wilson (1856–1939). At this time she published, in the *Bulletin of the Torrey Botanical Club*,[9] her first botanical paper, on the "pores" (i.e. pits) of the libriform fiber in plants. Later, she published a comprehensive work on the development of cork wings on certain trees.[10] Finding the position at Bryn Mawr not particularly satisfying, she left in 1888 to accept an appointment as a fellow at the University of Pennsylvania, to develop a botanical laboratory in their newly emerging Department of Botany. While in Philadelphia, she was invited by Nathaniel Lord Britton (1859–1934), the Professor of Botany at Columbia University, to present a paper at a meeting of the Torrey Botanical Club in New York City. Some had suggested that botany be taught as a subject at the newly formed Women's College of Columbia University, Barnard College, and after a time Gregory was approached by Britton to teach at Barnard.

Gregory[14] described the situation some years later:

> Dr. Britton told me that several of the ladies of the Trustees of Barnard wished to have him give some instruction in Botany. . . . He had told them it was impossible to undertake such work with any promise of success without a laboratory and the work he was doing in Columbia was entirely Systematic, and that there was no laboratory there for the men. . . . the proposed charter for Barnard College included a paragraph [indicating that] 'no person should teach in Barnard who was not officially connected with Columbia as a teacher.'

The requirement for Columbia status was a problem if Gregory was to be appointed. The problem was resolved by accepting her written plan for the establishment of a laboratory of botany without calling her a teacher. Gregory went again to Europe in the spring of 1889 with funds provided by the Torrey Botanical Club to purchase equipment and furniture for the laboratory at a cost of about $600. She taught the first year without salary or status, as for official purposes Britton was the person considered to be the teacher, and all assigned grades were approved and signed by him. In the second year, with all 16 students in the college taking botany, the charter of Barnard College was changed so that non-Columbia teachers could teach courses and Gregory was allowed to be called a lecturer. Starting with the summer of 1889, while making the equipment purchases for Barnard, and during the summers of 1893, 1894, and 1895, she returned to Europe to work in the laboratory of Professor Simon Schwendener in Berlin. During her third year at Barnard, she was given complete charge of the botany courses and became a professor there in 1896. Gregory assisted her pupils in preparing annual exhibits popularizing botany for the Progress of Science spring meetings of the New York Academy of Sciences in 1895, 1896, and 1897. She also gave evening classes to groups of men and women. Gregory was an active member of the Torrey Botanical Club beginning in 1887, and was associate editor of its *Bulletin* from 1890 until her death at age 55 on 21 April 1897. She was elected a member of the American Association for the Advancement of Science in 1892 and was an active member of the Society of American Naturalists.

What was the role of American women in the botanical community during Emily Gregory's lifetime? How available was the study of botany to young women growing up in New York State in the 1850s, 1860s, and 1870s? Few women of the previous generation had accomplished more than to become interested in the sentimental language of flowers, or perhaps make a collection of pressed plants or seaweeds. One notable exception had been Almira Hart Lincoln Phelps (1793–1884) who, inspired by the educator-naturalist Amos Eaton (1776–1842) and her educator sister Emma Willard (1787–1870), had written very successful introductory textbooks of botany that were widely used in schools, particularly in female seminaries during the period when Emily Gregory was in school.[17] She was, no doubt, familiar with Mrs. Lincoln's *Botany* (1829),[16] as it was known, and from it graduated to Asa Gray's more advanced *Botanical Text-book* (1842)[6] and *Manual* (1848),[7] both first editions of these books. In her own writings,[11] she referred to Asa Gray's first edition of another textbook, *First Lessons in Botany and Vegetable Physiology* (1857).[8]

Botany of the mid-nineteenth century in America was focused on the study of plant morphology and taxonomy. What was it that induced Emily Gregory to select botany as a subject of specialization when she attended Cornell University from 1876 to 1881? No doubt while there she learned more about the newly emerging studies of plant cytology and tissue anatomy, as well as plant physiology, from Professors Albert N. Prentiss (1836–1896) and William R. Dudley (1849–1911). In any case, the spark was strong enough to influence her to study in German-speaking laboratories, which were emerging in these new disciplines. She became well-trained and was apparently talented in laboratory procedures. Thus, she was called upon a number of times to establish plant anatomy and physiology laboratories at different institutions: Smith College, the Harvard University Annex, Bryn Mawr College, the University of Pennsylvania, and finally at Barnard College. Opportunities were improving during the last two decades of the nineteenth century for American women to become professionally involved in botany; however, they were still extremely limited. The newly emerging women's colleges provided some of the opportunities, and Gregory was able to capitalize on those opportunities. Not only did Emily Gregory become one of the few successful professional women botanists of the nineteenth century, but she also became one of the new breed of laboratory-oriented botanists who had been inspired by the progressive leadership in the German-speaking laboratories of Europe.

What was the place of Gregory's anatomy textbook in the annals of botanical endeavors? Although the anatomy of plants was studied as early as the seventeenth century, it was not until the second half of the nineteenth century that it became a major aspect of botanical research and teaching. The interest in cells in the mid-part of the century stimulated bright young German botanists to realize that not only did microscopic structural studies lead to a better understanding of gross morphology, but more importantly to a greater understanding of plant physiology. By 1877, Anton DeBary (1831–1888) was able to publish his major work summarizing plant anatomy.[4] It was into such heady Continental botanical laboratories, with bright enthusiastic young men, that Emily Gregory went to study and do research in the 1880s and 1890s. In 1893, she wrote:[11]

> As a science, botany may be divided in two branches, morphology and physiology, . . . or into two branches of such a character that one is and must be represented by systematic, the other by experimental and theoretical work.

At that time, the discipline was preoccupied with the external structure of botany, whereas later the discipline concerned anatomy and physiology. In the same paper, entitled "Anatomy as a Special Department of Botany," she[11] wrote:

> Fifteen years have passed since DeBary's book [*The Comparative Anatomy of Phanerogams and Higher Cryptogams*] was published, and during this time much progress has been made in this branch. Such progress in any department renders necessary a corresponding growth in text books . . . . In general, it is believed safe to say that the English written text books of the day are singularly deficient in this respect. Abstruse and difficult physiological problems are introduced to students illy-prepared to handle them, while they try to meet the deficiency by a partial description of anatomical characteristics, a knowledge of which is necessary even to a comprehension of the questions. Not only must the standard of text books be raised, but the methods of instruction must be improved if we expect to cope successfully with some of the most interesting problems of the day. . . . If any one doubts the necessity of training in anatomy as requisite to physiological study, let him attempt to explain the present theory of the ascent of water through the trunks of high trees, to a person ignorant of plant anatomy.

Believing that good botanical textbooks, particularly in anatomy, were important, Emily Gregory[19] wrote an introduction to anatomy that was published in 1895 for English language students.[12] This accomplishment came 92 years after the first American general botany textbook by Benjamin Smith Barton,[1] and ten years after the publication of George Lincoln Goodale's second volume of the sixth edition of Asa Gray's *Botanical Text-book* (1879),[5] which was devoted to physiological botany. The first five chapters of Goodale's volume, or about a third of it, was on the "outlines of the histology of phaenogamous plants." His treatment of anatomy is short on cell structure and, though brief, is comprehensive in its descriptive details about tissues. The approach is primarily comparative and descriptive.

Although Emily Gregory's *Elements of Plant Anatomy* (1895)[12] was not the first American textbook to consider plant anatomy, it has the distinction of being the first one completely devoted to plant anatomy. She wrote in its preface that the book resulted from

> . . . the substance of the lectures given to the classes in the last half of the second year's course in botany at Barnard College. . . . As a knowledge of the inner structure of plants is necessary to successful work in physiology, it is believed that a brief course in the elements of anatomy should precede the

> study of physiology. . . . There is only one other feature which serves as an excuse for the appearance of another text-book, that is, the effort to present the subject from the developmental point of view. The experience of several years has demonstrated the superiority of this method over that in common use. . . .It may also be found practicable, at a later date, to combine with the present volume an elementary treatise on plant physiology, and possibly to supplement both by a more advanced work on each subject separately. However this may be, it is quite certain that the measure of our progress in any science may be found in our ability to adapt the thought and experience of other nations to our special needs and resources.

Sadly, Gregory's plans for future volumes were not to materialize because of her death less than two years later.

The text of Gregory's *Elements of Plant Anatomy* (1895)[12] is arranged into two parts. The first part, about a third of the book, is devoted to the cell with consideration of the cell wall and its development separated from discussion of the cell contents. The second part is concerned with the anatomy of tissues (Fig. 21.2). Most of the pages are devoted to the tissues of angiosperms and gymnosperms, with only 15 pages in which the other groups of plants are considered. The book consists of 156 pages, and is, as Gregory wrote, "a brief outline of the elementary principles of anatomy. . . ." It drew heavily upon German sources, including almost all of its illustrations. The 23 authors cited were all Continental Europeans, and only one, A. P. DeCandolle (1778–1841), was non-German-speaking. Almost half of the cited authorities or figure sources were either Julius Weisner (1836–1916) or Theodor Hartig (1805–1880), both of whom are specifically thanked in the preface. Almost all the authorities cited were engaged in active anatomical research in the 1870s through the 1890s. Thus, it is fair to say that Gregory attempted to present American college-level students the fruits of

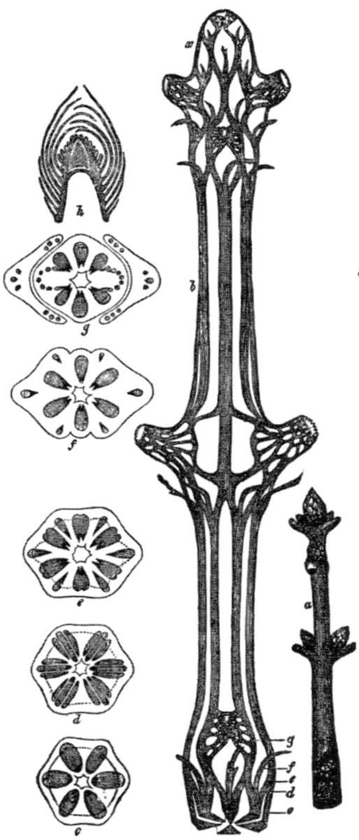

FIG. 48.

a tip of stem of *Atragene Alpina*; *b* skeleton of the same obtained by maceration. *b* an internode consisting of six bundles as seen in the cross-section, *c*. *h* the young bud with the leaf sheaths. *g* shows the cross-section through stem and leaf petiole at the point marked *g* on the skeleton figure. The leaf-traces of each leaf are separated into nine bundles. At *f* three simple leaf-traces are enclosed in the tissue of the stem, but still remain outside the circle of bundles. At *e* and *d* they enter into this circle and are fused with it.

Fig. 21.2. Illustrations of anatomical plant structures.

(From Emily L. Gregory. 1895. Elements of Plant Anatomy. Ginn & Co., Boston)

German anatomical research, of the recent past and contemporary period, in an introductory manner without presenting the complete arguments or details of conflicting interpretations of structure and development.

A review of the book appeared in the *Botanical Gazette* for 1896. It was probably written by the editor, John M. Coulter (1851–1928), and is not very complimentary. It said, in essence, "Amid much that is good, well-put, and correct, there is much that is crude, incorrect, or out of date."[3] Specific points, some picayune, were raised that included the "incompleteness of the account of mitosis," the definition of the term "bark," and the description of the development of interfascicular cambium. The review in an 1896 *Bulletin of the Torrey Botanical Club* by Smith Ely Jelliffe (1866–1945)[15] is more positive. It ended by saying:

> On the whole the work is timely, as there has been little published in this country which makes such an easy yet comprehensive introduction to the work of Sachs, DeBary and Vines. Hitherto, the student has had, in the main, to search through a mass of detail in these works to find the underlying principles, but in Dr. Gregory's work the main facts are outlined clearly, in excellent English, and the book loses nothing of its scientific value by being simple and concise.

Gregory answered the unfavorable comments in the *Botanical Gazette* review by stating that her aims were to present the material in a non-dry bones manner for beginning students, and that it was "in no way fitted for a reference book except for the learner." She[13] further answered each of the specific criticisms by pointing out that:

> In answer to several of these charges, it is only necessary to refer to numerous German, French and English text-books whose authority rests upon original monographs. . . . For one familiar with the facts of plant anatomy by years of study, not only of text-books but also of the plants themselves, it is hard to conceive how a candid critic could take exception to these statements. Indeed, if the author may be considered 'authority' on these subjects she has erred in company with the illustrious scientists of the present and past, with such men as Naegeli, Cramer, DeBary, Sanio, Vines, Reinke and Schwendener. The author heartily agrees with the reviewer in wishing the book were better and also in the hope that a revision, in the near future, may be made which will render it more useful.

Such a revision, of course, did not happen because Gregory died the next year, 21 April 1897.

Gregory, and her book, have long since faded into oblivion. An elementary textbook of a single edition is rarely cited in the scientific literature. Many of Gregory's papers were reviews of the contemporary anatomical and morphological research at German universities, or of German-language books, that served well the American botanists of that time, but are not of long-term significance. It is true that her book did not do justice to the details of nuclear division, nor did it anticipate the ecological approach to anatomy that was developing in German and American universities.

For the historian of American botany several points can be noted about Gregory and her anatomy textbook. First, Gregory was a pioneer in becoming a professional woman botanist at a time when very few women were entering the profession. Second, she was remarkable for being the first American woman to earn a European Ph.D. in botany. Third, she was, as were many of the ambitious young men of that period in America, well versed in the progressive botany being pursued in the German-speaking universities. Finally, her textbook of plant anatomy, the first in this country, was strongly based upon European Continental models rather than on British or American ones. Emily Gregory and her anatomy textbook were remarkable pioneers. As the resolution of the Torrey Botanical Club said upon her death, "We have lost in her an accomplished scientist [and] a devoted teacher."[2]

References Cited

1. Barton, Benjamin Smith. 1803. *Elements of Botany: or Outlines of the Natural History of Vegetables*. [The Author], Philadelphia. Pt. 1, xii, 302, 32 pp.; Pt. 2, 168, 38 pp.

2. Britton, Elizabeth G. 1897. Emily L. Gregory. *Bulletin of the Torrey Botanical Club* 24: [221]–228 + port. (unnumbered pl. opp. p. [221]). (Quote, p. 224). [The accompanying bibliography of 38 publications is incomplete. Most of these papers are reviews of German anatomical contributions. Schmid (1987)[18] provides a complete bibliography of the publications of Emily Gregory].

3. [Coulter, John M.]. 1896. [Book Review: Emily L. Gregory. *Elements of Plant Anatomy*. Ginn & Co., Boston. viii, 148 pp.]. *Botanical Gazette* 21: 239–240.

4. DeBary, Anton. 1877. *Vergleichende Anatomie der Vegetationsorgane der Phanerogamen und Farne. Handbuch der physiologischen Botanik,* Band 3. W. Engelmann, Leipzig. xvi, 558 pp.

5. Goodale, George Lincoln. 1885. *Physiological Botany. I. Outlines of the Histology of Phaenogamous Plants. II. Vegetable Physiology. Gray's Botanical Textbook.* 6th ed. Vol. II. Ivison, Blakeman, Taylor & Co., New York. xxi, 499 pp. + Practical Exercises, 36 pp.

6. Gray, Asa. 1842. *The Botanical Text-book.* Wiley and Putnam, New-York; Little and Brown, Boston. xii, [13]–413 pp.

7. _____. 1848. *A Manual of the Botany of the Northern United States*. James Munroe and Co., Boston and Cambridge. lxxii, 710 pp.

8. _____. 1857. *First Lessons in Botany and Vegetable Physiology*. Ivison & Phinney and G. P. Putnam & Co., New York. xii, 236 pp.

9. Gregory, Emily L. 1886. The pores of the libriform tissue. *Bulletin of the Torrey Botanical Club* 13: [197]–204, 233–244.

10. _____. 1888–1889. Development of cork-wings on certain trees. I-V. *Botanical Gazette* 13: [249]–258, [281]–287 + Pl. XXII, 312–318 + Pl. XXV; 14: 5–10, 37–44.

11. _____. 1893. Anatomy as a special department of botany. *Bulletin of the Torrey Botanical Club* 20: 100–107. (Quotes, pp. 104, 105, 107).

12. _____. 1895. *Elements of Plant Anatomy*. Ginn & Company, Boston and London. viii, 148 pp. (Quotes, pp. [iii], iv, iv-v).

13. _____. 1896. [Response] to the editors of the *Botanical Gazette*, pp. 72–73. *In*. Open letters. *Botanical Gazette* 22: 72–74.

14. _____. 1897. Statement Regarding the Origin and Development of the Botanical Department of Barnard College, New York, Jan. 25, 1897. [13] pp. (Quote, pp. [2–3]). [Handwritten manuscript housed in the Archives of Wollman Library, Barnard College, Columbia University].

15. Jelliffe, Smith Ely. 1896. [Book review: Emily L. Gregory. *Elements of Plant Anatomy*. Ginn & Co., Boston. viii, 148 pp.]. *Bulletin of the Torrey Botanical Club* 23: 62–63. (Quote, p. 63)

16. [Phelps], Almira Hart Lincoln. 1829. *Familiar Lectures on Botany*. H. and F. J. Huntington, Hartford; G. & C. & H. Carvill, New York; Richardson & Lord, Boston. x, [11]–335, [1], 4 pp. [At least 39 different printings and two major revisions (1845, 1852) were issued through 1869].

17. Rudolph, Emanuel D. 1984. Almira Hart Lincoln Phelps (1793–1884) and the spread of botany in nineteenth century America. *American Journal of Botany* 71: 1161–1167.

18. Schmid, Rudolf. 1987. Annotated bibliography of works by and about Emily Lovira Gregory (1841–1897). *Bulletin of the Torrey Botanical Club* 114: 319–324.

19. _____ and Dennis Wm. Stevenson. 1987. "Botanical text books," an unpublished manuscript (1897) by Emily Lovira Gregory (1841–1897) on plant anatomy textbooks. *Bulletin of the Torrey Botanical Club* 114: 307–318.

Publication Notes

Modified and edited from an unpublished manuscript.[a]

[a] 1984. "The First American Plant Anatomy Book and Its Author, Emily L. Gregory." Typewritten, 10 pp. [Manuscript].

[b] 1984. The first American plant anatomy book and its author, [Emily L. Gregory]. *American Journal of Botany* 71(5, pt. 2): 107–108. [Abstract].

[c] 1984. The first American plant anatomy book and its author, Emily L. Gregory. *The Ohio Journal of Science* 84(2): 6. [Abstract].

PART SEVEN

WRITING SHAW'S GARDEN HISTORY

In the late 1980s, William B. Faherty, Professor Emeritus of History at St. Louis University, prepared a manuscript on the history of the Missouri Botanical Garden, initially founded as Shaw's Garden. His manuscript was divided and published as two separate books, *Henry Shaw, His Life and Legacies* (1987)[1] and *A Gift to Glory In the First Hundred Years of the Missouri Botanical Garden (1859–1959)* (1989).[2] Peter Raven, Director of the Garden, envisioned a book that would encompass the history of the scientific investigations at the Garden from its inception. Ultimately Emanuel D. Rudolph was invited to write the Garden history, and the information he had prepared by June 1992 is organized and printed here in three papers (**Part Seven**). The first paper comments on the leadership of the Garden's six directors (**Paper No. 22**), the second one makes comparisons of the Missouri Garden with the New York Botanical Garden (**Paper No. 23**), and the third one discusses the founding of Shaw's Garden in St. Louis (**Paper No. 24**).

---

[1] William Barnaby Faherty. 1987. *Henry Shaw, His Life and Legacies*. University of Missouri Press, Columbia. [xi], 228 pp.

[2] William Barnaby Faherty. 1989. *A Gift to Glory in the First Hundred Years of the Missouri Botanical Garden (1859–1959)*. Harris & Friedrich, Ocean Park, Washington. xvi, 234 pp.

Introduction Nos. 22–24

Writing the Garden History (1988–1992)

Prior to the selection of Emanuel D. Rudolph as the writer of the Garden history Director Raven attempted to persuade Joseph Ewan, distinguished authority in the history of botany and research associate at the Garden since 1986, to write the desired definitive history, but Ewan declined to undertake the project. Raven also discussed his desire for the Garden's history with Frederick G. Meyer, a former graduate student affiliated with the Garden, who was supervisory botanist in the Herbarium of the National Arboretum, Washington, D.C. He, too, declined, and time passed without anyone becoming interested in the writing project.[1]

In the summer of 1988, the idea came to Marshall Crosby, then Assistant Director of the Garden, that Rudolph would be an excellent author for the definitive Garden history. When Marshall presented the idea to Director Raven, he also liked the idea, as did Rudolph when Marshall asked him in a telephone conversation. Rudolph appeared to be the ideal person for the project.[3] He had studied and worked at the Garden in the 1950s while a graduate student at Washington University in St. Louis, and he knew and worked with the Garden personnel at that time. As years passed, Rudolph became acquainted with many of the Garden's scientific alumni, and, of course, he was now nationally known for his studies in the history of botany, having presented papers at national meetings and published articles in recognized scholarly journals. Furthermore, his extensive private library would provide a rich source of reference material, and he was within two years of planned part-time retirement from University teaching and committee service work. The stage was set for the project.

Following correspondence between Raven[2] and Rudolph[3] with Crosby, Rudolph met with Raven and Crosby on 6 September 1988 at the Garden and accepted their official invitation to write the book-length history. The project was to cover the years from 1859 to 1971, the year when Raven became Director of the Garden. Rudolph confirmed the agreement in a letter of 12 September to Crosby. In this

[1] Ronald L. Stuckey. 1993. An Interview with Marshall Crosby at the Missouri Botanical Garden, St. Louis. 8 October. [Notes in possession of Ronald L. Stuckey, Columbus, Ohio].

[2] Peter Raven. 1988. Letter of 2 August to Marshall Crosby, Missouri Botanical Garden, St. Louis. 1 p.

[3] Emanuel D. Rudolph. 1988. Letter of 12 August to Marshall Crosby, Missouri Botanical Garden, St. Louis. 1 p.

letter, Rudolph[6] remarked that an ample amount of primary materials was available in the Garden's Library and Archives, and he was delighted that the manuscript and choice of publisher would belong to him, noting that his first choice was The Ohio State University Press. He also developed a tentative three-year budget, and agreed to present the after-dinner lecture at the 1989 Annual Systematics Symposium as part of the celebration of the 100th anniversary of the Garden. Rudolph[4] further provided a list of 13 individuals whom he would interview and record oral histories. On this list were Mrs. Bertha Dodge, recent widow of Carroll Dodge (1895–1988), who was Rudolph's major advisor, Frits Went (1903–1990), Mildred Mathias (1906–1995), Charles Heiser, Jr., Hugh Cutler, Hugh Iltis, Joseph Ewan (1909–1999), and Barry Commoner.

Rudolph ambitiously believed that the project could be accomplished in three years. He hoped "to tie the history of the Garden, including the Shaw School of Botany, the Program in Gardening Instruction, and the Arboretum, into some larger historical issues, such as: the civilizing of the Midwest; the rise of American philanthropy in American science; the parks and botanical gardens movement in America; the issues of smoke abatement in cities; as well as the rise of scientific botany involving the training of professional botanists and gardeners."[4] Enthusiasm for the project must have been at a high level, for in a letter two days earlier than the one to Crosby of 12 September 1988, Rudolph[5] acknowledged Mrs. Dodge's husband's death and told her of his Garden-history project. He inquired about his writing Carroll Dodge's obituary for *Mycologia*[6] and the whereabouts of his archival papers that might be useful to him.

Acquisition of Information

Rudolph worked assiduously to acquire data from the archives at the Missouri Botanical Garden, and to obtain information through interviews of former Garden scientists and associates. In October 1990, he made a trip to England to visit the Mill Hill School, where Garden founder and benefactor Henry Shaw was educated between 1812 and 1815. Earlier, on 26 February 1990, Rudolph[7] had written to Mr. A. C. Graham, the headmaster of the school, expressing his intention to make

[4] Emanuel D. Rudolph. 1988. Letter of 12 September to Marshall Crosby, Missouri Botanical Garden, St. Louis. 2 pp. + [List of] Candidates for Oral Histories. Typewritten, 11 September 1988. 1 p.

[5] Emanuel D. Rudolph. 1988. Letter of 10 September to Mrs. Bertha S. Dodge, Shelburne, Vermont. 1 p.

[6] Emanuel D. Rudolph. 1990. Carroll William Dodge, 1895–1988. *Mycologia* 82: 160–164.

[7] Emanuel D. Rudolph. 1990. Letter of 26 February to A. C. Graham, Headmaster, The Mill Hill School, Mill Hill Village, London, England. 1 p.

a visit there and learn of "costs to students, the faculty, the curriculum, and the special activities, both curricular and extracurricular, of that period. As Mr. Shaw wrote fondly about the Mill Hill School when he was in his eighties, particularly mentioning the pine trees and the student gardens, I would be pleased to see any sketches of the buildings and grounds."

The Mill Hill School in the northwest part of suburban London, according to Rudolph's report,[8] "is situated in a picturesque location in a somewhat village setting on the top of a hill with its slopes overlooking attractive rural views." Upon arrival Rudolph and his wife Ann were pleasantly welcomed by Headmaster Graham, who explained that "because of the tenuous financial condition of the School in its early period, and because there . . . [was] no archiving plan, no records survive for the early years. He showed me the published histories of the school, all of which I had already seen." One of the instructors at the school, Mr. Trevor Chilton, showed the Rudolphs the school grounds and the old trees, during which time Ann took photographs of the scenes, including one of Rudolph and Mr. Chilton together. (**Paper No. 24**).

Rudolph's[8] other mission in England at this time was his presentation on 28 September 1990 of a short progress report of his research on the history of the Missouri Botanical Garden. His report, titled "Henry Shaw, Millhillian, founder of the Missouri Botanical Garden," was delivered at a meeting of the Society for the History of Natural History held at Sussex University, near Brighton, England. The major component of the program was a symposium celebrating the bicentenary of the English geologist-paleontologist Gideon Mantell (1790–1851). Rudolph's manuscript or notes for the report given at this meeting have not been located.

Rudolph worked diligently to obtain both written records and oral recorded information about the Garden's history. Most of the raw data had been accumulated and were organized into various topics. Having finished his last class in the History of Biology on Friday, 12 June 1992, and having written the manuscript of the book's first chapter, Rudolph was now ready to write more chapters. The accident with his mini-auto-van occurred the next Friday, and he died on the following Monday, 22 June 1992. Upon the recommendation of estate executor Ronald L. Stuckey, all of the documents, papers, and notes[9] that Rudolph had assembled for the history of the Missouri Garden were given to the Garden's archives. In August 1992 Marshall Crosby took them to the Garden in St. Louis.

---

[8] Emanuel D. Rudolph. 1990. Report of a Trip to England, Sept. 25–Oct. 2, 1990. Typewritten, 1 p.

[9] Ronald L. Stuckey and John W. Frederick, compilers. 1992. Rudolph Papers: [List of] Materials for a Projected History of the Missouri Botanical Garden. Manuscript. Typewritten, [9] pp.

Introduction to the Individual Garden Papers

According to the agreement made at the time Rudolph was asked to write the history of the Missouri Botanical Garden, he was to present a special evening lecture, "One Hundred Years of the Missouri Botanical Garden." The lecture was delivered on 7 October 1989, as part of the program for the 36th Annual Systematics Symposium. The lecture, an excellent presentation, survives on a cassette tape which Stuckey recorded at the time of delivery. Rudolph's resulting paper was published in the *Annals of the Missouri Botanical Garden* (1991). A short summation[a,b] of this contribution is printed here. **(Paper No. 22)**.

Rudolph wrote on selected facets of the Garden history, which were the bases of three presentations he gave at scientific meetings. These meetings were the Historical Section of the Botanical Society of America on 7 August 1990 in Richmond, Virginia, where he spoke on "The Missouri and New York Botanical Gardens: Contrast and Similarity of Origins and Early Development," and on 6 August 1991 in San Antonio, Texas, where he gave the presentation, "Why Henry Shaw Established a Public Botanical Garden in St. Louis and How He Developed It." Earlier that same year, at the centennial meeting of The Ohio Academy of Science on 27 April 1991, on the day following his wife's death, he spoke, as scheduled, on "Henry Shaw's Path from Sheffield to Mill Hill School and the Founding of the Missouri Botanical Garden." In a report on his research for 1990, Rudolph[10] wrote concerning his presentation contrasting the histories of the two Botanical Gardens: "It went over so well that the people at New York want me to write a history of their institution when I finish Missouri." Left at his death was a handwritten version of this manuscript,[b] which was modified and edited for printing here. **(Paper No. 23)**.

Rudolph reworked his manuscripts prepared for the above-mentioned two presentations on Henry Shaw into a manuscript[a] he designated as the first chapter for his book on the history of the Missouri Botanical Garden. That manuscript,[a] with the permission of Marshall Crosby, has been further modified and edited for printing here. **(Paper No. 24)**.

[10] Emanuel D. Rudolph. 1990. Self-Evaluation Report for Teaching, Research, and Service at The Ohio State University, Columbus. 1 p.

Six Directors of the Missouri Botanical Garden

William Trelease
(1857–1945)

George T. Moore
(1871–1956)

Edgar Anderson
(1897–1969)

Building formerly housing Administration,
Herbarium, and Library,
the Missouri Botanical Garden

Frits Went
(1903–1990)

David M. Gates
(b. 1921)

Peter H. Raven
(b. 1936)

Paper No. 22 (1989)

The Missouri Botanical Garden, a Century of Leadership by Six Directors (1889–1989)

The Missouri Botanical Garden, now in operation for over 100 years since its establishment with a Board of Trustees, had its benefactor in its founder, Henry Shaw (1800–1889). Like a person, the Garden had a gestation phase and a youthful period strongly forming its character. Throughout the years various middle-age crises occurred which were weathered. Certain aspects of this remarkable history are presented here as a part of the celebration of the Garden's centennial history.[a]

In the hundred years since the Board of Trustees of the Missouri Botanical Garden implemented the stipulations of Henry Shaw's will, which prescribed functions for his public garden and provided an appropriate endowment, six directors have guided the garden's development. Henry Shaw's vision of the Garden was based upon his British background and the European botanical garden tradition. William Trelease (1857–1945), the first director from 1889 to 1912, set his highest priorities on building the herbarium and library and publishing substantial annual volumes, which he believed would make the Garden a major center for taxonomic research. George T. Moore (1871–1956), director for 40 years, from 1912 to 1953, moved the research into other areas of botany, applied botany, and horticulture. Keen on public display yet facing decreasing finances toward the latter part of his tenure, he began to seek private support for various programs, including a new arboretum located outside the city. Edgar Anderson (1897–1969, who had a short directorship (1954–1957), focused on Garden improvements with further attempts to raise funds and find support among garden clubs. Frits Went (1903–1990), the fourth director from 1958 to 1963, saw the Garden as a potential multifaceted cultural center for St.

Frontispiece Credits: Six Directors of the Missouri Botanical Garden. William Trelease (1857–1945), George T. Moore (1871–1956), Edgar Anderson (1897–1969), Frits Went (1903–1990). (From Rudolph, *Annals of the Missouri Botanical Garden* 78: 7, 9, 11, 12. 1991)

Administration, Herbarium, and Library Building, Missouri Botanical Garden. (Original source not known; in the Rudolph photograph collection)

David M. Gates (b. 1921). (From Gates, *Missouri Botanical Garden Bulletin* 59(5): 3. 1971)

Peter H. Raven (b. 1936). (From Rudolph, *Annals of the Missouri Botanical Garden*, 78: 14. 1991) (Permission to reproduce photographs from Garden publications, courtesy of the Missouri Botanical Garden, St. Louis, Missouri)

Louis. The geodesic-domed greenhouse, named the Climatron, opened in 1960, and brought local and national attention to the Garden. David Gates (b. 1921), the fifth director from 1965 to 1971, moved the Garden into ecological research and located support for a new research building that had been sought by Went. Peter H. Raven (b. 1936), the current director from 1971 to the present, has moved the Garden along on various fronts. He has expanded its horticultural displays and special gardens, widened the base of public support, enhanced the research in systematic botany, and projected the role of the Garden into international efforts to understand our biosphere and protect it. Each director has guided the Missouri Botanical Garden in certain directions during the last hundred years to make it the major botanical garden that it is now.[b]

Upon concluding this brief history of the Missouri Botanical Garden, it might be appropriate to quote the eccentric essayist-historian Lytton Strachey (1880–1932) who once wrote that ". . . ignorance is the first requisite of the historian,"[1] because the historian must be selective and interpretive and must ignore much. Perhaps it would be more accurate to say that the historian is more like a maker of myths, those myths that have one or more morals. What then is the mythology of the Missouri Botanical garden? One moral is that an institution with a divided administrative structure, Board and Director, can thrive, though at times strong conflicts may exist. Another moral is that a botanical garden, with strong traditions, can be an effective force for good in a community and in society. The Missouri Botanical Garden, for example, has helped to beautify St. Louis, and helped to clean up its air, has educated Missourians about their natural and horticultural heritage, and has helped expand people's view of the importance of vegetation for human survival. It has shown that botanical and horticultural research can be made meaningful to the general public.[a,b]

The founder and six directors, each with differing views of its mission but all concerned with education and research, have molded the Garden. The ordinary citizen and the scientific community have been educated in the importance of the Garden's research, whether it be through a flora of Missouri or Panama, a new hybrid water lily, an understanding of the genetics of significant crop plants, an experiment with an ecologically designed greenhouse, or a race to know the plants of the disappearing tropical rainforests. The "mythology" of the Missouri Botanical Garden is one that is real, perhaps more real than fact or history.[a,b]

Reference Cited

1. Strachey, Lytton. [ca. 1918]. *Eminent Victorians. The Star Series.* Garden City Publishing Co., Inc., Garden City, New York. ix, 351 pp. (Quote, p. v).

Publication Notes

Compiled and edited from the presentation,[a] and the resulting paper[b] published 1991 in the *Annals of the Missouri Botanical Garden*; delivered 7 October 1989 at the 36th Annual Systematics Symposium of the Missouri Botanical Garden Celebrating Its 100th Anniversary, at the Garden, St. Louis.

[a] 1989. "One Hundred Years of the Missouri Botanical Garden." [Cassette tape made by Ronald L. Stuckey; presentation delivered 7 October 1989 in St. Louis, Missouri]. [Manuscript at the Missouri Botanical Garden].

[b] 1991. One hundred years of the Missouri Botanical Garden. *Annals of the Missouri Botanical Garden* 78: 1–18. (See pp. 1, 14–15).

Fig. 22.1. Shaw's summer home at Shaw's Garden, the Missouri Botanical Garden.

(Taken by K. U. Kramer, March 1955; in the Rudolph photograph collection)

First Directors and Buildings: Missouri and the New York Botanical Gardens

Henry Shaw's Museum Building, Missouri Botanical Garden

Nathaniel Lord Britton (1859–1934)

William Trelease (1857–1945)

Construction of the First Building at The New York Botanical Garden, housing the Administration, Herbarium, and Library

Paper No. 23 (1990)

The Missouri and New York Botanical Gardens: Contrast and Similarity of Origins and Early Development

William Trelease and Nathaniel Lord Britton, both noted and notable botanists during the period just before the arrival of the twentieth century, became directors of botanical gardens that were in their formative stages. Trelease[6,15,18,23,29,38] became the first director of the Missouri Botanical Garden in St. Louis where businessman Henry Shaw[1,2,17,19] had founded and operated a public garden in 1859, and endowed and designed its organization in his will of 1885.[44] It was connected by overlapping staff with the newly emerging Washington University, and administered by a self-perpetuating Board of Directors. The New York Botanical Garden was the inspiration of Britton[20,27,35,47] and his wife Elizabeth G. Britton.[5,26] The Garden was incorporated in 1891 with the understanding that if $250,000 could be subscribed, the City of New York would provide land for the establishment of a public botanical garden. This accomplishment became a reality in four years by a distinguished board of wealthy and influential businessmen and Columbia College.[a]

During the last decade of the nineteenth century and the first decade of the twentieth century, American botany was rapidly changing. William Trelease and Nathaniel Lord Britton made the institutions they directed important vehicles for these changes. American botany was maturing because of the many new professional botanists, the improved departments at American universities, and the new botanical

---

Frontispiece Credits: First Directors and Buildings: Missouri and the New York Botanical Gardens. Shaw's Museum, Missouri Botanical Garden, St. Louis. (Photographed by E. L. Evinger, from anonymous, *Missouri Botanical Garden Bulletin* 49(6): 96. 1961; permission to reproduce photograph courtesy of the Missouri Botanical Garden)

Nathaniel Lord Britton (1859–1934). (Photograph by Underwood and Underwood, July 1907, from Howe, *Journal of the New York Botanical Garden* 35: portrait opp. p. 169. 1934)

William Trelease (1857–1945). (From *Transactions of the Illinois State Academy of Science* 38: 131. (1945) [1946]; permission to reproduce photograph courtesy of the Illinois Academy of Science)

Construction of New York Botanical Garden Building. (Photograph taken November 1901, from Gamdee, *Garden Journal* 23(2): 38. 1974) (Permission to reproduce photographs in Garden publications courtesy of the New York Botanical Garden, New York ©1934, ©1974)

institutions which were becoming world class in quality. These factors made the earlier desirability of a German doctorate for American students less necessary. These men received varied evaluations of their scientific and administrative abilities. For example, of William Trelease, as noted in Lenz (1986), Marcus E. Jones (1852–1934) wrote, "... in my judgment he has no scientific judgment and is utterly unfit to monograph any genus,"[31] while another, Louis Otto Kunkel (1961) wrote "he rendered a great service to the pioneers of the middle west and to botanical science in America and throughout the world."[29] Of Nathaniel Lord Britton, Lloyd Shinners (1962) wrote "taxonomically speaking [the two editions of his *Illustrated Flora of the Northeastern United States*] represent the most backward steps ever taken in American botany,"[45] while Elmer D. Merrill (1938) wrote "one marvels at his energy, his ceaseless mental activity and at the diverse productive results."[35] Little doubt exists, however, that each of these administrators was a dominant force in shaping and perpetuating his own particular botanical garden.

The Missouri Botanical Garden

William Trelease (1857–1945), a native of upstate New York, received his undergraduate degree in 1880 from Cornell University.[6,15,18,23,29,38] There he studied insects under John H. Comstock (1849–1931), publishing several papers about insect pollination. From Cornell he went to the University of Wisconsin as an instructor and then to Harvard University for graduate studies. His doctorate in 1884 under William Gilson Farlow (1844–1919) is the first in America on bacteria. After several earlier teaching positions, he returned to the University of Wisconsin in 1884 as professor of botany and head of the department, only to leave a year later to accept the first Engelmann Professorship of Botany at the newly established Henry Shaw School of Botany at Washington University, St. Louis. Asa Gray (1810–1888) and others had recommended him for the position at the University.

Henry Shaw (1800–1889), a bachelor of English birth immigrated to the United States in 1819, settling in St. Louis where he established a very successful hardware business.[1,2,17,19,42] Because of family connections in his native Sheffield, and astute business practices that included importing by ships through the port of New Orleans, he quickly became wealthy and was able to retire from active business while investing wisely in sound real estate. Having a long-held interest in gardens, a desire to repay the city for his success, and a longing for the perpetuation of his memory, Shaw consulted Sir William Jackson Hooker (1785–1865) at Kew about founding a public garden. Through Hooker, Shaw was put into contact with Asa Gray at Harvard University and George Engelmann (1809–1884) in St. Louis. In 1859, the Missouri State Legislature chartered an open-to-the-public, tax-free garden, the Missouri Botanical Garden, surrounding Shaw's country home, Tower Grove.[42]

When Trelease arrived in 1885, the Garden had grown under Shaw's guidance into a considerable establishment with many planted trees, ornamental herbs, and greenhouses of plants. With Engelmann's assistance and Gray's advice, a small library, herbarium, and museum collection were acquired and then housed in an especially-built Museum Building. Upon Shaw's death in 1889, William Trelease at age 32 was appointed the Director of the Garden by the Board of Trustees. Shaw's will,[44] which Asa Gray helped draft, contained specific instructions for how the Missouri Botanical Garden was to be governed and organized. William Trelease was to be its director for the next 23 years.

The New York Botanical Garden

Nathaniel Lord Britton (1859–1934), two years Trelease's junior, also had a diverse background.[20,27,35,47] Born on Staten Island, the County of Richmond of the City of New York, into a family that had been living there since 1695, he was a well established member of New York society. Britton entered Columbia College to study mining, engineering, and geology, obtaining a mining engineer's degree in 1879 at age 20. Britton's long held interest in plants and local botany led him to join the Torrey Botanical Club in 1877 and to publish the first of his many papers, in its *Bulletin*[7] that year. The year of his graduation from Columbia also was one in which his *Flora of Richmond County* (1879),[25] co-authored with Charles Arthur Hollick (1857–1933), was published. Immediately he became an assistant to John Strong Newberry (1822–1892), the professor of geology and botany at Columbia. Britton accepted the assignment of teaching botany as well as some other subjects, and also became an assistant for the Geological Survey of New Jersey, which in 1881 published his *A Preliminary Catalogue of the Flora of New Jersey*.[8] Apparently this study served as his doctoral dissertation, the same year that he received his Ph.D. from Columbia College. Four years later he married the ardent bryologist Elizabeth Gertrude Knight (1858–1934), and became a Professor of Botany in a newly established botany department at Columbia College.

With the failure of the Elgin Botanical Garden in 1810, established by David Hosack (1769–1835) in 1801, it was given to the Board of Trustees of Columbia College in 1814.[40,41] Since then the City of New York had no Botanical Garden. While the Brittons were staying at the Royal Botanic Gardens, Kew, in the summer of 1888 studying South American plant specimens, Mrs. Britton was led to ask anew, "Why couldn't we have something like this in New York?"[26,27] Britton, always one to follow his decisions with rapid and decisive actions, began a campaign for a botanical garden. At the regular meeting of the Torrey Botanical Club, 24 October 1888, Mrs. Britton gave an illustrated description of the Botanical Gardens at Kew.[49] A committee of the Torrey Botanical Club, consisting of Timothy F. Allen (1837–1902), N. L. Britton, Addison Brown (1830–1913), Thomas Hogg (1820–1892), J. S. Newberry, Henry H. Rusby (1855–1940), and E. E. Sterns (1846–1926), chairman, was formed at a meeting in November of 1888.[3] Soon thereafter the proper city officials and prominent citizens were contacted for their support. A charter was obtained in 1891 and amended in 1894,[14] for the establishment of a city botanical garden in Bronx Park, provided an endowment of $250,000 could be raised by subscription within seven years. By 18 June 1895 the endowment had been realized with $25,000 contributions each from J. Pierpont Morgan (1867–1943), Andrew Carnegie (1835–1919), Cornelius Vanderbilt (1843–1899), John D. Rockefeller I. (1839–1937), Darius O. Mills (1825–1910), and Judge Addison Brown (1830–1913), as well as sizeable contributions from others including Britton.[22] By the following year the city had appropriated $500,000 for the construction of a Museum,[10,48] herbarium,[46] library,[39,54] laboratory building,[32] greenhouses,[24] roads, and a few years later a Conservatory.[30] Thus, the New York Botanical Garden was underway in 1896 when Britton was named its Director-in-Chief, a position he was to hold for the next 31 years.[4]

The Two Gardens Compared

In comparing the two botanical gardens, it is well to begin with a comparison of their governance. The Missouri Botanical Garden is governed by a self-perpetuating board of trustees, the original members of which were enumerated in Henry Shaw's will probated in 1889.[44] Although three members of the scientific community, Asa Gray, George Engelmann and Spencer F. Baird (1832–1887) were named as honorary trustees in the will, each one had predeceased Shaw and were not replaced by the board with other scientists. The remaining trustees were business colleagues or friends of Henry Shaw, with five ex-officio trustees: the mayor of the city, the president of the school board, the chancellor of Washington University, the president of the St. Louis Academy of Science, and the Episcopal Bishop of Missouri. This Board was the beneficiary of Shaw's properties and was empowered to appoint the Director and to oversee the daily operation of the Garden. From the beginning, the mayor of St. Louis refused to be an active participant on the Board, thus removing the Garden from most political pressures.

By contrast, the Board of Managers of the New York Botanical Garden is divided into Scientific Directors: the President of Columbia College, its professors of botany, geology, and chemistry, the President of the Torrey Botanical Club, and an ex-officio member, the President of the Board of Education; with other managers to be the Mayor of the city, the President of the Board of Commissioners of the Department of Public Parks, and nine other managers elected by the board. Thus, it has a scientific component as well as a political one. Further, the financing of the Garden's activities were private in the Missouri case and mostly public in the New York case.

Given these differences, the initial development of both gardens was very similar. Both had young active directors who believed that scientific activities were of paramount importance. They built strong libraries and herbaria on firm foundations. In Missouri Henry Shaw's and George Engelmann's libraries,[50] supplemented the pre-Linnean library of Edward Louis Sturtevant (1842–1898).[28,51] In New York, the Garden early acquired the botanical library of Columbia College,[54] and later over 5,000 duplicates from the Conservatoire Botanique of the University in Geneva,

Switzerland, were purchased for $13,000 by the Garden.[39] The Herbarium at Missouri was built on those of J. Johann Bernhardi (1774–1850), George Engelmann, William Trelease, and collections from Washington University;[52] while at New York, herbaria came from Columbia College,[9,46] including the vascular plant collections of John Torrey (1796–1873)[47a], John J. Crooke (1824–1911), and the Torrey Botanical Club; in addition to the private collections of Job B. Ellis (1829–1905)[53] fungi, William Mitten (1819–1906)[21,26] mosses, and Thomas Morong (1827–1894)[11] vascular plants.

Both gardens initiated active publishing programs which resulted in important publication exchanges. Each added rapidly to its library and herbarium. Both institutions attempted to establish strong research programs. Britton by his position as Secretary of the Board of Managers, and by his skillful administrative and persuasive powers, was able to expand his scientific staff rapidly and to keep it intact producing numerous floras and monographs. By contrast, Trelease was less successful in expanding his scientific staff who produced mostly monographs rather than floras. It is true that in time Trelease hoped that the monographs, mostly of southwestern genera, would aid in the production of a national flora. Britton, in opposition, believed that floras should be written first and that the monographs could be produced simultaneously or later. Quick and useful publications were Britton's ideal. He himself with considerable assistance from Addison Brown,[12,36] published the first completely illustrated American regional flora, *The Illustrated Flora of the Northern United States and the British Possessions* (3 volumes, 1896–1898).[13,43] Britton used it, as well as other New York-generated publications, including West Indian, southeastern United States and also South American floras to further the Rochester (1892), or American (1907) Codes of Botanical Nomenclature, which he had devised.[16,33,34,37,55] These Codes led to a serious split in the American botanical community with the Harvard botanists leading the group who adhered to a more International Code, while those at New York remained loyal to the American Code. Britton, because of his independence from other institutions or cities, was able to be firm in his resolution. Others, including Trelease, had varied allegiances, including one to Harvard.

The personalities of the two directors were different, though each one was a highly qualified botanist. Trelease was respected and loved, whereas Britton was respected and feared. When Trelease resigned from the directorship of the Missouri Botanical Garden in 1912, because of serious disagreements with the Board of Trustees over the role of scientific activities relative to the purely horticultural activities of the Garden, he had built a strong foundation, upon which others were to build. Although Alfred Russel Wallace (1823–1913) in his April 1887 visit to St. Louis called the Garden the "Trelease Botanical Garden," it was the Shaw Botanical Garden to those in St. Louis. Its directors were following Henry Shaw's vision as interpreted by the trustees. By contrast, the New York Botanical Garden was the vision of Nathaniel Lord Britton. It was his brainchild, and he molded it to his liking. By 1912 it was Britton's Botanical Garden and belonged to no one else.

References Cited

1. A[nderson], E[dgar]. 1954. Henry Shaw—A pictorial biography. *Missouri Botanical Garden Bulletin* 42: 103–118.

2. _____ and Dorothy Brockhoff. 1967. Henry Shaw: A pictorial biography. *Missouri Botanical Garden Bulletin* 55(6): 1–16.

3. [Anonymous]. 1889. A million-dollar garden. *American Art Journal* 52: 84–85.

4. [_____]. 1906. Descriptive guide to the grounds, buildings and collections. *Bulletin of the New York Botanical Garden* 5: 107–194. [Reprinted, New York, 1907].

5. Barnhart, John Hendley. 1935. The published work of Elizabeth Gertrude Britton. *Bulletin of the Torrey Botanical Club* 62: 1–17.

6. Bay, J. Christian. 1945. *William Trelease, 1857–1945: Personal reminiscences.* Privately Printed, Chicago. 12 pp.

7. Britton, N. L. 1877. Staten Island plants. *Bulletin of the Torrey Botanical Club* 6: [177]–179.

8. _____. (comp.), with the assistance of eminent botanists. 1881. *A Preliminary Catalogue of the Flora of New Jersey*. Office of the [Geological] Survey [of New Jersey], Rutgers College, New Brunswick. xiii pp., 223 leaves, pp. 225–233.

9. _____. 1896. The Herbarium of Columbia University, and the Herbarium of the New York Botanical Garden. *Bulletin of the Torrey Botanical Club* 23: 253.

10. _____. 1900. The Museum Building. *Journal of the New York Botanical Garden* 1: 1–5 + photograph opp. p. 1.

11. _____. 1901. The Morong Herbarium. *Journal of the New York Botanical Garden* 2: 35–37.

12. _____. 1913. Addison Brown. *Journal of the New York Botanical Garden* 14: 119–121.

13. _____ and Addison Brown. 1896–1898. *An Illustrated Flora of the Northern United States, Canada and the British Possessions*. Charles Scribner's Sons, New York. 3 vols. (1: xii, 612 pp., 2: iv, 643 pp., 3: xiv, 588 pp.).

14. [Brown, Addison, Charles P. Daly, N. L. Britton, and others]. 1896. Act of Incorporation, as Amended by Chapter 103 of the Laws of 1894, Approved March 7, 1894. *Bulletin of the New York Botanical Garden* 1: [1]–61.

15. Buchholz, J. T. 1945. William Trelease 1857–1945. *Science, n. ser.* 101: 192–193.

16. Committee on Nomenclature. 1893–1894. List of Pteridophyta and Spermatophyta growing without cultivation in northeastern North America. *Memoirs of the Torrey Botanical Club* 5: 1–377.

17. Dimmock, Thomas. 1890. Henry Shaw: A biographical sketch, pp. 7–25 + port. (frontis.). *In* [*First Annual Report*], *Missouri Botanical Garden*, St. Louis. 165 pp. + 5 folded maps.

18. Ewan, Joseph. 1976. William Trelease, p. 456. *In* C. C. Gillispie (ed.), *Dictionary of Scientific Biography*, Vol. 13. Charles Scribner's Sons, New York.

19. Faherty, William Barnaby. 1987. *Henry Shaw: His Life and Legacies*. University of Missouri Press, Columbia. x, 228 pp.

20. Fernald, M. L. 1936. Nathaniel Lord Britton, 1859–1934. *Proceedings of the American Academy of Arts and Sciences* 70: 504–505.

21. Fleming, Mary and Rupert C. Barneby. 1964. Treasures of the Garden's Herbarium III. The Bryophytes of William Mitten. *Garden Journal* 14: 146–148.

22. Gambee, Eleanor Brown. 1974. The founding fathers: The Garden, too, was just an idea once. *Garden Journal* 24: 34–37.

23. Greenman, J. M. 1945. William Trelease. *Missouri Botanical Garden Bulletin* 33: 71–72 + port. (pl. 12) opp. p. 71.

24. Henshaw, Samuel. 1900. The range of horticultural houses. *Journal of the New York Botanical Garden* 1: 81–84 + photograph (pl. V) opp. p. 81.

25. Hollick, [Charles] Arthur and Nathaniel L. Britton. 1879. *The Flora of Richmond County, New York*. Staten Island, New York. 36 pp.

26. Howe, Marshall A. 1934. Elizabeth Gertrude Britton. *Journal of the New York Botanical Garden* 35: 97–103. (Quote, p. 102).

27. _____. 1934. Nathaniel Lord Britton 1859–1934. *Journal of The New York Botanical Garden* 35: 169–180 + port. opp. p. 169. (Quote, p. 172).

28. Hutchings, C. E. 1903. A supplementary catalogue of the Sturtevant Prelinnean Library, pp. 233–316. *In Fourteenth Annual Report, Missouri Botanical Garden, St. Louis*. 316 pp.

29. Kunkel, Louis Otto. 1961. William Trelease, February 22, 1857–January 1, 1945. *Biographical Memoirs, National Academy of Sciences* 35: [307]–332 + port. opp. p. [307]. (Quote, p. 313).

30. Lees, Carlton B. 1974. The Conservatory: Past, present, and future. *Garden Journal* 24: [40]–45.

31. Lenz, Lee W. 1986. *Marcus E. Jones: Western Geologist, Mining Engineer and Botanist*. Rancho Santa Ana Botanic Garden, Claremont, California. xv, 486 pp. (Quote, p. 205).

32. MacDougal, D. T. 1900. The Laboratories. *Journal of the New York Botanical Garden* 1: 65–68 + (pl. IV) opp. p. 65.

33. Members and Alternates of the Nomenclature Commission. 1904. Code of Botanical Nomenclature. *Bulletin of the Torrey Botanical Club* 31: 249–261.

34. _____. 1907. American Code of Botanical Nomenclature. *Bulletin of the Torrey Botanical Club* 34: 167–178.

35. Merrill, Elmer D. 1938. Biographical memoir of Nathaniel Lord Britton 1859–1934. With bibliography by John Hendley Barnhart. *Biographical Memoirs, National Academy of Sciences* 19: 147–202 + port. (unnumbered pl. opp. title p.) (Quote, p. 151).

36. Montague, Gilbert H. 1954. The extraordinary Judge Addison Brown. *Garden Journal* 4: 1–4, 26.

37. Nicholson, Dan H. 1991. A history of botanical nomenclature. *Annals of the Missouri Botanical Garden* 78: 33–56.

38. Pammel, Louis H. 1927. *Prominent Men I Have Met, III, Dr. William Trelease*. Published by The Author, Ames, Iowa. 84 pp.

39. Reed, John F. 1969. The Library of The New York Botanical Garden. *Garden Journal* 19: 77–88.

40. Robbins, Christine Chapman. 1964. David Hosack: Citizen of New York. *Memoirs of the American Philosophical Society* 62: v–vii, 1–246.

41. Robbins, William J. 1952. The founding of The New York Botanical Garden. *Garden Journal* 2: 12–14, 27.

42. Rudolph, Emanuel D. 1991. One hundred years of the Missouri Botanical Garden. *Annals of the Missouri Botanical Garden* 78: 1–18.

43. Rusby, H. H. 1896. [Review of] An Illustrated Flora of the Northern United States, Canada and the British Possessions, . . . By Nathaniel Lord Britton . . . , and Hon. Addison Brown, . . . , Vol. 1, 1896. *Bulletin of the Torrey Botanical Club* 23: 371–378.

44. Shaw, Henry. [undated]. *Henry Shaw's Will [of 1885] Establishing the Missouri Botanical Garden: Admitted to Probate at St. Louis, Missouri, September 2, 1889*. Pamphlet Published by the Missouri Botanical Garden, St. Louis. 3 pp.

45. Shinners, Lloyd H. 1962. Evolution of the Gray's and Small's manual ranges. *Sida* 1: [1]–31. (Quote, p. 8).

46. Small, J. K. 1900. The Herbarium. *Journal of the New York Botanical Garden* 1: 33–38 + photograph (pl. II) opp. p. 33.

47. Sprague, T. A. 1934. Nathaniel Lord Britton. *Kew Bulletin* 1934: 275–279.

47a. Steere, William C. 1967. John Torrey and the New York Botanical Garden. *Garden Journal* 17: 138–143.

48. _____. 1974. The Museum Building: An idea crystallized. *Garden Journal* 24: [38]–39.

49. [Sterns, E. E.]. 1888. Proceedings of the Club, 24 October 1888. *Bulletin of the Torrey Botanical Club* 15: 301–302.

50. Trelease, William. 1892. Third annual report of the director, for 1891, pp. 11–20. *In Third Annual Report, Missouri Botanical Garden*, St. Louis. 170 pp.

51. _____. 1896. The Sturtevant Prelinnean Library, pp. 123–209. *In Seventh Annual Report, Missouri Botanical Garden*, St. Louis. 209 pp.

52. _____. 1905. Sixteenth annual report of the director, pp. 13–30. *In Sixteenth Annual Report, Missouri Botanical Garden*, St. Louis. 257 pp.

53. Underwood, Lucien M. 1900. The Ellis collection of fungi. *Journal of the New York Botanical Garden* 1: 38–40.

54. Vail, Anna Murray. 1900. The Library. *Journal of the New York Botanical Garden* 1: 17–22.

55. Weatherby, C. A. 1949. [Symposium on botanical nomenclature]. II. Botanical nomenclature since 1867. *American Journal of Botany* 36: 5–7.

Publication Notes

Modified and edited from the abstract[a] and unpublished manuscript[b] read 7 August 1990 at the meeting of the Historical Section, Botanical Society of America, Richmond Centre, Richmond, Virginia. The first paragraph of the revised paper printed here is taken from the meeting abstract.

[a] 1990. The Missouri and New York Botanical Gardens; Contrast and similarity of origins and early development. *American Journal of Botany* 77(6, supplement): 80–81. [Abstract].

[b] [1990]. "The Missouri and New York Botanical Gardens: Contrast and Similarity of Origins and Early Development." Handwritten, 21 pp. [Manuscript].

Post card views of Shaw's Garden, the Missouri Botanical Garden, St. Louis, about 1900 (In the collection of RLS) ↕

Post card view of the main building at the New York Botanical Garden, Bronx, about 1910 (In the collection of RLS)

Entrance to Shaw's Museum Building. At the time of the photograph this building housed the Library of Joseph and Nesta Ewan. l. to r. Nesta Ewan, unidentified, Joseph Ewan, Emanuel D. Rudolph (Taken by Ronald L. Stuckey, 6 October 1989; in the collection of RLS)

Emanuel D. Rudolph with Trevor Chilton, instructor at Mill Hill School (Taken by Ann Rudolph, 1 October 1990; in the Rudolph photograph collection)

Mill Hill School and the Missouri Botanical Garden

Henry Shaw (1800–1889)

The Museum at Shaw's Garden,
the Missouri Botanical Garden

Celebrating 100 Years of the Founding of the Missouri Botanical Garden. The following speakers made presentations at the 36th annual Systematics Symposium, held at the Garden, 6–7 October 1989. (l. to r.) Shoichi Kawano, Marshall R. Crosby, Peter H. Raven, Dan H. Nicolson, James L. Reveal, F. Nigel Hepper, Barbara G. Briggs, Joseph Ewan, and Emanuel D. Rudolph, who gave the keynote address, "One Hundred Years of the Missouri Botanical Garden."

Paper No. 24 (1991)

Henry Shaw's Path from Sheffield to Mill Hill School and the Founding of the Missouri Botanical Garden

Why would Henry Shaw, a businessman, retiring at age 39 in 1840 from an active retail and wholesale hardware business in St. Louis, 19 years later decide to establish a public botanical garden? A letter to Sir William Jackson Hooker (1785–1865), Director of the Royal Botanic Gardens, Kew, in 1856 simply states that, "being the proprietor of lands in the vicinity of this town—and having a desire to found and endow a public botanic garden and in commencing such a considerable undertaking I wish to obtain such hints and information as may be useful to me . . ."[59] Neither this initial inquiry, nor later ones, tell the reasons for Shaw's "desire." To determine the probable reasons it is necessary to consider Shaw's early life and those times.

Sheffield

Born in the first year of the nineteenth century in Sheffield, England, Henry Shaw (1800–1889) was initiated to social consciousness as a youth during the Napoleonic Wars.[2,16,19,20] The son of a middle class merchant-factor, of Baptist persuasion, in an important industrial town, he was instilled with certain attitudes and behavior. Youths are indoctrinated by their parents, and by older relatives and friends, with attitudes that these elders themselves learned in youth and modified through time. Thus, each generation inherits baggage of the past which it molds for itself as it matures. Henry Shaw was brought up in a merchant's household with late eighteenth century British Dissenter values. Among these would have been a respect for hard work, initiative, individuality, and wealth, as well as a highly religious moral tone. The town of Sheffield had certain characteristics that also influenced a growing boy.

Sheffield, noted since Renaissance times as a center for cutlery making, became an important eighteenth century manufacturing center because of ample nearby ore

Frontispiece Credits: Mill Hill School and the Missouri Botanical Garden. Henry Shaw (1800–1889). (Photograph taken in 1885, from anonymous, *Missouri Botanical Garden Bulletin* 49(6): 85. 1961; reproduced with permission)

The Museum at Shaw's Garden, the Missouri Botanical Garden. (From Rudolph, *Annals of the Missouri Botanical Garden* 78: 5. 1991; reproduced with permission)

Celebrating 100 Years of the Founding of the Missouri Botanical Garden. (Taken by Ronald L. Stuckey, 7 October 1989; in the collection of RLS)

and coal fields.[4,30,31,32,33,44,46,56,57,71] Characterized by small-sized factories and piece-work finishing in many homes in town and nearby, it lacked the strong separation between workers, craftsmen, and entrepreneurs so common in some other British industrialized towns. The *Gentlemen's Magazine* of 1764 notes that Sheffield, with a population of about 20,000, has as its principal manufactures, "knives, forks, scissors, razors, lancets, phleams, files, edge tools for carpenters, shears, etc."[25]. As a metal- producing and manufacturing town, it was high on dirt, and low on distinguished buildings, about which eighteenth and early nineteenth century visitors commented.[4,11,25,32,54,61,65,71] In particular, the smoke and dirty appearance of the buildings and people were noted. Mary Howitt remembered her visit to Sheffield in the second decade of the nineteenth century for its ". . . broken up ground and half-built, ruinous houses . . . [and] . . . some haunt of squalor."[34] In 1800, the year before Henry Shaw was born, the population was over 31,000, by 1811 it had increased to over 35,000, and by the time he left in 1819, it was about 40,000.[31,56,71] A German governmental visitor in 1814 described Sheffield as "the capital of the English iron and steel industries."[48] Unlike Manchester, but like Birmingham, Sheffield had many small businesses in which little capital expenditure was needed for their beginning. During the two decades Shaw was there, the spread of wealth was not extreme, and few businessmen accumulated fortunes. The first large suburban house, for example, was not built until 1773.[43] Shaw's father, Joseph, and his mother's family, the Hooles, were in the cutlery and related trades. It was a town in which foreign trade was important and in which ". . . French, German, and other foreign languages,. . . [were] much studied by the opulent traders of Sheffield, and considered by many as necessary branches of education."[4]

Attractive open spaces still existed then outside the town. A volume of 1812 reports that "Wharncliffe park, the seat of the Honourable James Archibald Stuart Wortley, . . . six miles to the north-west of Sheffield, excites the admiration of tourists by its beautiful and picturesque scenery."[4] The hilly nature of the terrain promoted romantic visions. Several large old oaks, the Brincliffe oaks, were particularly notorious.[33] Even so, some remarked on the spoiling of the hillside beauty by the presence of numerous coal pits.[65]

At the beginning of the nineteenth century the town, while developing the newer technologies, such as still crucibles and steam power, had reminders of earlier times.[64,71] It was poorly lighted with oil lamps until 1819 when the first gas lamps were installed, and a public gibbet-post existed with remnants of a clothed body hanging on it until 1817.[43] Open spaces were available and ". . . common land where youths and men could have taken exercise at cricket, quoits, football and other exercises . . . " and "perhaps as beautiful botanical gardens as any in the kingdom, but these are opened only once or twice a year to the poorer classes, . . . and hermetically sealed on a Sunday . . ., the only day when members of the working classes have the leisure to enjoy them."[11,31,41,53,54(quote)] Many citizens commonly cultivated beautiful flowers, and avenues of sycamores grew along the Sheaf River.[32,43] It was said that many working men had a garden, which they cultivated for pleasure. Ample views of nature and flowers were available to the young Henry Shaw, perhaps more appealing because of their contrast to the drab townscape.

Sheffield was in some ways a typical provincial conservative town, but in other ways a center for national radicalism. In the late eighteenth century it had been a strong radical area, fanned by Thomas Paine's *Rights of Man*, but by the time of Shaw's youth the flame had waned only to reignite later. Dissenter Protestant sects and Catholics were well represented in the population. Non-conformist attendance at religious services was said to be over fifty percent of those attending all services by mid-century.[9,71] Joseph Shaw (ca. 1766–1844), Henry's father, was an active Baptist, although other sects were more prominent in the town. Baptists, who had first met in another sect's chapel in 1806, had their own chapel with a minister by 1814.[71] Baptists were usually associated with lower-middle and working classes, while Unitarians and Quakers were likely to be associated with the upper-middle levels.[15]

The nature of "classes" in British society during the nineteenth century is a matter of considerable interest to historians and sociologists.[8,12,15,28,35, 45,49–51,55,57,62,67,69,70] Individuals undoubtedly recognized the "classes" to which they felt they belonged, and to which others thought they belonged, although it is not always clear now how the divisions should be made. Instability and mobility, particularly in a place like Sheffield, was a fact of the times. It is clear that Henry Shaw's family was neither of the "working class," nor of the aristocracy. The Shaws and the Hooles were not involved in the various scientific activities in Sheffield in the first decade of the century. In truth, these activities were not very strong in that period,[35,36,37,38,39,40] although between 1799 and 1820, 69 itinerant scientific lecturers are known to have visited. Shaw certainly must have known about Jonathan Salt (1759–1810), a Sheffield cutler, who studied the local flora and was "engaged in this interesting pursuit [botany] with such patient ardour and unalloyed delight, that he not only acquired a correct and comprehensive knowledge of plants, but also was regarded by the first professors of his day as an ornament and benefactor to the science, . . . "[17,31,32(quote)]

By their religious and economic activities, the Shaws fit into the "lower ranks of the middle class" as defined by Davidoff and Hall,[15] which may be misleading for Sheffield. Considerable mobility existed, and one commentator wrote about the 1820s, that there then "was the aristocracy of mere craft and position, but one generation deep, and vulgar though affected, looking back with horror and contempt at the democratic base whence they had just sprung."[70] Also, the class structure in Sheffield was less rigid than that of some of the other manufacturing towns.[55,57] One way to achieve upward mobility was through better education. Perhaps that was the aim of Joseph Shaw in sending his son Henry to the newly established Dissenters' public school at Mill Hill, called the Mill Hill School, about ten miles north of London.

Mill Hill School

When Mary Howitt of near Sheffield later remembered how in 1809, at age 10, she and her older sister went off to boarding school outside of London, she wrote: "The next day we were in London-London! How the very thought transported us with joy and astonishment! . . . We felt ourselves in a new world at [school]."[34] The eleven-year-old Henry Shaw in 1811 must have had similar feelings when he also arrived at Mill Hill School. The School, founded late in 1807, was young and the first nonconformist public school to be modeled on the established older Anglican types such as Eton, Winchester, Westminster, and Harrow.[1,3,6,68] A British public school is defined as "a 'Boarding Academy for Young Gentlemen,' which draws its pupils from all parts of the Country, and is not a Private Adventure School."[52] The nonconformist groups had previously depended upon boarding academies operated by one individual to provide secondary-level education.[52] Mill Hill School was one of the earliest to be governed by a committee, then consisting of six ministers and 21 laymen. In the earliest years, the committee dominated the headmaster, causing some friction.[6] When Henry Shaw came to the School in 1811, the Reverend Maurice Phillips had just become headmaster and he remained there through Shaw's four years of school.

Mill Hill School was established in the village of Mill Hill in Middlesex on the former estate of the prominent eighteenth century Quaker merchant Peter Collinson (1693–1768).[7,13,22,24,66] This estate was inherited from Collinson's wife's father, Michael Russell, with a building called Ridgeway House, said to be built before 1524 and located on pleasant high ground overlooking long rustic vistas. Collinson corresponded extensively with Benjamin Franklin (1706–1790) and John Bartram (1699–1777), as well as with others of scientific bent in America. His stimulus is credited with interesting Franklin in electricity–whom he aided in establishing the Library Company in Philadelphia, and who visited him at Mill Hill in 1757.[10,23] Collinson was aggressive in introducing exotic plants, particularly from North America, first into his garden at Peckham, and with increased industry into his Mill Hill garden after inheriting it sometime after 1749. A number of garden plants and exotic trees are credited to Collinson's first introductions into Britain. A manuscript catalogue of plants in his Mill Hill garden was published many years after Collinson's death.[18,42] Collinson's son then tried to maintain the garden, but other later owners and renters were not as careful, and many herbaceous and woody plants were removed or lost.[13,63] When the School began in the Ridgeway House in 1808 many large exotic trees still existed on the grounds, including three cedars of Lebanon said to have been planted there by Linnaeus.[5,21,63]

Collinson was proud of his garden and its many rarities. He wrote to Linnaeus in 1759 that his garden "is enriched with many rare plants and trees not to be seen elsewhere."[63] Seven years later he noted, "I survey my garden with raptures, to see the infinite variety with which the great Creator has enriched the vegetable world."[29, 64(quote)]

Henry Shaw[60] in his last years wrote about his remembrance of Mill Hill School and its gardens:

> Between 1750 and 1760 he [Collinson] made a collection of all the pines and firs that could be procured in his ground at Mill Hill, and several pines, cedars, and spruces still remained in 1812. . . . Collinson corresponded with Linnaeus, who when in England visited him at Mill Hill (where in the garden there stood in 1812 a venerable upright Cypress tree said to have been planted by Linnaeus). . . . Collinson's large greenhouse had been altered into a classroom, at one end of which the window still remained, looking from his library into the greenhouse below. In 1812 many fine specimens of rare plants still grew in the garden, with the Cypress of Linnaeus; but the most admired trees, were the noble cedars of Lebanon, that grew near the play ground, protected from injury by a high board fence.

An American visitor there in 1832 wrote, "The grounds, encircled by groves and plantations, and studded with single trees, descend beautifully towards the west, and command a charming prospect."[65] The noted landscape gardener John Loudon (1783–1843), upon a visit to Mill Hill School in 1835, commented on the many fine specimen trees and shrubs, some dating from Collinson's time.[47] Surely, the young Henry Shaw was very much impressed in 1812 by the fine plantings on the school grounds, a mental seed that was to grow many years later. Shaw thought Linnaeus had planted a cypress, others thought the cedars of Lebanon; what is the truth matters little, it is what people perceive that matters, and the Linnaean trees became a tradition at Mill Hill.[42,47] Large specimen trees were part of the School's grounds from its beginning, and have been part of its tradition ever since. The Ridgeway House, with some additions, served as the school until it was replaced in 1826 with a neo-classical structure that still stands.[47]

The Mill Hill School is a Protestant Dissenters' school, but even in its early days it had Anglican, and even Catholic, pupils. When Shaw entered in 1811–1812, it had 73 boys.[6] The newly rewritten plan included instruction in religious education; the Greek, Latin, and French languages; arithmetic; geography; mathematics; history; and drawing.[6] Thus, unlike most public schools, French and other non-classical subjects, though in Shaw's time no sciences, were a regular part of the curriculum.[26,52,60] Winter and summer, the boys had relatively little time for recreation. Team sports were not then an important part of the public school activities as they were to become later,[14, 26] and Shaw chose to cultivate a little garden rather than play games. He[60] remembered:

> Nearby [to the playground] were the boys' little gardens of about 25 feet square each, well fenced to guard against the intrusions of the football, for those that preferred gardening for a recreation, of which number was—The writer of this Guide—who from 1812 to 1815 thus relieved fugging at Latin, Greek and Euclid, by growing pinks and geraniums, and retiring to a seat on a corner of his little domain, shaded by a broad trained Irish yew, purchased from a neighboring nursery at the price of ten shillings.

The fees were fixed at three guineas a quarter for the first three classes and two guineas for the three lower, thus ten shillings for a pupil was a considerable cost, and illustrates Shaw's dedication to gardens and exotic plants at a young age by rendering that amount for a tree.[6,26,27] Shaw ended his formal education just as the Napoleonic Wars ended. He was not to continue his education, but rather to be involved with business. The Mill Hill experience was to make Shaw a lifelong Francophile.[60]

The economy of Sheffield, particularly that based upon steel and cutlery, went through several depressions in the early years of the nineteenth century. Markets for cutlery, which often is stock-piled and can be produced in large quantities by price-cutting little masters, journeymen, and home finishers, is particularly subject to strongly fluctuating prices.[33,46,58] As a result of these fluctuating prices and the incurrence of a debt from a bad partnership, Joseph Shaw decided to investigate American markets in 1818.[20] With his son Henry, he sailed to Canada whence Henry was to travel to the United States for new markets, and a new life.

Henry Shaw's Sheffield background provided several attitudes that influenced his later activity in St. Louis. The rapid and frequent fluctuations of the economy in his youth had resulted in unemployment, debt and hardships, and made him a wary businessman who wanted to have a financial cushion. The drabness of the city as compared with the rustic yet controlled settings of several gardens outside of the city made him appreciate the latter. These gardens and particularly the traditions of Peter Collinson's garden at Mill Hill School probably stimulated his desire for having a recreational garden there from 1811 to 1815. There too, he learned the French language and developed an appreciation for French people, something that was to be beneficial to him in his new life in America.

References Cited

1. A[nderson], E[dgar]. 1931. Henry Shaw's alma mater and Peter Collinson. *Missouri Botanical Garden Bulletin* 19: 88–94.

2. _____ and Dorothy Brockhoff. 1967. Henry Shaw, a pictorial biography. *Missouri Botanical Garden Bulletin* 55(6): 1–16. [Reprint].

3. [Anonymous]. 1894. *An undenominational public school, Mill Hill School: Past and present.* St. James Budget, February 1894: 22–24.

4. Bigland, John. 1812. Yorkshire. *In The Beauties of England and Wales; or, Delineations, Topographical, Historical, and Descriptive, of Each County, Vol. 16.* J. Harris, etc., London. viii, 938, [22] pp. [Sheffield, pp. 816–827]. (Quotes, pp. 825, 827).

5. Blunt, Wilfrid, with the assistance of William T. Stearn. 1971. *The Compleat Naturalist: A Life of Linnaeus. A Studio Book.* The Viking Press, New York. 256 pp. [See pp. 111–115].

6. Brett-James, Norman G. [ca. 1909]. *The History of Mill Hill School 1807–1907.* Andrew Melrose, London. xii, 415 pp. [See pp. 15, 41–42, 46–48].

7. _____. [ca. 1925]. *The Life of Peter Collinson, F.R.S., F.S.A.* Published for the Author by Edgar G. Dunstan & Co., [London]. 287, [5] pp.

8. Briggs, Asa. 1960. The language of 'class' in early nineteenth-century England, pp. 43–73. *In* Asa Briggs and John Saville (eds.), *Essays in Labour History: In Memory of G. D. H. Cole, 25 September 1889–14 January 1959. With recollections of G. D. H. Cole by Ivor Brown [and others].* Macmillan, London; St. Martin's Press, New York. vii, 363 pp. [See pp. 43–73].

9. _____. 1965. *Victorian Cities.* Harper and Row, New York and Evanston. 424 pp. [See p. 66].

10. Clark, Ronald W. 1983. *Benjamin Franklin, a Biography.* Random House, New York. viii, 530 pp.

11. Colman, Henry. 1850. European life and manners, p. 53. *In Familiar Letters to Friends, Vol. 1.* Charles C. Little and James Brown, Boston. 2 vols. [See Vol. 1, p. 53].

12. Crossick, Geoffrey. 1984. The petite bourgeoisie in nineteenth-century Britain: The urban and liberal case, pp. 62–94. *In* Geoffrey Crossick and Heinz-Gerhard Haupt (eds.), *Shopkeepers and Master Artisans in Nineteenth-century Europe.* Methuen, London and New York. xii, 283 pp.

13. Darlington, William. 1967. *Memorials of John Bartram and Humphry Marshall.* Facsimile of the 1849 ed., with introduction by Joseph Ewan. *Classica Botanica Americana. Supplement* I. Hafner Publishing Co., New York. lii, 585 pp. [See p. 460].

14. Darwin, Bernard. 1929. *The English Public School.* Longmans, Green, London, New York. xv, 175 pp. [See p. 75].

15. Davidoff, Leonore and Catherine Hall. 1987. *Family Fortunes, Men and Women of the English Middle Class, 1780–1850.* Hutchinson Education, London. 576 pp. [See p. 24].

16. Denken, Jean and James R. Reed. 1977. *Henry Shaw, His Life and Legacy.* [Missouri Botanical Garden, St. Louis]. [16] pp.

17. Desmond, Ray. 1977. *Dictionary of British and Irish Botanists and Horticulturists, Including Plant Collectors and Botanical Artists.* With historical introduction by William T. Stearn. Taylor and Francis, London. xxvi, 747 pp. [See p. 539].

18. Dillwyn, L. W. 1843. *Hortus Collinsonianus: An Account of the Plants Cultivated by the Late Peter Collinson, Esq., F.R.S., Arranged Alphabetically According to Their Modern Names, from the Catalogue of His Garden, and Other Manuscripts, Not Published.* W. C. Murray and D. Rees, Swansea. vii, 64 pp.

19. Dimmock, Thomas. 1890. Henry Shaw, a biographical sketch, pp. 7–28 + portrait (frontis.). *In* [*First Annual Report*], *Missouri Botanical Garden*, St. Louis. 165 pp. + 5 folded maps. [Reprinted as a pamphlet, late 1930s].

20. Faherty, William Barnaby. 1987. *Henry Shaw, His Life and Legacies*. University of Missouri Press, Columbia. [xi], 228 pp. [See p. 7].

21. [Flanagan, Mark]. 1988. *The Grounds at Mill Hill School, an Advisory Report*. [Royal Botanic Gardens, Kew]. 16 pp.

22. Fox, R. Hingston. 1919. *Dr. John Fothergill and His Friends: Chapters in Eighteenth Century Life*. Macmillan and Co., London. xxiv, 439 pp.

23. Franklin, Benjamin. 1840. *The Works of Benjamin Franklin: Containing Several Political and Historical Tracts Not Included in Any Former Edition, and Many Letters, Official and Private, Not Hitherto Published, with Notes and a Life of the Author*. By Jared Sparks. Hillard, Gray, and Co., Boston. 10 vols. [see vol. 1, p. 30].

24. Frick, George F. 1971. Collinson, Peter, pp. 349–351. *In* Charles Coulston Gillispie (ed.), *Dictionary of Scientific Biography, Vol. 3*. Charles Scribner's Sons, New York.

25. G., E. 1764. Natural history of Sheffield. *Gentlemen's Magazine, and Historical Chronicle* 34(April): 157–161. (Quote, p. 159).

26. Gathorne-Hardy, Jonathan. 1977. *The Public School Phenomenon*. Hodder and Stoughton, London. 478 pp. [See pp. 52–53, 276, and Appendix A].

27. Gibbins, H. de B. 1910. *Industry in England: Historical Outlines*. 6th ed. Charles Scribner's Sons Ltd., New York. viii, 483 pp. [See p. 424].

28. Gretton, R. H. 1917. *The English Middle Classes*. G. Bell and Sons, London. xii, 238 pp.

29. Harvey, John H. 1981. A Scottish botanist in London in 1766. *Garden History* 9(1): [40]–75. ["Peter Collinson's Garden, Mill Hill, 13 September 1766," pp. 59–60].

30. Hey, David. 1972. The Rural Metalworkers of the Sheffield Region: A Study of Rural Industry before the Industrial Revolution. *Department of English Local History Occasional Papers, 2nd. ser., No 5*. Leicester University Press, [Leicester]. 60 pp.

31. Holland, G. Calvert. 1843. *The Vital Statistics of Sheffield*. Robert Tyas, London; J. H. Greaves, Sheffield. xii, 263. [See pp. 13, 27, 39].

32. Holland, John. [ca. 1860]. *Sheffield and Its Neighbourhood*. Photographically illustrated by Theophilus Smith. Pawson & Brailsford, Sheffield; Robert White, Worksop, Park-street. [ix], 132 pp. [See pp. 23–24, 58].

33. _____ and James Everett. 1854–1856. *Memoirs of the Life and Writings of James Montgomery, Including Selections from His Correspondence, Remains in Prose and Verse, and Conversations on Various Subjects*. Longman, Brown, Green, and Longmans, London. 7 vols. [See vol. 2, p. 328; vol. 3, p. 317].

34. Howitt, Mary. 1889. *Mary Howitt, An Autobiography.* Edited by her daughter Margaret Howitt. Wm. Isbister Limited, London. 2 Vols. (Quotes, Vol. 1, pp. 78, 85).

35. Inkster, Ian. 1973. The development of a scientific community in Sheffield, 1790–1850: A network of people and interests. *Transactions of the Hunter Archaeological Society* 10: 99–131.

36. _____. 1973. A phase in middle class culture: Phrenology in Sheffield, 1824–1850. *Transactions of the Hunter Archaeological Society* 10: 273–279.

37. _____. 1975. Science and the mechanics' institutes, 1820–1850: The case of Sheffield. *Annals of Science* 32: 451–474.

38. _____. 1976. Culture, institutions and urbanity: The itinerant science lecturer in Sheffield 1790–1850, pp. 218–232. *In* Sidney Pollard and Colin Holmes (eds.), *Essays in the Economic and Social History of South Yorkshire*. South Yorkshire County Council, Recreation, Culture and Health Department, [Barnesley]. 308 pp.

39. _____. 1977. Marginal men: Aspects of the social role of the medical community in Sheffield 1790–1850, pp. 128–163. *In* John Woodward and David Richards (eds.), *Health Care and Popular Medicine in Nineteenth Century England: Essays in the Social History of Medicine*. Croom, Helm, London. 195 pp.

40. _____ and Jack Morrell (eds.). 1983. *Metropolis and Province: Science in British Culture, 1790–1850*. Hutchinson and Co., London. 288 pp.

41. Lamb, J. P. (comp. and ed.). [1951]. *Sheffield: A Handbook Prepared for the Festival of Britain*. Published for the City Council by the Civic Information Service, Sheffield. 67 pp. [See p. 61].

42. Lambert, Aylmer Bourke. 1811. Notes relating to botany, collected from the manuscripts of the late Peter Collinson, Esq., F.R.S. *Transactions of the Linnean Society of London* 10: 270–282.

43. Leader, Robert Eadon. 1876. *Reminiscences of Old Sheffield*. Leader and Sons, Sheffield. 144 pp. [See pp. 117, 145, 164, 195].

44. _____. 1901. *Sheffield in the Eighteenth Century*. The Sheffield Independent Press, Sheffield. [iv], 362 pp.

45. Lewis, Roy and Angus Maude. 1949. *The English Middle Classes*. Phoenix House, London. 320 pp.

46. Lloyd, G. I. H. 1913. The *Cutlery Trades: An Historical Essay in the Economics of Small-scale Production*. A. M. Kelley, New York. xvi, 493 pp. [Reprinted 1968 by Thomas Nelson, London; Augustus M. Kelley, Publishers, New York]. [See pp. 192–193].

47. Loudon, J. C. 1854. *Arboretum et Fruiticetum Britannicum; or, The Trees and Shrubs of Britain.* 2nd ed. Henry G. Bohn, London. 8 vols. [See vol. 1, p. 56].

48. May, Johann Georg. 1968. Chapter IV. Report on a journey to England by Factory Commissioner J. G. May in 1814, pp. 131–169. *In* W. O. Henderson (ed.), *Industrial Britain under the Regency: The Diaries of Escher, Bodmer, May, and de Gallois, 1814–18.* Reprints of Economic Classics. Augustus M. Kelley, Bookseller, New York. xi, 188 pp. (Quote, p. 152).

49. Morris, R. J. (ed.). 1968. *Class, Power and Social Structure in British Nineteenth-century Towns.* Themes in Urban History. Leicester University Press, Leicester. xiii, 222 pp.

50. Neale, R. S. 1972. *Class and Ideology in the Nineteenth Century.* Routledge & Kegan Paul, London and Boston. viii, 200 pp.

51. _____. 1981. *Class in English History 1680–1850.* Basil Blackwell, Oxford. vi, 250 pp.

52. Ogilvie, Vivian. 1957. *The English Public School.* The Macmillan Co., New York. 228 pp. (Quote, p. 7). [See pp. 117, 119].

53. Olive, Martin. 1991. Letter of 28 January 1991. [Copy of letter in The Ohio State University Herbarium Archives, Columbus].

54. Pike, E. Royston. 1966. *Human Documents of the Industrial Revolution in Britain.* Allen & Unwin, Ltd., London. 368 pp. (Quote, p. 359). [Quote from letter of John Wardle, 1813].

55. Pollard, Sidney. 1959. *A History of Labour in Sheffield.* Liverpool University Press, Liverpool. xix, 372 pp. [See p. 3].

56. Rees, Abraham. [no date]. Sheffield. *In The Cyclopaedia; or, Universal Dictionary of Arts, Sciences, and Literature,* Vol. 33. lst American ed., revised, corrected, and adapted to this country. Samuel F. Bradford; and Murray, Fairman and Co., Philadelphia. 41 vols.

57. Reid, Caroline. 1976. Middle class values and working class culture in nineteenth century Sheffield ... the pursuit of respectability, pp. 275–295. *In* Sidney Pollard and Colin Holmes (eds.), *Essays in the Economic and Social History of South Yorkshire.* With an introduction by Sidney Pollard. South Yorkshire County Council, Recreation, Culture and Health Department, [Barnesley]. 308 pp. [See p. 275].

58. Rostow, W. W. 1948. *British Economy of the Nineteenth Century: Essays.* At the Clarendon Press, Oxford. 240 pp. [See pp. 33, 37].

59. Shaw, Henry. 1856. Letter from Henry Shaw to Sir William Jackson Hooker. [Letter in the correspondence archives of Sir. W. J. Hooker, The Royal Botanic Gardens Library, Kew, Richmond, Surrey, England]. [not seen].

60. _____. [no date]. [The quoted passage from Rudolph's manuscript, probably from a document in Henry Shaw's archives at the Missouri Botanical Garden Library in St. Louis, has not been verified].

61. [Simond, Louis]. 1815. *Journal of a Tour and Residence in Great Britain, during the Years 1810 and 1811: With Remarks on the Country, Its Arts, Literature, and Politics, and on the Manners and Customs of Its Inhabitants.* By a French Traveller. Published for the Author: T. & W. Mercein, Printers, [New York]. 2 vols. [See vol. 2, p. 78].

62. Smith, Dennis. 1982. *Conflict and Compromise: Class Formation in English Society 1830–1914: A Comparative Study of Birmingham and Sheffield.* Routledge and Kegan Paul, London. xiii, 338 pp.

63. Smith, James Edward, Sir. 1821. *A Selection of the Correspondence of Linnaeus and Other Naturalists from the Original Manuscripts.* Longman, Hurst, Rees, Orme and Brown, London. 2 vols. (Quote, Vol. 1, pp. 51, 72). [See also Vol. 1, p. 4].

64. Smith, Wilfred. [ca. 1948]. *An Economic Geography of Great Britain.* E. P. Dutton and Co., Inc., New York. 747 pp. [See p. 115].

65. Stewart, C. S. 1834. *Sketches of Society in Great Britain and Ireland.* Carey, Lea, and Blanchard, Philadelphia. 2 vols. (Quote, vol. 1, p. 174). [See also vol. 2, pp. 42–43].

66. Swem, E. G. 1949. Brothers of the spade, correspondence of Peter Collinson, of London, and of John Custis, of Williamsburg, Virginia, 1734–1746. *Proceedings of the American Antiquarian Society* 58: 17–190.

67. Thompson, E. P. 1964. *The Making of the English Working Class.* The Classic Modern Study. Pantheon Books, New York. 848 pp.

68. Timpson, George F. [ca. 1958]. *Et Virtutem—Essays on Mill Hill.* [The Author], [Printed for Private Circulation . . . by the K. R. Webb Printing Co., Gloucester, England]. 96 pp.

69. Wade, John. 1835. *History of the Middle and Working Classes: With a Popular Exposition of the Economical and Political Principles Which Have Influenced the Past and Present Condition of the Industrious Orders. . . .* 3rd. ed. Effingham Wilson, London. xx, 604 pp.

70. White, Alan. 1988. Class, culture, and control: The Sheffield Athenaeum Movement and the middle class 1847–64, pp. 85–115. *In* Janet Wolff and John Seed (eds.), *The Culture of Capital: Art, Power and the Nineteenth Century Middle Class.* Manchester University Press, Manchester. vii, 236 pp. (Quote, p. 86).

71. Wickham, E. R. 1957. *Church and People in an Industrial City.* Lutterworth Press, London. 292 pp. [See pp. 20, 22, 25, 74]

Publication Notes

Modified and edited from an unpublished manuscript[a] prepared as Chapter 1 for Rudolph's book on "The History of the Missouri Botanical Garden." This topic[b,c] was presented 27 April 1991 at the Annual Meeting of The Ohio Academy of Science, The Ohio State University, Columbus; and 6 August 1991 at the meeting of the Historical Section, Botanical Society of America, Convention Center, San Antonio, Texas.

[a] 1991. "Henry Shaw's Path from Sheffield to Mill Hill School and the Founding of the Missouri Botanical Garden." Handwritten, 5 pp.; typewritten, 13 pp. Typewritten, 25 pp., as Chapter 1. [Manuscript].

[b] 1991. Henry Shaw's path from Sheffield to Mill Hill School and the founding of the Missouri Botanical Garden. *The Ohio Journal of Science* 91(2): 28. [Abstract].

[c] 1991. Why Henry Shaw established a public botanical garden in St. Louis and how he developed it. *American Journal of Botany* 78(6, suppl.): 105. [Abstract].

PART EIGHT

WRITING BOTANICAL HISTORY

Emanuel D. Rudolph's six papers in this group (**Part Eight**) are topically heterogeneous, but they have in common the theme of writing the history of American botany. The first paper describes the accomplishments of Andrew Denny Rodgers III, a celebrated writer of botanical history (**Paper No. 25**), the second one discusses the introduction of the natural system of plant classification into the United States (**Paper No. 26**) and the third one provides a commentary on writing the history of botany (**Paper No. 27**). The remaining three papers record botanical achievements in a particular region covering a specific time period. The fourth one discusses trans-Atlantic connections for classifying plants (**Paper No. 28**), the fifth paper is a commentary on biological history in Antarctica (**Paper No. 29**), and the sixth one summarizes one hundred years of plant science in the United States (**Paper No. 30**).

Introduction No. 25

From Lawyer to Botanical Historian (1965–1981)

Rudolph was a great admirer of Andrew Denny Rodgers III (1900–1981), and the seven scholarly biographies of botanists that Rodgers wrote. These classic books provide a solid foundation for the study of botanical history. Rodgers, born in Columbus, Ohio, lived most of his life in Bexley, the adjacent town to the east, where in the 1940s and early 1950s he devoted full time to the writing of these books. After that time, he seemed to have disappeared from the scene of scholarship.

When Rudolph came to Columbus in 1961, he sought Rodgers for his friendship and to learn more about his writing in botanical history. Their first meeting may have occurred through the efforts of Adolph E. Waller (1892–1975), associate professor of botany and plant pathology at The Ohio State University. Waller, who taught courses in economic botany and the history of botany, was Rodgers' neighbor, and in the living room of the Waller home Rodgers learned much about plants and botanists. On various occasions Rodgers, along with other University friends and faculty who shared some interest in the history of science, was an evening-dinner guest at the Rudolph home. I had the privilege of several invitations to these events and heard after-dinner stories that Rodgers told about writing his books.

Born 19 January 1900, Andrew Denny Rodgers III was the son of Andrew Denny, Jr. and Mary Louise (Price) Rodgers of Columbus. His paternal grandmother was the daughter of William Starling Sullivant (1803–1873), who became world-known for his studies in bryology during the mid-nineteenth century.[1,2] William's father, Lucas Sullivant (1765–1823), was a deputy government surveyor in the Virginia Military District. The elder Sullivant founded the towns of Franklinton and Columbus at the junction of the Olentangy and Scioto Rivers in central Ohio.

---

[1] Andrew Denny Rodgers III. 1940. *"Noble Fellow." William Starling Sullivant.* With a compilation of the new species of mosses and liverworts described by William S. Sullivant and those described by William S. Sullivant and Leo Lesquereux, prepared by Richard T. Wareham of The Ohio State University. With a foreword by Dr. Adolph E. Waller. G. P. Putnam's Sons, New York. xxii, 361 pp. [Facsimile ed., published by Hafner Publishing Co., New York, 1968].

[2] Ronald L. Stuckey and Marvin L. Roberts. 1991. Frontier Botanist: William Starling Sullivant's Flowering-Plant Botany of Ohio (1830–1850). *Sida, Botanical Miscellany* No. 6. x, 65 pp.

Rodgers received his A.B. degree from nearby Ohio Wesleyan University (1922), attained his L.L.B. degree from The Ohio State University (1925), and took graduate courses in speech at Northwestern University (1933–1935). He taught history at Harding High School in Marion, Ohio (1924–1925) and then practiced law in the firm of Henderson, Burr, Randall, and Porter of Columbus (1926–1933) before turning to writing poetry and articles about his great-great grandfather, Lucas Sullivant.[3,4]

In the after-dinner talk sessions, Rodgers related how during his days as a practicing lawyer, he would steal time from his work while at his desk writing poetry and stories concerning the history of Columbus. At that time, he began to realize that the legal profession was not for him, and that what he really wanted to be was a writer. Upon leaving the law firm, Rodgers enrolled in graduate courses at Northwestern University to study speech, drama, English, and writing. There he received much encouragement from his professors for a practical career in that field of endeavor. Still not satisfied with his life and work there, he left the University and traveled throughout the world.

In the mid-1930s Rodgers was beset with various personal problems, among them being difficulties with his marriage and major depression followed by hospitalization for some months. While recovering from these situations, Rodgers began to study the family history and discovered that his great grandfather, William Starling Sullivant, was considered by botanists as the world's foremost authority on the science of bryology for the nineteenth century. Sullivant had been associated with the great American botanists John Torrey and Asa Gray. For Rodgers to write a definitive biography of his relative and the botanist Sullivant, he had to become acquainted with the current-day botanists who could provide him with information. He obtained considerable help from his botanist-neighbor, Adolph Waller; ecologist-algologist Edgar N. Transeau (1895–1960), who was chairman of the Department of Botany and Plant Pathology at The Ohio State University; and Richard T. Wareham (b. 1905), the Department's bryologist. Later, as Rodgers began to write his various books, he sought the acquaintance of botanists, librarians, and others throughout the country. Writing these botanical history books was a full-time endeavor, and so the unanswered question was, how did he make a living during those years of writing? He occasionally mentioned that the Sullivant-Rodgers descendants depleted the family wealth and spent their inheritance income. Rodgers

[3] Ronald L. Stuckey and Emanuel D. Rudolph. 1981. Andrew Denny Rodgers III (1900–1981). *The Ohio Journal of Science* 81: 190–191. [See also: *Taxon* 30: 876–877. 1981].

[4] Ronald L. Stuckey, Emanuel D. Rudolph, [and others]. 1987. [Andrew Denny Rodgers III], pp. iv–v, 65–77 *In* Ronald L. Stuckey, comp., *First Botanists of Columbus: A Sullivant Family Scrapbook*. Compiled by Ronald L. Stuckey for Field Trip to Green Lawn Cemetery. 38th Annual AIBS Meeting, Columbus, Ohio, 9 August 1987. *Ronald L. Stuckey Compiled Publication Series*, No. 2. Zip Services, Columbus, Ohio. xiv, 86 pp.

apparently shared in that family wealth and could live the life of a full-time writer without income from employment.

National recognition of Rodgers' writing was soon to come following publication in 1952 of his seventh book. In 1956, he was awarded one of 50 certificates of merit presented to living botanists at the Golden Jubilee Celebration of the Botanical Society of America. The citation described him as a "unique figure on the American Literary Scene," and his biographies as "readable, scholarly, and authentic."[5] At the time of the Botanical Society's 50th anniversary, several of the merit award recipients wrote papers about their disciplines. These essays were published under the heading of "Special Papers" in the *American Journal of Botany* (1956, 1957) and collected in a book, *Fifty Years of Botany* (1958),[6] edited by bryologist William Campbell Steere (1907–1989). Rodgers[7] as author of one of these papers, wrote what he considered to be "a fair summary of . . . [his] practical experience in this field." He described what constitutes the history of science, maintained that it is to be factually written, and stated that the need continues to exist "for full-scale 'summaries and syntheses' predicated in science and in history, and in major instances more than biographies in presentation." Rodgers also was the recipient of three Ohioana Book Awards for outstanding biographical scholarship. These awards were for his books on *Sullivant* in 1941, *John Torrey* in 1942, and *Bailey* in 1950. All of them were given by the Martha Kinney Cooper Ohioana Library, Columbus. An honorary Doctor of Letters degree was conferred on him by Ohio Wesleyan University in 1946.

Rudolph was one of several members of the botany faculty who suggested that Rodgers be nominated for an Honorary Doctor of Letters by The Ohio State University. During 1965 Bernard S. Meyer (1901–1987), chairman of the Department of Botany and Plant Pathology, appointed a committee of Emanuel D. Rudolph, as chairman, Clarence E. Taft (1906–1986), and Adolph E. Waller, all members of the Department, to prepare the formal nomination and documentation to the Committee on Honorary Degrees. At the commencement of the University on 14 June 1966, Rodgers was awarded the honorary degree. The biographical

---

[5] Bernard S. Meyer. 1956. Awards of certificates of merit at the fiftieth Anniversary meeting [of the Botanical Society of America]. *American Journal of Botany* 43: 859–861.

[6] William Campbell Steere, ed. 1958. *Fifty Years of Botany: Golden Jubilee Volume of the Botanical Society of America*. McGraw-Hill Book Co., Inc., New York, Toronto, London. 1958. xiii, 638 pp. (See pp. 14–19).

[7] Andrew Denny Rodgers III. 1957. Botanical history. American Journal of Botany 44: 186–188. [Reprinted, pp. 467–470 *In* William Campbell Steere, ed., *Fifty Years of Botany: Golden Jubilee Volume of the Botanical Society of America*. McGraw-Hill Book Co., Inc., New York, Toronto, London. 1958. xiii, 638 pp.].

account of his achievements that appeared in the commencement program was written by Rudolph.

Rodgers was always enthusiastically helpful to others in their historical studies, and he was genuinely interested in learning what other potential writers were contemplating. In later years, he reluctantly would speak at informal meetings with students in Rudolph's history of biology class. These informal sessions were held in the evenings at the Rudolph home and library,[8] where Rodgers would intensively quiz each student to learn of his/her interest in the history of science. At these sessions he often shared some of his personal experiences in seeking historical material for his own writing. He would not speak to groups in public or present formal papers at professional meetings.

As part of the documentation for Rodgers' honorary degree, Rudolph[9] compiled a list of published reviews on his seven books. Following Rodgers' death at age 81 on 7 January 1981, Stuckey and Rudolph (1981)[3] published his obituary. Using information from some of the book reviews from Rudolph's files and data from Rodgers' personal papers and letters deposited at The Ohio Historical Society, Columbus, Stuckey prepared an evaluation of the reviews. He also described the methodology and philosophy that Rodgers used in writing his books. Rudolph contributed ideas and text to this unpublished manuscript,[b] and his portion is printed here. **(Paper No. 25)**.

[8] Ronald L. Stuckey 1995. Emanuel David Rudolph (1927–1992): Book collector and library friend. *The Michigan Botanist* 34: 4–23.

[9] Emanuel D. Rudolph 1965. ["Index to and Passages from reviews of Andrew Denny Rodgers III Seven Books in American Botany History"]. Handwritten, 12 pp. + Typewritten, 44 pp. & 3 additional pages. [Manuscript].

Andrew Denny Rodgers III, Historian of Botany

Andrew Denny Rodgers III
(1900–1981)

Richard A. Popham (1913–1988),
Andrew Denny Rodgers III,
Emanuel D. Rudolph (1927–1992),
in the Rudolph Home

Biographical Sketch, below

LIFE OF ANDREW DENNY RODGERS III

RODGERS, Andrew Denny, III (Litt.D.), Author, 30 S. Dawson Ave.. Bexley, Columbus 9, Ohio. BIOGRAPHY IN AMERICAN SCIENCES. b. Columbus, Ohio, Jan. 19, 1900. s. Andrew Denny, Jr. and Mary Louise (Price) R. Edn.: A.B., Ohio Wesleyan Univ., 1922; LL.B., Ohio State Univ., 1925; Grad. Study, Northwestern Univ., 1933-35. Hon. Deg.: Litt.D., Ohio Wesleyan Univ., 1946. m. Elinor Chase Rogers, Oct. 15, 1934. Exp.: Practice of Law, Henderson, Burr, Randall and Porter, Columbus, Ohio, 1926-33; Author and Teacher; Fellow, Internatl. Inst. of Arts and Letters. Mem.: Am. Assn. Adv. Sci.; Hist. of Sci. Soc.; Sigma Alpha Epsilon; Delta Sigma Rho; Phi Delta Phi; Toastmasters Soc. Author: "Noble Fellow" William Starling Sullivant, 1940; John Torrey, A Story of North Am. Botany, 1942; John Merle Coulter Missionary in Science, 1944; American Botany, 1873-1892, Decades of Transition, 1944; Liberty Hyde Bailey, A Story of American Plant Sciences, 1949; Bernhard Eduard Fernow. A Story of North American Forestry, 1951; Erwin Frink Smith, A Story of North American Plant Pathology, 1952; etc. Contbr. to: Am. Jour. of Botany; Various Historical Soc. Publications. Serv. Rec.: Students Army Trng. Corps, 1918. Gen. Int.: Recd. Cert. of Merit, Bot. Soc. of Am., 1956. Areas of Rsch.: Am. Botanical History; Reclamation and Irrigation Development of Lands West of 98th Meridian; Ohio History.

Paper No. 25 (1965–1981)

Andrew Denny Rodgers III (1900–1981), Focussing on His Seven Books in the History of American Botany

Biographies and bibliographies of individual botanists have long been a major form of recording botanical history. Botanical biographies and bibliographies appear in books or as small articles in botanical or other journals. In the 1940s and early 1950s, Andrew Denny Rodgers III (1900–1980), of Columbus, Ohio, a lawyer by profession, left the practice and later devoted full time to the writing of biographies of botanists. Seven biographies appeared from his pen.[a]

Initially, Rodgers' interest in family history prevailed, and his first book-length biography, "*Noble Fellow*" *William Starling Sullivant* (1940),[2] was described as "a remarkable book, centered about the accomplishments of a remarkable man" (Caldwell, 1945).[1] It presented in colorful accurate detail the life and accomplishments of his great grandfather. In addition, the book provided a narrative of family life and a history of Columbus with central Ohio serving as the background (Fig. 25.1). With encouragement and help from many botanists, Rodgers continued his writing, producing six additional biographically-centered books in the history of botany: *John Torrey: A Story of North American Botany* (1942),[3] *American Botany 1873–1892: Decades of Transition* (1944),[4] *John Merle Coulter: Missionary in Science* (1944),[5] *Liberty Hyde Bailey: A Story of American Plant Sciences* (1949),[6] *Bernhard Eduard Fernow: A Story of North American Forestry* (1951),[7] and *Erwin Frink Smith: A Story of North American Plant Pathology* (1952).[8,a]

Rodgers' books are unique in that not only does each volume focus on one individual, except for the *American Botany* book which centers on three, but also each develops the history of that facet of botany for which the individual was a leader. Presenting these life stories of American botanists has also provided a detailed account of the development of botanical science in the United States. As a result of Rodgers'

Frontispiece Credits: Andrew Denny Rodgers III, Historian of Botany. Andrew Denny Rodgers III. (Taken 14 June 1966, from Photo Archives, The Ohio State University, Columbus)

Richard A. Popham (1913–1988), Andrew Denny Rodgers III (1900–1981), and Emanuel D. Rudolph (1927–1992). (Taken in the Rudolph home by Ann Rudolph, 19 December 1971; in the Rudolph photograph collection)

Photograph of Andrew Denny Rodgers III, (1900–1981). (From *Le Bijou* [Yearbook of Ohio Wesleyan University, Delaware]. 1923) Biographical sketch. (From Robert C. Cook, ed., *Leaders in American Science* 5(1962–1963): 585. 1963)

“NOBLE FELLOW”

William Starling Sullivant

BY

ANDREW DENNY RODGERS III

WITH A COMPILATION OF THE NEW SPECIES OF MOSSES AND LIVERWORTS DESCRIBED BY WILLIAM S. SULLIVANT AND THOSE DESCRIBED BY WILLIAM S. SULLIVANT AND LEO LESQUEREUX, PREPARED BY RICHARD T. WAREHAM OF THE OHIO STATE UNIVERSITY

Andrew Denny Rodgers III

With a Foreword by

DR. ADOLPH E. WALLER

ILLUSTRATED

G. P. PUTNAM'S SONS
NEW YORK
1940

Fig. 25.1. Autographed title page of *“Noble Fellow” William Starling Sullivant* by Andrew Denny Rodgers III (1940).

(From the copy in the library of RLS)

William Starling Sullivant (1803–1873)

(In the Rudolph photograph collection)

writings more is known about American botanists than about any other group of American scientists. His books are readable, scholarly, and authentic, and they have become "classics" for their significant documentation and evaluation of American botanical history. Since 1965, five of Rodgers' books have been issued in facsimile editions by the Hafner Publishing Company, New York, and one of these five, the *Fernow* book, was reprinted again. The reprinting of his books attests to their value and their current need as reference works in the history of science. Rodgers' pioneering research laid the foundation for disciplined, scholarly study of the history of American botany.[a]

From discussions with Rodgers and examination of his working manuscripts and data books, some of which are in the Rare Books and Special Collections of The Ohio State University Libraries, his methods became clearer. He believed that a biographer of scientists had a duty to explain the person's involvement in scientific endeavors. Strictly personal situations or unsettled controversial scientific matters were to be avoided because they were believed to have little to do with the scientific accomplishment which was at the heart of the biography. Rodgers chose only major figures in various disciplines of plant science for study. He said that plant ecology as a discipline was not selected because it was still too controversial, and a biographical treatment of a major figure should wait until scientists agreed upon the "correct" approaches in that field. Thus he declined to write a biography of Frederic Clements (1874–1945) after doing some preliminary study on him.[b]

Rodgers worked in a legalistic manner. He gathered published papers and comments, unpublished manuscripts and letters, and interviewed individuals throughout the country. He first organized his materials in loose-leaf notebooks in a chronological sequence. When he was ready to put into words what he wanted to say, particularly in his later works, he wrote what might be considered an extended legal brief. Being careful to exclude his own evaluations, he began with the assumption that the scientist under consideration was truly outstanding and a giant in his field.[b]

Rodgers discovered and helped to preserve important collections of manuscript source materials which might otherwise have been lost. His work and efforts in this endeavor have set a fine example for the preservation of scientific material useful for historical purposes. His own archives are preserved at the Ohio Historical Society, Columbus.[b]

References Cited

1. Caldwell, Frank C. 1945. [Review of Andrew Denny Rodgers III's] *William S. Sullivant. The Ohio State Archaeological and Historical Quarterly* 54: 395–408. (Quote, p. 395).

2. Rodgers, Andrew Denny III. 1940. *"Noble Fellow." William Starling Sullivant.* With a compilation of the new species of mosses and liverworts described by William S. Sullivant and those described by William S. Sullivant and Leo Lesquereux, prepared by Richard T. Wareham of The Ohio State University. With a foreword by Dr. Adolph E. Waller. G. P. Putnam's Sons, New York. xxii, 361 pp. [Facsimile ed., published by Hafner Publishing Co., New York, 1968].

3. _____. 1942. *John Torrey. A Story of North American Botany.* Princeton University Press, Princeton, New Jersey; Humphrey Milford, London; Oxford University Press, [Oxford, England]. 352 pp. [Facsimile ed. published by Hafner Publishing Co., New York, 1965].

4. _____. 1944. *American Botany 1873–1892. Decades of Transition.* Princeton University Press, Princeton, New Jersey; Humphrey Milford, London; Oxford University Press, [Oxford, England]. 340 pp. [Facsimile ed. published by Hafner Publishing Co., New York, 1968].

5. _____. 1944. *John Merle Coulter. Missionary in Science.* Princeton University Press, Princeton, New Jersey. viii, 321 pp.

6. _____. 1949. *Liberty Hyde Bailey. A Story of American Plant Sciences.* Princeton University Press, Princeton, New Jersey. 506 pp. [Facsimile ed. published by Hafner Publishing Co., New York, 1965].

7. _____. 1951. *Bernhard Eduard Fernow. A Story of North American Forestry.* Princeton University Press, Princeton, New Jersey. 623 pp. [Facsimile ed. published by Hafner Publishing Co., New York, 1968; also reprinted by Forest History Society, Durham, North Carolina, 1991, with xii, 623 pp.].

8. _____. 1952. *Erwin Frink Smith. A Story of North American Plant Pathology. Memoirs of the American Philosophical Society* 31: [1]–x, 1–675.

Publication Notes

Compiled and edited from a paragraph[a] published 1974 in a paper by Stuckey and Rudolph, and a 1981 manuscript[b] on Andrew Denny Rodgers III written by Stuckey, of which three paragraphs were prepared by Rudolph.

[a] Stuckey, Ronald L. and Emanuel D. Rudolph. 1974. History of botany 1947–1972 with a bibliographic appendix, pp. [237]–261. *In* 25 years of botany. *Annals of the Missouri Botanical Garden* 61: [1]–261. [See pp. 242–243, 259].

[b] _____ and _____. 1981. "Andrew Denny Rodgers III (1900–1981): Historian of American Botany." Typewritten, by Stuckey, 12 pp.; handwritten, by Rudolph, 3 pp., typewritten, 1 p. [Manuscript].

Introduction No. 26

Scientific "Revolution" or "Skirmish?" (1980–1982)

In 1962, historian of science Thomas S. Kuhn proposed in his book *The Structure of Scientific Revolutions*[1] that some changes in science are the result of "revolutions" in thinking which are relatively sudden and disruptive of the "normal science" at that time. Since then, historians of science have been inclined to search for "revolutions" in all conceptual changes. An important influence on botanical science and upon students who were to become future botanists was a change that occurred in English speaking countries in the nineteenth century. That change was the replacement of the Linnaean artificial Sexual System of plant classification with Jussieu's Natural System.

Rudolph always seemed impressed with this "revolution" approach used to investigate the history of science. His attempt at applying the idea to the change from the artificial to the natural system of plant classification was not as "revolutionary" as expected. He[c] wrote:

> Two of America's major botanists of the early and mid-century, John Torrey and Asa Gray, were prime movers in accomplishing this [plant classification] change in the United States. Although the change perhaps is not a true "revolution" in the Kuhnian sense, it is, at least, an interesting and important "skirmish" in the progress toward a new professional field of botany in America: a change in approach that strongly influenced students who were to become botanists.

An abridged version of Rudolph's published paper[c] is printed here. **(Paper No. 26)**.

[1] Thomas S. Kuhn. 1962. *The Structure of Scientific Revolutions*. University of Chicago Press, Chicago, Illinois. xv, 172 pp.

John Torrey and the Natural System of Plant Classification

John Torrey (1796–1873)

AN

INTRODUCTION

TO THE

NATURAL SYSTEM OF BOTANY:

OR,

A SYSTEMATIC VIEW

OF THE

ORGANIZATION, NATURAL AFFINITIES, AND GEOGRAPHICAL DISTRIBUTION

OF THE WHOLE

VEGETABLE KINGDOM;

TOGETHER WITH THE USES OF THE MOST IMPORTANT SPECIES IN MEDICINE, THE ARTS, AND RURAL OR DOMESTIC ECONOMY.

BY JOHN LINDLEY, F. R. S., L. S., G. S.

MEMBER OF THE IMPERIAL ACADEMY NATURÆ CURIOSORUM;
OF THE BOTANICAL SOCIETY OF RATISBON; OF THE PHYSIOGRAPHICAL SOCIETY OF LUND;
OF THE HORTICULTURAL SOCIETY OF BERLIN;
HONORARY MEMBER OF THE LYCEUM OF NATURAL HISTORY OF NEW YORK, &c. &c.
AND PROFESSOR OF BOTANY IN THE UNIVERSITY OF LONDON.

FIRST AMERICAN EDITION, WITH AN APPENDIX.

BY JOHN TORREY, M. D.,

Professor of Chemistry and Botany in the College of Physicians and Surgeons in the City of New York, Member of the Wernerian Society of Edinburgh, Fellow of the Mineralogical Society of Jena, Member of the Physiographical Society of Lund, Sweden, &c., &c.

" C'est ainsi que sont formées les familles trés naturelles et généralement avouées. On extrait de tous les genres qui composent chacune d'elles les caractères communs à tous, sans excepter ceux qui n'appartiennent pas à la fructification, et la réunion de ces caractères communs constitue celui de la famille. Plus les resemblances sont nombreuses, plus les familles sont naturelles, et par suite le caractère général est plus chargé. En procédant ainsi, on parvient plus sûrement au but principal de la Science, qui est, non de nommer une plante, mais de connoître sa nature et son organization entière."—JUSSIEU.

NEW YORK:

G. & C. & H. CARVILL, 108, BROADWAY.

1831.

Title page from *An Introduction to the Natural System of Botany*, by John Lindley; first American edition by John Torrey (1831)

Paper No. 26 (1980–1982)

Introducing the Natural System of Plant Classification to Nineteenth Century American Students

The shift from the Sexual System of Linnaeus[4,5] to the Natural System of Jussieu[2] for classifying plants was a gradual one in the United States. In 1831, John Torrey (1796–1873) introduced the Natural System by preparing an American edition of John Lindley's *An Introduction to the Natural System of Botany*[3] issued in London the previous year. Torrey's young friend Asa Gray wrote the first American botanical textbook, *Elements of Botany* (1836),[1] that consistently followed the Natural System. Some other popular botanical textbook writers, Almira Hart Lincoln Phelps (1793–1884), Samuel Griswold Goodrich (1793–1860), and John Lee Comstock (1789–1858), continued to employ the artificial Linnaean Sexual System of classification well past the middle of the century. Alphonso Wood (1810–1881) became a serious competitor of Asa Gray (1810–1888) for the botanical student audience during the mid-century, partly because he supported the Natural System while continuing in his early editions to give the Linnaean System in outline. The supporters of the Sexual System contended, as had the prominent American teacher-naturalist Amos Eaton (1776–1842), that students learned to know plants easier and with greater confidence by using it.[a]

By the end of the century teaching the Sexual System of Linnaeus was all but forgotten. The earlier conflict for replacing it by the Natural System was accomplished by the previous generation and was no longer an issue. The newer issues were different ones, and they tended to depreciate taxonomy. In the period when the Natural System was being promoted, John Torrey had the support of The New York Lyceum of Natural History and its leader Dr. Samuel Latham Mitchill (1764–1831), and Torrey's disciple Asa Gray had the prestige of Harvard University and the strength of his popular textbooks and manual. Even before the advent of evolutionary theory, various Natural Systems were surely displacing the Artificial Systems and the skirmish for control of the young minds had been all but won.[b,c]

Frontispiece Credits: John Torrey and the Natural System of Classification of Plants. John Torrey (1796–1873). (From frontispiece, *Bulletin of the Torrey Botanical Club* 4(3): opp. p. 9. 1873)

Title page of *An Introduction to the Natural System of Botany*. (From the copy in the library of The Ohio State University)

References Cited

1. Gray, Asa. 1836. *Elements of Botany*. G. & C. Carvill & Co., New-York. xiv, 428 pp.

2. Jussieu, Antoine Laurent de. 1789. *Genera Plantarum Secundum Ordines Naturales Deposita*. Apud Viduam Herissant, Typographum, . . . et Theophilum Barrois, Parisii. 24, lxxii, 498 pp.

3. Lindley, John. 1831. *An Introduction to the Natural System of Botany: or, A Systematic View of the Organization, Natural Affinities, and Geographical Distribution of the Whole Vegetable Kingdom* . . . 1st American ed., with an appendix by John Torrey. G. & C. & H. Carvill, New York. lxxix, 393 pp.

4. Linnaeus, Carolus. 1737. *Genera Plantarum*. Apud Conradum Wishoff, Lugduni Batavorum. 384 pp.

5. _____. 1753. *Species Plantarum*. Impensis Laurentii Salvii, Holmiae. 2 vols. (1: 560 pp.; 2: pp. 561–1200 + index).

Publication Notes

Compiled and edited from an abstract[a] and published portions[c] of a manuscript[b] read 14 July 1980 at the meeting of the Historical Section, Botanical Society of America, University of British Columbia, Vancouver.

[a] 1980. The introduction of the natural system of classification to American students in the nineteenth century. *Botanical Society of America, Miscellaneous Series, Publication* 158: 98. [Abstract].

[b] 1980. "The Introduction of the Natural System of Classification to Nineteenth Century American Students." Typewritten, 15 pp. [Manuscript].

[c] 1982. The introduction of the natural system of classification of plants to nineteenth century American students. *Archives of Natural History* 10: 461–468. (See pp. 461, 465–466).

Introduction No. 27

Writing Botanical History (1947–1972)

Rudolph and Stuckey (Fig. 27.1) were invited to present a paper on the development of the history of botany as part of a symposium on "25 Years of Botany" for the years 1947 to 1972, sponsored by the Botanical Society of America at its 1972 annual meeting at The University of Minnesota, Minneapolis-St. Paul. The main portion of the paper with examples and the bibliographical citations was prepared by Stuckey, who read the paper at the meeting. The first, second, and fourth paragraphs, in addition to the paragraph preceding the subtitle, "The Future," were composed by Rudolph.

The symposium was held not only to commemorate the close interactions during the preceding 25 years between the Botanical Society of America (BSA) and the American Institute of Biological Sciences (AIBS), but also to present an historical summary of the major progress in botanical science during the period since the end of World War II. The entire symposium was published in 1974 with an introduction by Warren H. Wagner, Jr. (1920–2000). The contribution by Stuckey and Rudolph (1974),[b] titled "History of botany 1947–1972 with a bibliographic appendix," is the last of 11 papers in the compilation. Those paragraphs written by Rudolph have been identified and assembled into a separate contribution printed here. **(Paper No. 27)**.

Fig. 27.1. Ronald L. Stuckey (l.) and Emanuel D. Rudolph (r.) at a meeting of the Historical Section, Botanical Society of America, Richmond, Virginia.

(Taken 7 August 1990 by Dale Johnson for RLS; in the collection of RLS)

List of Journals with Papers on Botanical History Founded from 1951 to 1968

TABLE 1. Scientific journals that have been established within the past 25 years that contain papers or information on American botanical history—evaluations, perspectives, biographical sketches, exploration, book reviews, and bibliographies of individuals and/or disciplines.

Taxon International Association for Plant Taxonomy, Utrecht, Netherlands. J. Lanjouw, original editor; now edited by Frans A. Stafleu.	1951
Regnum Vegetabile: A Series of Publications for the use of Plant Taxonomists International Association for Plant Taxonomy, Utrecht, Netherlands. J. Lanjouw, original editor; now edited by Frans A. Stafleu.	1953
Plant Science Bulletin Botanical Society of America, Tampa, Florida. Harry J. Fuller, original editor; now edited by Robert W. Long.	1955
British Journal for the History of Science The British Society for the History of Science, 393 Cowby Road, Oxford. M. P. Crosland, editor.	1962
History of Science: An Annual Review of Literature, Research, and Teaching W. Heffer & Sons Ltd., Cambridge, England. A. D. Crombie and M. A. Hoskin, editors.	1962
The Michigan Botanist The Michigan Botanical Club, Ann Arbor, Michigan. Edward G. Voss, editor.	1962
Sida: Contributions to Botany Privately published, Dallas, Texas. The late Lloyd Shinners founder and editor; now edited by William F. Mahler and John W. Thieret.	1962
Huntia: A Yearbook of Botanical and Horticultural Bibliography The Hunt Botanical Library and Carnegie Institute of Technology [now called Hunt Institute for Botanical Documentation], Carnegie-Mellon Universiy, Pittsburgh, Pennsylvania. George H. M. Lawrence, original editor.	1964
Bibliography of the History of Medicine National Library of Medicine, Bethesda, Maryland.	1966
The Journal of the History of Biology Belknap Press of Harvard University, Cambridge, Massachusetts. Everett Mendelsohn, editor.	1968

Paper No. 27 (1972)

Commentary on Writing the History of Science, Including Botany

In a broad context, the endeavor of science, by its nature, is progressive and directive; building on what came before, discarding what is unsound in the light of new information, and finding new approaches or new theories of explanation. History on the other hand, is reportive and analytical; the good and the bad, the successful and the unsuccessful, as well as the progressive and the retrogressive all deserve consideration. The historian of science faces a dilemma. Should he make value judgments about the course of his history saying this line was "good" or that one "bad" or "retrogressive?" Too often it has been shown that the bias of a particular time changes, and what was thought "good" at one time is thought later, or by someone else, to be not as "good."

Further complications or interpretation arise in the breadth of approach made to the study of the history of science or to a particular science. For example, the history of botany, or mycology, or radiation botany, could be considered by its students as only consisting of the events and concepts within that science—what is termed internal history. On the other hand, the student might consider the events in the science and its concepts in the broader stage of human activities as they relate directly or indirectly to the history of his science— what is termed external history. How should the historian of botany interpret the changes that have taken place in his science? Perhaps these changes can be interpreted as a series of sudden overthrows of established theory, procedure, and persons by the new challengers; or as a series of revolutions as proposed by Thomas Kuhn.[6] Or should he interpret the historical framework as a more gradual evolutionary series of changes in conceptualization as is the more traditional opinion? These are the kinds of questions that concern the historian of science today. Many of these concerns have emerged during the past 25 years (1947–1972) during which time the history of science as a discipline has grown and become a viable independent field of study in this country. As bio-historians and botanists, we should be prepared to be examined by historians and philosophers of science, to be asked different questions about why we approach our studies in the manner we do, and to learn to appreciate the value of critical external as well as self examination. Our work can be interesting and exciting and at the same time it can be scholarly and enlightening.

Frontispiece Credits: List of Journals with Papers on Botanical History Founded from 1951 to 1968. Table 1. (From Stuckey and Rudolph, *Annals of the Missouri Botanical Garden* 6(1): 245. 1974)

Why should botany students and teachers, or for that matter any botanists, be concerned with the history of their science? Two important reasons suggest themselves: insight and understanding. Insight is the quality brought into the methods of explanation and discovery that comes only from the perspective of history. By looking to the past, one can develop a better sense of the present. For example, the very detailed historical analysis of cellular respiration written by Keilin,[5] one of the major workers on that problem, provides an exciting perspective for understanding the present and suggests ways that science develops. His approach makes valuable contributions to the teacher who must explain modern cellular respiration. The second reason, understanding, is particularly relevant to those who teach and learn. If one needs to understand some aspect of botany, one way that is useful is to trace the history of the development and understanding of that subject. It will start at the beginning in a situation where the facts are assembled and a hypothesis suggested. Upon this simple base, new experiments and observations are made, and step by step the superstructure is built. False starts, dead ends, and bursts of understanding occur, but all of these add to the excitement of understanding the current solution to the problem.

This "case history" method was used successfully by Conant.[1] His approach still has much to recommend to us today. Science students are frequently surprised at the difficulties that appear along the way to current understanding as much as they are with the bursts of genius. It is valuable for them to realize both of these aspects. People and how they think are interesting to other people. Thus, history can inject humanity into beginning botany that can be obtained in no other way. For example, in two beginning botany texts,[2,4] persons are introduced as individuals. The historians and philosophers of science add another reason for studying the history of botany. Their approach is to attempt to understand the way concepts have developed in relation to the basic philosophical framework of the science of botany and its broader relationship with the biological and physical sciences. Thus, the history of botany has an intrinsic interest all of its own. Perhaps more botanists should develop this sort of interest!

Within the 25-year period, 1947–1972, an amazing proliferation of materials—papers, journals, and books—for studying the history of botany has come into existence. These items, as well as newspapers, manuscripts, letters, or herbarium labels, may be reproduced easily by photocopy or microfilm. The use of the computer in storing and retrieving information should develop [and so it has by 2000] into an asset for the study of the history of botany. Today, facsimiles and reprints of classical works, or collections of reprints of important papers are available, often in reasonably priced paperback editions. Many of these editions contain valuable interpretative introductions and supplements. Another development that is no less striking is the appearance of inexpensive methods of fine color reproduction which makes possible better graphic presentation of original drawings of the past, such as those of Ewan[3] on William Bartram. The "tools" and literature available to the historian of botany today are indeed greater than those of the 1940s and 1950s.

References Cited

1. Conant, James Bryant (ed.). 1957. *Harvard Case Histories in Experimental Science.* Harvard University Press, Cambridge, Massachusetts. 2 vols. (xvi, 639 pp.).

2. Cronquist, Arthur. 1971. *Introductory Botany.* 2nd ed. Harper & Row, Publishers, New York. ix, 885 pp.

3. Ewan, Joseph. (ed.). 1968. William Bartram: Botanical and zoological drawings, 1756–1788, reproduced from the Fothergill album in the British Museum (Natural History). *Memoirs of the American Philosophical Society* 74: [i]–x, 1–180.

4. Jensen, William A. and Frank B. Salisbury. 1972. *Botany: An Ecological Approach.* Wadsworth Publishing Co., Belmont, California. xi, 748 pp.

5. Keilin, David. 1966. *The History of Cell Respiration and Cytochrome.* Prepared for publication by Joan Keilin. Cambridge University Press, Cambridge, [England]. xix, 416 pp.

6. Kuhn, Thomas. 1970. *The Structure of Scientific Revolutions.* 2nd ed. *International Encyclopedia of Unified Science*, 2, No. 2. University of Chicago Press, Chicago. xiii, 210 pp.

Publication Notes

Prepared from those portions written by Rudolph in an invited manuscript[a] authored by Stuckey and Rudolph and read 28 August 1972 by Stuckey in a Symposium on 25 Years of Botany (1947–1972), at the annual meeting of the Botanical Society of America, University of Minnesota, Minneapolis-St. Paul; paper published 1974.[b]

[a] Ronald L. Stuckey and Emanuel D. Rudolph. 1972. "Twenty-five Years of the History of Botany (1947–1972), with a Bibliographic Appendix." Typewritten, 18 pp. + bibliography and notes of many pages; also an earlier incomplete version. Handwritten and typewritten, 39 pp. [Manuscript].

[b] _____ and _____. 1974. History of botany 1947–1972 with a bibliographic appendix. *Annals of the Missouri Botanical Garden* 61(1): 237–261.

Introduction No. 28

The Manasseh Cutler Lecture (1985)

Rudolph was the second of five individuals invited to present the Manasseh Cutler Lecture at Marietta College in Ohio. The lecture series was a memorial to the Reverend Manasseh Cutler (1742–1823), clergyman, botanist, and colonizer, who throughout his life was minister of the Congregational Church in Ipswich, now Hamilton, Massachusetts. Active in many endeavors, Cutler may have been the principal author of the Ordinance of 1787, which provided for the orderly annexation of new territories and states and was therefore second in importance only to the Constitution itself in setting the direction of the new American nation. In botany, he was the first to write a flora of New England. Supported by the Dawes family, descendants of Cutler who had a long association with Marietta College, the lecture series recognized Cutler's colonizing efforts in the community where the College is located and his accomplishments in religion, botany, politics, education, and history. The series also was part of the bicentennial commemoration of the Northwest Ordinance.

Rudolph's Manasseh Cutler Lecture was presented 24 September 1985 and titled "From God's Order to Ordered Inspiration: Trans-Atlantic Connections for Classifying Plants." The lecture was summarized in *The Marcolian*, the campus newspaper, of which selected portions follow (Owens, 1985):[1]

> 'The way people look at nature is a microcosm of the way they look at their own world.' That is what Emanuel D. Rudolph . . . told about 65 people . . . at the second lecture of the Manasseh Cutler lecture series. Rudolph focussed on four time periods describing how the conditions during those periods affected the study of nature.

The four periods on which Rudolph spoke were: (1) the late 18th century of "ordered nature," with God's order there, it remained for humans to discover its intricacies; (2) the mid-19th century of "changing nature," when order and stability were challenged; (3) the early 20th century of "controlled nature," when people believed an increase in science and technology would lead to prosperity and better health, and with the development of genetics, understanding natural development helped illuminate relationships of organisms; and (4) late 20th century of "perplexing

[1] Dave Owens. 1985. Nature, society related. *The Marcolian* (Marietta College, Ohio), 27 September, p. 1.

Fig. 28.2. Manasseh Cutler (1742–1823).

(From Cutler, *Bulletin of the Lloyd Library of Botany, Pharmacy, and Materia Medica*, No. 7 *Reproduction Series*, No. 4. 1903; permission to reproduce photograph courtesy of the Lloyd Library, Cincinnati, Ohio)

The Manasseh Cutler
Lecture Series
Thomas Hall
September 24, 1985
7:30 p.m.

Welcome
Dr. Sherrill Cleland
President of the College

Remarks
C. Burr Dawes

Introduction of Speaker
Dr. Sherrill Cleland

Lecture
From God's Order to Ordered Inspiration: Trans-Atlantic Connections for Classifying Plants
Emanuel D. Rudolph
Professor of Botany
The Ohio State University

Reception
Public reception follows at the president's home.

Fig. 28.1. Program of the Manasseh Cutler Lecture Series Marietta College, Ohio, 24 September 1985.

(In the Rudolph collection)

nature," when considerable uncertainty existed due to political and social problems in a heavily populated world.

Rudolph stated that mankind has covered "a whole spectrum of ideas of how we shall look at diversity in nature." The dominant organization during each era "reflected how we looked at the world in general."[2] He intended to publish his manuscript[a] with additional documentation. Selected additional references have been added to the edited manuscript printed here. **(Paper No. 28)**.

[2] Anonymous. 1985. Scholar lectures on ABCs of Order. *Marietta World* (Marietta College, Ohio), Fall 1985, p. 2.

Two European Philosophers of Botany

John Ray (1627–1705)

Jean-Baptiste Lamarck (1744–1829)

Paper No. 28 (1985)

From God's Order to Ordered Inspiration: Trans-Atlantic Connections for Classifying Plants

I am delighted to have been asked to present the Second Manasseh Cutler Memorial Lecture at Marietta College. It gives me the opportunity to share with you some insights about how the world of nature has been perceived from the time of our honoree, Manasseh Cutler (1742–1823) in the eighteenth century, to the present. The way people look at the world of nature is a microcosm of the way they look at their own world and at the universe. Thus, my lecture is about nature and particularly the vegetable part, but nature as it is set in a human historical frame.

Carl Sandburg (1878–1967) once wrote, "The past is a bucket of ashes,"[17] which is another way of saying that history is dead stuff. It is the duty of those who study the past to rekindle the fire that made the ashes and to render those who are deceased vital, interactive individuals who participated in shaping their world. That is an exacting, yet very exciting exercise.

Two themes are interwoven in this discussion. The first is how people at different times attempted to understand the complexity of the kinds of organisms; the second examines interactions of people across space, particularly between America and Europe. This examination concentrates on four periods of time: The eighteenth century (the period of "ordered nature"), the mid-nineteenth century (the period of "changing nature"), the early twentieth century (the period of "controlled nature"), and contemporary time (the period of "perplexing nature").

Eighteenth Century: Ordered Nature

Let us imagine ourselves living in the eighteenth century. For most people it was a time of relative calm and stability. The majority of people belonged to a particular social class into which they had been born. But gradually changes were taking place in economic and social domains with the beginnings of industrialization, the consequent rise of city factories, and bourgeois industrialists. And, some dramatic

---

Frontispiece Credits: Two European Philosophers of Botany. John Ray (1627–1705). (Frontispiece opposite title page to Ray's work, *The Wisdom of God Manifested in the Works of the Creation. In Two Parts*. London. 1704; from Rudolph's copy in the Rare Book Room, Herbarium, The Ohio State University)

Jean-Baptiste Lamarck (1744–1829). (From a published photograph in an unidentified source; in the Rudolph photograph collection)

political changes were taking place, especially toward the end of the century with the American and French revolutions.

The way people understood the natural world was much influenced by the Newtonian world view, which at the beginning of the century had made the physical world subject to precise mathematically-defined laws, which had indicated to Isaac Newton a clock-work universe ordered by God. God's order was there; it remained for human beings to discover its intricacies. For those interested in living organisms, it was not so easy to fit them into an orderly law-governed system. New and different forms were continually being discovered in newly explored regions, and the systems used for describing and classifying the limited number of European organisms were strained beyond hope by these discoveries.

The world was viewed as God's world and all must be right with it. John Ray (1627–1705), the important British naturalist at the end of the seventeenth century, attempted to place all plants and animals into orderly systems. They were quite complicated and difficult for most people to use. He also wrote books on various subjects, including one entitled *The Wisdom of God Manifested in the Works of the Creation* (1704).[16] In it, he reviewed the complexities of nature from the stars to lowly plants, finding ample evidence of God's plan. For example, after considering the various wonderful structures of plants from their whole bodies to their microscopic parts, he declared: "There seems to be necessary some intelligent *plastick Nature* which may understand and regulate the whole Oeconomy of the Plant . . . ."[16] Thus, for Ray an intelligent God had arranged the world and everything's place in it. On the level of the working naturalists, who in those days were leisurely gentlemen of independent means, including physicians, or perhaps, clergymen, like Ray, a need existed for a better, more workable system for classifying organisms. It was a system that would allow for the ready identification of plants and animals, and the easy incorporation of new forms into it.

The Swede Carl Linnaeus (1707–1778) was the individual who was able to provide God's ordered system of nature for organisms. In 1735, while in Holland, where he had journeyed to obtain a medical degree, he published his master plan—his *Systema Naturae* (1964, reprint),[13] whose English translated title is, *The System of Nature, or, the Three Kingdoms of Nature, Systematically Proposed in Classes, Orders, Genera, and Species*. His classification of the major groups of animals did not differ greatly from those of earlier naturalists, particularly that of John Ray, because most of the animal groups have quite obvious distinguishing characteristics. On the other hand, his classification of the groups of plants, based as it was on the numbers of stamens and pistils in their flowers, was very new and effective. It separated plants into easily-determined groups even when they did not have other easily-observable differences.

The system of Linnaeus was that established by God at the Creation and was indeed an orderly one that allowed the placement of those forms created by God, which human beings were just finding in distant parts of the globe. One of his postulates stated:[13]

> As there are no new species; as like always gives birth to like; as one in each species was at the beginning of the progeny, it is necessary to attribute this progenitorial unity to some Omnipotent and Omniscient Being, namely *God*, whose work is called *Creation*. This is confirmed by the mechanism, the laws, principles, constitutions and sensations in every living individual.

Further he noted that, "Natural bodies are divided into *three kingdoms of nature*: viz. the mineral, vegetable, and animal kingdoms. *Minerals* grow; *Plants* grow and live; *Animals* grow, live, and have feeling. Thus the limits between these kingdoms are constituted."[13] He had to admit that his system for ordering plants, called the "Sexual System," because it was based upon the reproductive structures in flowers, was an artificial one and not the God-given completely natural system that yet remained to be discovered once all the plants of the world had been found. Linnaeus' Sexual System for plants had the great virtue that it was easy to learn and use, and allowed for the inclusion of new species as they were identified. In time, he devised detailed rules for describing and naming new organisms, with a binomial of genus and species names, and he published catalogues, which he kept revising, of all the known plants and animals placed into his systematic arrangements. Linnaeus' ordering of nature provided an additional stimulus for naturalists and others to find new organisms and to learn the names of organisms found in the places where they were living.

By the middle of the eighteenth century, Linnaeus had become the world's foremost naturalist. He was a popular teacher at Uppsala University where students came from all parts of the civilized world to study with him. He wrote numerous books, many of which were quickly translated from Latin or Swedish into various other languages including English. He received specimens of plants and animals from distant places, many collected by former students who explored the ends of the earth for him, quite a few losing their lives in the process. Never had one person been the namer and describer of so many new organisms. No wonder that Linnaeus was called the "Prince of Botanists." He was not a modest person, and in fact he seems to have relished the accolades that were showered upon him. In an autobiographical sketch written in the third person, he wrote of himself, "No person has ever had a more solid knowledge of all the three kingdoms of nature. [No person has ever] proved himself a greater botanist or zoologist."[15] By the time of his death in 1778, Linnaeus had become a Swedish national hero and an honored naturalist visited by many interested in nature, including several heads of state. His disciples and followers continued to use his classification systems and his rules. Laudatory biographies, even some for children, were written.

Now, let us move to America where visitors and residents were learning about their natural surroundings. The settlers in the southern colonies were the earliest to try to name the plants and animals, and in the eighteenth century it is not surprising that they used Linnaeus as their guide. By the mid-eighteenth century, John Clayton (1686–1773) had collected enough plants so that the Dutch botanist John Frederick Gronovius (1686–1762) was able to use them to publish a flora of Virginia, arranging them in the Linnean Sexual System. Many plants collected by Alexander Garden (1730–1792) in Carolina were sent to Linnaeus who described them in his writings. Thus were the beginnings of trans-Atlantic connections between native Americans and Europeans. In Philadelphia, John Bartram (1699–1777) whose curiosity led him to collect plants locally, and later to travel to other parts from northern New York and southern Canada to the tip of Florida, sent his plants to British correspondents who wanted them for their gardens; he also sent some of them to Linnaeus. In New York, the Governor of the colony, Cadwallader Colden (1688–1776), shipped plant and animal specimens to Linnaeus, and he conferred with Linnaeus' student Peter Kalm (1716–1779), who had been sent to the colonies to collect specimens and learn about natural products. Colden[5] wrote on 1 February 1750 to Linnaeus:

> Mr. Kalm being now on his return home, I cannot suffer him to go without acknowledging the favours which I have received from you, a person so highly esteemed and distinguished in the republic of letters, and to whom the world owes so much, for the vast acquisitions it has gained in Botany and all parts of Natural History, by your wonderful skill and indefatigable labours.

Native New Englanders were somewhat slower to collect and study their plants. However, in 1785, Manasseh Cutler brought forth the first flora of the region, "An account of some of the vegetable Productions, naturally growing in this Part of America, botanically arranged" (1785),[6] the first such flora published by an American in this country. He noted in his introduction:

> We have, perhaps, as great a variety of indigenous plants, as any country produces, in a similar climate. But a great part of them have never been so far noticed as to receive even a trivial name [that is the species name, for example]. *Canada* and the southern states, beside the attention paid to their productions by some of their own inhabitants, have been visited by eminent botanists from *Europe*. But a great part of that extensive tract of country, which lies between them, including several degrees of latitude, and exceedingly diversified in its surface and soil, seems still to remain unexplored.

Even though Linnaeus was by now dead, Cutler used the accepted Linnaean Sexual System in the arrangement of his own plants. He was in correspondence about botanical matters with the English physician, Jonathan Stokes (1755–1831), a friend of Linnaeus' son.

In 1787, Cutler, with several friends, visited the aged John Bartram at his home and garden south of Philadelphia. He[7] described the visit in this manner:

> I instantly entered on the subject of botany with as much familiarity as possible, and inquired after some rare plants which I had heard that he had. He presently got rid of his embarrassment, and soon became very sociable, which was more than I expected, from the character I had heard of the man. I found him to be a practical botanist, though he seemed to understand little of the theory.

Cutler continued to have a strong interest in natural history, and it is evident from his diaries and letters that he greatly regretted not being able to take more time for botanical studies from his very busy life as minister, lawyer, physician, statesman, and citizen. He wished to prepare for publication more about native plants and their medicinal virtues, as is evident from his notes and letters, some of which he sent to the Pennsylvanians, the Reverend Henry Muhlenberg (1753–1815) and Doctor Benjamin Smith Barton (1766–1815).

The eighteenth century pioneer naturalists of this country were anxious to expand the knowledge of the natural ordered world of God as it was on this continent. It almost became a religious mission, and certainly a patriotic duty, for them to discover and name the new organisms and to help fill in blanks in God's great chain of being. Discovering the natural, orderly, and stable world of nature was a noble eighteenth century mission.

Mid-nineteenth Century: Changing Nature

Moving ahead a half century, let us examine the world view of the middle of the nineteenth century. Order and stability were being seriously challenged. The Industrial Revolution had accelerated, resulting in greater social turmoil. As the British author Phyllis Deane[10] described it:

> Out of about 3 1/2 million people living in London and the main English and Welsh towns in 1851, only about a third had been born where they were then living. These early Victorian towns, impersonal, unwholesome and harshly competitive, were growing faster than the ability of the municipalities to cope with the sheer physical and social problems of urbanization. In every spare corner of the towns another house was squeezed. With largely unpaved streets and open sewers the living conditions of the poorer, most overcrowded sections of the main townships were appalling: epidemics of cholera or typhoid were uncomfortably frequent. Nor were they confined to the poor. Albert the Prince Consort died of typhoid fever in 1861.

The revolutionary spirit of 1848 on the European continent had further challenged the social and political order. As in the macrocosm of human life, so too there were disturbing challenges to the order of nature. Early in the century, the French naturalist, Jean Baptiste Lamarck (1744–1829) in his book on *Zoological Philosophy* (1809, reprint 1984),[12] had proposed that species had changed and were changing into new and different ones to fit into new environments. Almost all naturalists accepted his idea of evolution. However, the seeds of doubt had been sown, and as more fossils were discovered and studied, it became more difficult to account for the major changes of life that were shown in these past ages.

Charles Darwin (1809–1882), after his five year voyage around the world as naturalist on the British Royal Navy's ship *Beagle*, for 20 years gathered data to help explain the past and current variations in organisms. In 1858, as he was in the process of writing a manuscript of a massive several-volume work on his findings, he received from the naturalist Alfred Russel Wallace (1823–1913), a correspondent of his in the East Indies, the draft of a paper for comment and possible publication. It described independently, in brief, the outline of Darwin's Theory of Evolution concluding that changes in environment had acted and selected for survival only those individuals within a species from all the variations in its populations. In time, by sexual and geographical isolation of these surviving populations, new species would gradually arise. Wallace[22] wrote it this way:

> . . . There is a general principle in nature which will cause many *varieties* to survive the parent species, and to give rise to successive variations departing further and further from the original type . . .

Darwin had called the process natural selection. Wallace's paper, together with one of Darwin's unpublished essays written fully 19 years earlier and rewritten 14 years earlier, with part of a letter written to the American botanist Asa Gray two years before, was given at a meeting of the Linnean Society of London. Within a year, Darwin (1979, reprint) prepared a shorter book-version than he had originally planned, *On the Origin of Species by Means of Natural Selection, or the Preservation of Favoured Races in the Struggle for Life*.[9] Unlike a number of other works proposing evolution rather than stability of species and published since the time of Lamarck, Darwin's book was readily accepted as valid by many important naturalists. Among the earliest was Darwin's good friend and confidant, the British botanist Joseph Dalton Hooker (1817–1911) (Fig. 28.3), and his American correspondent, Asa Gray. Darwin[8] wrote to Gray thanking him for his favorable comments on the *Origin of Species* in a letter of 28 January 1860:

> Hooker has forwarded to me your letter to him; and I cannot express how deeply it has gratified me. To receive the approval of a man whom one has long sincerely respected, and whose judgment and knowledge are most universally admitted, is the highest reward an author can possibly wish for, and I thank you heartily for your most kind expressions.

In these ways the new evolutionary view of nature replaced the stable ordered view that had been the earlier standard. The new view was one of "changing nature." In his book Darwin[9] had eloquently expressed it thus:

> It is interesting to contemplate a[n] entangled bank, clothed with many plants of many kinds, with birds singing on the bushes, with various insects flitting about, and with worms crawling through the damp earth, and to reflect that

> these elaborately constructed forms, so different from each other, and dependent on each other in so complex a manner, have all been produced by laws acting around us. These laws, taken in the largest sense, being Growth with Reproduction; Inheritance which is almost implied by reproduction; Variability from the indirect and direct action of the external conditions of life, and from use and disuse; a Ratio of Increase so high as to lead to a Struggle for Life, and as a consequence to Natural Selection, entailing Divergence of Character and the Extinction of less-improved forms. Thus, from the war of nature, from famine and death, the most exalted object which we are capable of conceiving, namely, the production of the higher animals, directly follows. There is grandeur in this view of life, with its several powers, having been originally breathed into a few forms or into one; and that, whilst this planet has gone cycling on according to the fixed law of gravity, from so simple a beginning endless forms most beautiful and most wonderful have been, and are being, evolved.

The fashion soon began to look at classification of organisms, which had previously been imagined to be in stable and unchanging order, as fitting into a changing evolutionary system. Some botanists, such as Asa Gray, considered the evolutionary order one planned by a creator; however, the order could no longer be considered stationary. As Gray[11] wrote in 1879:

> . . . species are not "simple curiosities of nature," to be catalogued and described merely, but . . . they have a history, ... The fixity of species under this view is not absolute and universal, but relative.

Fig. 28.3. Joseph Dalton Hooker (1817–1911).

(From Mea Allen. 1967. *The Hookers of Kew, 1785–1911*. Michael Joseph, London. opp. page 192; permission to reproduce photograph courtesy of the Linnean Society of London)

Not only did evolution help to explain the great diversity of life, its distribution on earth, its past history, and its current variability, but for some it also helped to explain the human condition. "Social Darwinism" claimed that those people who were the most successful were so because they were the best fit in an evolutionary sense. This view tended to justify prevalent social structures. On the other hand, some claimed that by changing the human environment for the better, the human race would be evolutionarily improved. It is clear that the mid-nineteenth century was a period of radical change in thinking about nature and the classification of organisms. No longer was it possible to ignore the past history of life. No longer was it possible to believe that species were stable. Now systems of classification were said to be evolutionary, to reflect the descent of the organisms and their real natural relationships. From morphological evidence it was thought to be possible to reconstruct family trees with primitive types arranged below the higher advanced ones. Evidence also existed that strong trans-Atlantic connections occurred at this time among naturalists who were more nearly equals, rather than, as earlier, between leaders in Europe and followers in America.

Toward the end of the nineteenth century Darwinism had a serious flaw in its explanation for the mechanism of changes in organisms that become more and more evident. The problem was: How did inheritance produce the changes from one type into another, when it was believed that heredity was a blending of characteristics from both parents and any changes would be diluted by back crossing to an unchanged population. Darwin had postulated that numerous invisible gemmules existed regulating inheritance. These gemmules were formed in various parts of organisms all the time and they moved to the reproductive structures where they were to be passed on in the sex cells to the next generation. In his view, called pangenesis, these gemmules were influenced by the environment and somewhat changed by it. Gradually the changed gemmules were thought to cause changes in the inheritance of the organisms in which they were located. It was a hypothesis that had few advocates. To many it seemed to be another form of the Lamarckian notion that characteristics acquired during the life of an organism were passed on to their changed offspring. Progressive biologists followed August Weismann (1893)[23] who claimed that the environment could not effect any changes in the sex cells of an organism but could only effect changes in the body cells. Thus, bodily changes could not be passed on to offspring by any known mechanism. A mechanism for change was published by some, for example, by Hugo de Vries (1901–1903),[20] a Dutch botanist. His claim was that one species could change suddenly, or mutate, into a new self-perpetuating species in one generation. This sudden change went counter to the Darwinian natural-selection theory with its emphasis on very slow change. The end of the nineteenth century was a period of questioning the mechanism of heredity.

Early Twentieth Century: Controlled Nature

When the twentieth century arrived, the general feeling was one of optimism. The Industrial Revolution and the great technological and scientific accomplishments of the nineteenth century had improved the lot of more people. Alfred Russel Wallace (1970, reprint)[21] wrote in his book, *The Wonderful Century, Its Successes and Its Failures*, that the important scientific discoveries of the nineteenth century were greater than all those that had been made in all of the previous centuries. He did point out some failures, such as continued poverty and militarism. Most people found great hope for a better future. Increased science and technology would surely lead to prosperity and better health and general welfare for more people. Change was becoming viewed as normal.

While the evolutionary explanation was generally accepted, the nagging problem of explaining the hereditary mechanisms for evolution remained unsolved. The mutation proponent[20] Hugo de Vries (1848–1935), one of three researchers studying inheritance of plant characteristics in 1900, discovered a long ignored paper written in 1865 by an unknown monk, Gregor Mendel (1822–1884)[14] whose work explained de Vries' own experimental results. What Mendel had shown in his study of garden peas—not sweet peas as some textbooks have stated—was that hereditary characteristics were not blended in offspring, but rather they remained unmixed like discrete particles which could be masked by other particles; however, they would show up in mathematically predictable ratios in future generations. The controlling factors or particles were to be termed "genes." The mutation notion could be combined with Mendelian genetics by showing that these genes could somehow change suddenly. In a relatively few years, better basic genetic understanding and successful use of it in breeding plants and animals were possible.

A major proponent of the new Mendelian genetics was the English biologist William Bateson (1861–1926). He took on the task of actively defending the new genetics much in the manner that earlier Thomas Henry Huxley (1825–1895) had defended the new evolution theory. In England, Bateson had a special problem because the entrenched group of researchers on heredity, who were called the biometricians, led by Walter F. R. Weldon (1860–1906), were strongly committed to the blending explanation. Bateson wrote a slim book, *Mendel's Principles of Heredity*; *a Defence* (1902).[3] It was a strident call to arms for studying Mendelian principles. The book stated:

> In the study of Evolution progress had well-nigh stopped. The more vigorous, perhaps also the more prudent, had left this field of science to labour in others where the harvest is less precarious or the yield more immediate. Of those who remained some still struggled to push towards truth through the jungle of phenomena: Most were content supinely to rest on the great clearing Darwin made long since. Such was our state when two years ago it was suddenly discovered that an unknown man, Gregor Johann Mendel, had, alone, and unheeded, broken off from the rest—in the moment that Darwin was at work—and cut a way through. This is no mere metaphor, it is simple fact. Each of us who now looks at his own patch of work sees Mendel's clue running through it: whither that clue will lead, we dare not yet surmise. It was a moment of rejoicing, and they who had heard the news hastened to spread them and take the instant way. In this work, I am proud to have borne my little part. But every gospel must be preached to all alike. It will be heard by the Scribes, by the Pharisees, by Demetrius the Silversmith, and the rest. Not lightly do men let their occupation go; small, then, would be our wonder, did we find the established prophet unconvinced. Yet, is it from misgiving that Mendel had the truth, or merely from indifference, that no naturalist of repute, save Professor Weldon, has risen against him?

• • •

> In the world of knowledge we are accustomed to look for some strenuous effort to understand a new truth even in those who are indisposed to believe. It was therefore with a regret approaching to indignation that I read Professor Weldon's criticism. Were such a piece from the hand of a junior it might safely be neglected; but coming from Professor Weldon there was the danger—almost the certainty —that the small band of younger men who are thinking of research in this field would take it they had learnt the gist of Mendel, would imagine his teaching exposed by Professor Weldon, and look elsewhere for lines of work. . . . I venture to express the conviction, that if the facts now before us are carefully studied, it will become evident that the experimental study of heredity, pursued on the lines Mendel has made possible, is second to no branch of science in the certainty and magnitude of the results it offers.

Who was this fiery author, William Bateson? As a young zoologist, he had come from England to the United States in the summers of 1883 and 1884 to study embryology with Johns Hopkins University Professor William K. Brooks (1848–1908) at a field station on Chesapeake Bay. It was there that he became interested in variation and discontinuous evolution. After some travels to investigate variation in nature, he began to breed plants and animals to examine in detail what he termed discontinuous variation. A major book (1894)[2] described his and others' experimental results. Thus, in 1900 when Bateson read a paper by de Vries in which Mendel's explanation of inheritance was described, he was instantly ready to accept its implications. The discontinuity of variation was due to the re-emergence of traits after one or more generations which had remained unchanged, though masked, during that time.

Two young American investigators, George Harrison Shull (1874–1954) with the Carnegie Institution at Cold Spring Harbor in New York, and Edward Murray East (1879–1938) at the Illinois Agricultural Experiment Station in Urbana and subsequently at the Connecticut Agricultural Experiment Station in New Haven, independently were to take up the study of Mendelian inheritance in crop plants. Both inbred maize to get pure lines for hybridizing. When they crossed these poor quality, low-yielding inbreds, they were amazed to discover that the hybrid offspring were more robust than even the original varieties from which the inbreeds had been obtained. After much further research by Shull, East, and others, modern types of hybrid corn were developed which dramatically changed American agriculture in this century. Experimental genetics with its understanding of gene control proved to be applicable to improving the human condition. The success of the new genetics with its emphasis on the importance of gene mutation undercut the older Darwinian explanation of the mechanism of evolution change by gradual genetic change. The problem was resolved by showing that mutations in breeding populations would serve as raw materials for natural selection and could result in gradual changes in whole populations over time.

Because genetic control was the key to understanding natural variation, it is not surprising that it was soon used to illuminate the relationships of organisms in revised classification schemes. Genetic relationships were determined through breeding experiments, and in turn they became important in determining how closely related were different plants. The term used for this close relationship is "phylogenetic." Delimiting a species was now based upon genetic similarity and breeding success among individuals, and accordingly a species was now termed "biological." As Shull[18] put it:

> The intensive and extensive investigations which have been made in genetics and cytology have made it obvious that the reality of distinctions between different taxonomic forms rests almost exclusively on the reality and permanence of differences in the number and the constitutions of the chromosomes.

The chromosomes were the carriers of the genes in cells. In this way breeding experiments and chromosome analyses became essential in plant and animal taxonomy and systematics. These data were vital in the classifying and natural evolutionary arrangement of organisms.

The optimism of the beginning of the century seemed justified by the accomplishments of genetics in improving domesticated plants and animals. Some even started a eugenic movement to genetically improve humans. Nature was controlled by knowable genes which were the clues to understanding relationships. It was also a period of much trans-Atlantic scientific exchange. Now the scientists on both sides of the ocean were truly equals. It might be said that the genetic studies of Thomas Hunt Morgan and his school, as well as those of others such as George Harrison Shull and Edward Murray East, were to give the United States an edge in leadership.

When the Britisher William Bateson came to lecture in America in 1902, he was impressed with the keen interest in matters relating to genetics, something not true at home. Bateson[4] noted:

> At the train yesterday, many of the party arrived with their 'Mendel's Principles' in their hands! It has been 'Mendel, Mendel all the way' and I think a boom is beginning at last!

By the time of his next visit in 1907, he[4] could write to his wife:

> I never before felt what an exhilarating thing it is to speak to a really large and enthusiastic audience. It is dreadfully intoxicating! . . . The Americans are rather absurd in their hero-worship and one has continually to remember that they keep a constant procession of heroes on the march. But after years of snubbing, it is rather pleasant to get appreciation even though in an overdose. I go about like a queen-bee on a comb, and I should not be surprised to see some admirer backing out of my presence . . . .

Hugo de Vries also lectured in the United states in 1904, 1906, and 1912. He, too, was lionized. In such an atmosphere it is not surprising that genetics became the dominant approach to interpreting plant and animal relationships.

Contemporary Time: Perplexing Nature

In the 1950s and 1960s, several biologists, primarily Robert Sokal of the University of Kansas and Peter H. A. Sneath of the University of Leicester in England, originated another approach to classifying organisms. They used computers that placed equal values on all measurable characteristics of organisms and separated different species by finding gaps among groups of individuals with very similar characteristics. In this way numerical taxonomy was born. It claimed that the similarities of characteristics merely reflected gene expression in the phenotype or body cells of individuals, and that their approach did not necessarily indicate the past history or phylogeny of the group. Sokal and Sneath[19] stated it this way:

> The points we wish to make are that phylogenetic speculations should not be involved in the classificatory process and that any such speculations should be based upon a sound phenetic classification.

They also claimed that this phenetic classification, based on currently observable characteristics, could not yet be related to phylogeny. According to them, the existing phylogenetic systems were qualitative and intuitive, not quantitative and deductive, and therefore they were not good science. Almost all of those making classification systems at that time were phylogenetic classifiers. It is not surprising that a strong divergence of opinion developed.

In the past 20 years, those numerical taxonomists who maintain that it is really not possible to determine the history of living organisms and their evolutionary relationships from observable characteristics of living organisms have continued to make phenetic classifications. They claim that their classifications are solely artificial, but also useful for purposes of identification. On the other end of the spectrum of classifiers are those who continue to believe that the phylogenetic approach can be used and can result in systems that reflect evolutionary relationships. Their techniques of genetic experimentation have become more refined. For example, they have used computer programs to measure genetic differences and similarities and have matched gene molecules among species. They believe that their natural classifications composed of three dimensional trees are more and more precise reflections of evolutionary history. It might even be possible to include the chemicals from fossils in the information used in the phylogenetic systems. For the numerical taxonomists reading the direction of evolution still requires an intimate knowledge of the group of organisms for the phylogenetic system to be meaningful.

Somewhere between the two extremes of phenetic and phylogenetic approaches to classification is a group of taxonomists who follow a cladistic approach. A cladogram is a two-dimensional branching diagram that measures relationships among natural groups of individuals or "clades." It often uses many of the computer techniques of the numerical taxonomists, but changed so that different characteristics are given different weighting. Those characteristics thought to be much more evolutionarily significant are heavily weighted. The cladists claim to be able to generate cladograms which are true evolutionary diagrams because of their rules for determining the direction of evolution.

Because a continuum of opinion exists, at least three or more groups of taxonomists and systematists think they have the best system of classification. All of these investigators say that organisms are the result of evolutionary processes; but differ on how to use and interpret the available information about organisms. Taxonomy and systematics are now exciting fields precisely because of the heated discussions among the different schools of thought. An air of uncertainty exists about what is the true or correct path to understanding. "Perplexing nature" is a way of describing the current uncertainty about the truth of classification.

What about the trans-Atlantic connections in the contemporary scene? The phenetic taxonomists claim that they are the descendants of an eighteenth century French naturalist, Michel Adanson (1763; 1966, reprint),[1] who tried to use all kinds of equally weighted characteristics of organisms in his elaborate classification schemes. His very complicated scheme for the plant groups was ignored in his own time. His phenetic followers are now located on both sides of the Atlantic Ocean.

The phylogenetic, or traditional taxonomists can be viewed as the intellectual descendants of Darwin, by way of his Mendelian rebirth as so-called Neo-Darwinism earlier in this century. They too, are located on both sides of the Atlantic and include the largest number of taxonomists.

The major group of cladistic taxonomists has taken as its mentor a German taxonomist, Willi Hennig (1913–1976). In the 1950s and 1960s Hennig proposed methods and principles designed to help determine the direction of evolutionary changes within groups of organisms. He claimed to be able to construct truly evolutionary cladograms. Several years ago the cladists in this country founded the Willi Hennig Society which now has a worldwide membership and exerts an important influence on the field of systematics.

The attempts to make sense out of the diversity of life and to put living individuals into some meaningful order continue to be a challenge to biologists. This brief history has followed the tortuous path of these attempts since the eighteenth century. The God's "ordered nature" of Manasseh Cutler's time, became the "changing nature" of Darwin's time, which became "controlled nature" because of Mendelian genetics acted on by natural and human selection early in this century. These past discoveries have now contributed to the "perplexing nature" of our times because of differing views of how to interpret data, whether they be phenetic, phylogenetic, or cladistic. Ideas and understandings develop within a human context. The tenor of the times and the personalities of the individuals who generated them are crucial for their interpretation. A lily or a seaweed would be placed differently in the various schemes of classification over these two and a half centuries. Each scheme was devised by thinking, concerned persons who were trying to order their natural worlds. It is a quest that yet continues to beckon thoughtful scientists.

References Cited

1. Adanson, Michel. 1966. *Familles des Plantes*. Reprint [of the 1763 Paris ed.]. With an introduction by Frans A. Stafleu. *Historiae Naturalis Classica*, Tomus XLVI. J. Cramer, Lehre; Wheldon & Wesley, Codicote, Herts.; Stechert-Hafner Service Agency, Inc., New York. [xv], cccxxv, 640 pp.

2. Bateson, William. 1894. *Materials for the Study of Variation Treated with Especial Regard to Discontinuity in the Origin of Species*. Macmillan and Co., London and New York. xv, 598 pp.

3. _____. 1902. *Mendel's Principles of Heredity; A Defence*. With a Translation of Mendel's Original Papers on Hybridisation. At the University Press, Cambridge, [England]. xiv, 212 pp. (Quotes, pp. [v]–vi, ix).

4. _____. 1984. *William Bateson, F.R.S.: His Essays & Addresses*. With a memoir by Beatrice Bateson. [Reprint of the 1928 ed.]. *The History of Hereditarian Thought*. Garland Publishing, Inc., New York, London. ix, 473. (Quotes, pp. 81, 109–110).

5. Colden, Cadwallader. 1821. His Excellency Cadwallader Colden, Governor of New York, to Linnaeus, pp. 451–459. *In* Sir James Edward Smith, *A Selection of the Correspondence of Linnaeus and Other Naturalists from the Original Manuscripts, Vol. II*. Longman, Hurst, Rees, Orme & Brown, London. II vols. (Quote, Vol. II, p. 457). [Reprinted in 1978 by Arno Press, New York, in the series, *Biologists and Their World*].

6. Cutler, Manasseh. 1785. An account of some of the vegetable Productions, naturally growing in this Part of America, botanically arranged. *Memoirs of the American Academy of Arts and Sciences* 1: 396–493. (Quote, p. 396).

7. Cutler, William Parker and Julia Perkins Cutler. 1888. *Life Journals and Correspondence of Rev. Manasseh Cutler, LL.D.* Robert Clarke & Co., Cincinnati. II vols. (I: xii, 524 pp.; II: iv, 495 pp.). (Quote, vol. I, p. 273).

8. Darwin, Charles. 1958. *The Autobiography of Charles Darwin and Selected Letters*. Edited by Francis Darwin. [Facsimile of the 1892 American ed.]. Dover Publications Inc., New York. 365 pp. (Quote, p. 239).

9. _____. 1979. *On the Origin of Species*. A facsimile of the first edition with an introduction by Ernst Mayr. 5th printing. Harvard University Press, Cambridge, Massachusetts, and London. xxvii, ix, 513 pp. (Quote, pp. 489–490).

10. Deane, Phyllis. 1977. *The First Industrial Revolution.* [1977 printing of the 1965 ed.]. Cambridge University Press, Cambridge, London, New York, Melbourne. vii, 295 pp. (Quote, p. 260).

11. Gray, Asa. 1879. *Gray's Botanical Text-book.* 6th ed. Vol. I., *Structural Botany or Organography on the Basis of Morphology. to Which Is Added the Principles of Taxonomy and Phytogeography, and a Glossary of Botanical Terms.* Ivison, Blakeman & Co., New York and Chicago. (Quote, pp. 330–331).

12. Lamarck, Jean. 1984. *Zoological Philosophy: An Exposition with Regard to the Natural History of Animals.* Translated by Hugh Elliot. With introductory essays by David L. Hull and Richard W. Burkhardt, Jr. The University of Chicago Press, Chicago and London. lxvi, 453 pp. [Translation of Philosophie zoologique, published in Paris, 1809].

13. Linnaeus, Carolus. 1964. *Systema Naturae.* Facsimile of the first edition. With an introduction and a first English translation of the "Observationes" by Dr. M. S. J. Engel-Ledeboer and Dr. H. Engel. *History of the Dutch Classics on Science* VIII. B. De Graaf, Nieuwkoop, [The Netherlands]. 30, [19] pp. (Quotes, pp. 18, 19).

14. Mendel, Gregor. 1930. Appendix: Experiments in Plant-Hybridisation. Harvard University Press, Cambridge, Massachusetts. pp. 313–353. [Reprint of the translation made by the Royal Horticultural Society of London, 1901, with minor changes by William Bateson of Gregor Mendel's paper, Versuche über Pflanzenhybriden, published in *Verhandlungen des Naturforschenden Vereins in Brünn*, 4. 1865 (appeared in 1866)].

15. Pulteney, Richard. 1805. *A General View of the Writings of Linnaeus.* 2nd ed., with corrections, considerable additions, and memoirs of the author, by William George Maton. To Which Is Annexed the Diary of Linnaeus, Written by Himself. J. Mawman, London. (Quote, p. 565).

16. Ray, John. 1704. *The Wisdom of God Manifested in the Works of the Creation.* In Two Parts . . . 4th ed. corrected, and very much enlarg'd. Printed by J. B. for Sam. Smith, London. 2 Parts. (Quote, part I: p. 118).

17. Sandburg, Carl. 1946. Prairie, pp. 157–164. *In Poems of the Midwest: Containing Two Complete Volumes, Chicago Poems and Cornhuskers.* Illustrated by photographs selected by Elizabeth McCausland. Introduction by Lloyd Lewis. The Living Library. The World Publishing Co., Cleveland and New York. 267 pp. (Quote, p. 164).

18. Shull, George H. 1929. Point of view of genetics, pp. 1578–1586. *In* B. M. Duggar (ed.), *Proceedings of the International Congress of Plant Sciences, Ithaca, New York, August 16–23, 1926,* Vol. II. The Collegiate Press, George Banta Publishing Co., Menasha, Wisconsin. II vols. (I: viii, 951 pp.; II: vi pp. + pp. 953–1799). (Quote, vol. II: p. 1580).

19. Sokal, Robert R. and Peter H. A. Sneath. 1963. *Principles of Numerical Taxonomy.* W. H. Freeman and Co., San Francisco and London. xvi, 359 pp. (Quote, p. 8).

20. de Vries, Hugo. 1901–1903. *Die Mutationstheorie: Versuche und Beobachtungen über die Entstehung von Arten im Pflanzenreich*. Verlag von Veit & Comp., Leipzig. 2 vols. (1: xii, 648; 2: xiv, 752 pp.).

21. Wallace, Alfred Russel. 1970. *The Wonderful Century: Its Successes and Its Failures*. [Facsimile of the 1st (1898) London ed.]. Gregg International Publishers Limited, Westmead, Farnborough, England. xii, 400 pp.

22. _____1971. On the tendency of varieties to depart indefinitely from the original type, pp. 89–98. *In* Alfred Russel Wallace and Charles Darwin on species and varieties, pp. 83–98. *In* H. Lewis McKinney (ed.), *Lamarck to Darwin: Contributions to Evolutionary Biology 1809–1859*. Coronado Press, Lawrence, Kansas. 122 pp. [From: "Charles Darwin and Alfred Russel Wallace, 'On the Tendency of Species to Form Varieties; and on the Perpetuation of Varieties and Species by Natural Means of Selection,' in *Journal of the Proceedings of the Linnean Society, Zoology*, 3 (1859), 46–62. Read July 1, 1858"]. (Quote, p. 90).

23. Weismann, August. 1893. *The Germ-plasm: A Theory of Heredity*. Translated by W. Newton Parker and Harriet Rönnfeldt. *The Contemporary Science Series*. Charles Scribner's Sons, New York. xxiii, 477 pp.

Publication Notes

Revised and edited from an unpublished manuscript[a] for the invited presentation, 24 September 1985 as part of the Manasseh Cutler Lecture Series, Marietta College, Marietta, Ohio.

[a] 1985. "From God's Order to Ordered Inspiration: Trans-Atlantic Connections for Classifying Plants." Handwritten, outline, 2 pp., title page, 1 p., text, 67 pp.; typewritten, 22 pp. [Manuscript].

Introduction No. 29

Lichenology on the Antarctic Continent (1961–1969)

As a highly motivated research scientist, Rudolph[1] was the first botanist to undertake extensive field investigations on the vegetation of the Antarctic Continent during five field expeditions in the 1960s. His interest in the vegetation, particularly the lichens, was fostered while he was a graduate student under the advisorship of Carroll W. Dodge (1895–1988) at Washington University and the Missouri Botanical Garden, St. Louis. At Wellesley College he continued to develop research projects on Antarctic lichens. Because he had a non-tenure track appointment there and with no prospect of financial support or release time for research, Rudolph was encouraged by George Llano to seek a position at a large university committed to active on-going research.

Llano, a lichenologist by profession, was then with the administration of the National Science Foundation, Washington, D.C. (Fig. 29.1). He wanted to initiate a biology research program in Antarctica, and invited Rudolph to submit a research proposal on the taxonomy and ecology of lichens. Because of Rudolph's uncertain future at Wellesley College, he corresponded in the spring of 1960 with Richard P. Goldthwait (1911–1992), Director of the newly formed Institute of Polar Studies at The Ohio State University. By April 1961, Rudolph accepted an appointment at the University, with a three-quarter-time commitment in the Institute of Polar Studies and a one-quarter assignment in the Department of Botany and Plant Pathology, under the chairmanship of Bernard S. Meyer (1901–1987). With financial support from the National Science Foundation through the Institute, Rudolph began studies of the cryptogamic plants in Antarctica.

Rudolph went to Antarctica on five research expeditions, from one to three months each time during the austral summers of 1961–1962, 1962–1963, and 1963–1964. His headquarters were at Hallett Station in the Ross Sea Sector about 300 miles north of McMurdo Station (Figs. 29.2, 29.3, 29.4, 29.5). In 1966–1967, Rudolph supervised the Byrd Land Botanical Survey whose investigation collected cryptogamic plants along the Marie Byrd Land coast, a seven-mile portion of that region which previously had never been explored for plants. In 1968–1969, he had his headquarters at the McMurdo Station, where he studied the ecology and

[1] Ronald L. Stuckey. 1994. Emanuel D. Rudolph (1927–1992): Polar lichenologist and historian of botany. *The Bryologist* 97: 437–446.

Fig. 29.1. Lichenologists in Washington, D.C., l. to r. Rolf Santesson, Mason E. Hale, Jr. (1928–1990), Emanuel D. Rudolph (1927–1992), George A. Llano, Henry A. Imshaug. The event was organized by Llano in honor of Santesson, visiting from the Institute of Systematics, Uppsala University, Sweden.

(Taken 9 May 1966, by John A. Stevenson; in the Rudolph photograph collection. The photograph is reproduced in reverse in *The Bryologist* 69: 391. 1966)

distribution of lichens and obtained microclimatic data and information on substrate types for a comprehensive biological analysis of the lichens (Fig. 29.4).

The three contributions selected for this memorial volume are: (1) an unpublished manuscript on the ecology of lichens,[a] (2)information contained in a grant proposal for studies on the history of Antarctic biology and botany,[d] and (3) selections from two papers Rudolph published that are cited in the publication notes[b,c] **(Paper No. 29)**.

Fig. 29.2. Emanuel D. Rudolph at his desk, Hallett Station, Antarctica.

(Permission to reproduce photograph courtesy of Clifford Wetmore, taken by him during season 1961–1962; in the collection of RLS)

Fig. 29.3. Emanuel D. Rudolph at left end of table with scientists and support staff having Thanksgiving dinner at Hallett Station, Antarctica, 22 November 1962.

(From the Rudolph photograph collection)

Fig. 29.4. Emanuel D. Rudolph studying lichens using microclimatic equipment in the field on the Antarctic Continent (1963).

(Taken 1963 at Hallett Station, Antarctica; in the Rudolph photograph collection)

(Above, l.) Emanuel D. Rudolph conducting experiments on lichens in the field at Hallett Station, Antarctica (Taken January 1963; in the Rudolph photograph collection)

(Above, r.) At Christchurch, New Zealand, leaving for Antarctica; l. to r. Edmund Schofield, Vernon Ahmadjian, Clifford Wetmore (Taken October 1963; in the Rudolph photograph collection)

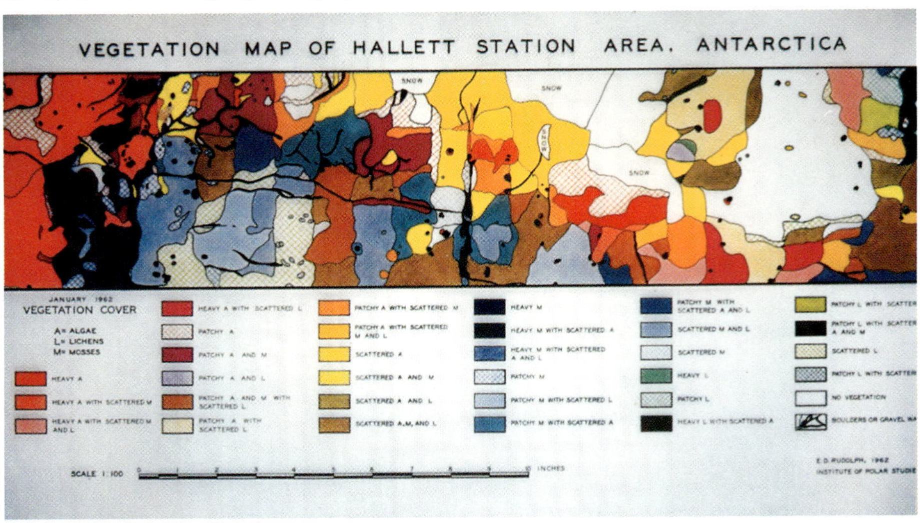

Vegetation map of Hallett Station area, Antarctica. Drawn 1962 in color by Emanuel D. Rudolph; the first vegetation map prepared for this region of the Antarctic Continent (Color photograph of the original map; in the Rudolph photograph collection)

At Right) Emanuel D. Rudolph among the polar books (Taken January 1988; in the Rudolph photograph collection)

(Above and Below) The room with polar books on the second floor of the Rudolph home at 384 Arcadia Avenue (Two photographs, taken by A. E. Spreitzer, August 1992; in the collection of RLS)

Two Antarctic Lichenologists, Dodge and Lamb

Carroll William Dodge (1895–1988)

lichen flora
of the ANTARCTIC CONTINENT and ADJACENT ISLANDS

CARROLL W. DODGE
Department of Botany
University of Vermont

PHOENIX PUBLISHING
Canaan, New Hampshire 03741

Title page from *Lichen Flora of the Antarctic Continent and Adjacent Islands*, by Carroll W. Dodge (1973)

Ivan Mackenzie Lamb (1911–1990)

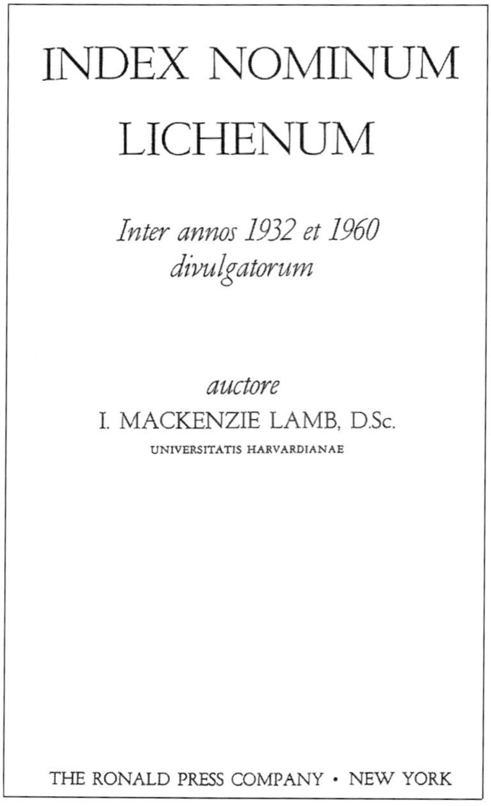

INDEX NOMINUM
LICHENUM

Inter annos 1932 et 1960
divulgatorum

auctore
I. MACKENZIE LAMB, D.Sc.
UNIVERSITATIS HARVARDIANAE

THE RONALD PRESS COMPANY • NEW YORK

Title page from
***Index Nominum Lichenum*,**
by I. Mackenzie Lamb (1963)

Paper No. 29 (1961–1969)

Commentary on the History of Biology and Lichenology in Antarctica

Antarctica, the only continent that has never had a native human population, is a large continent, the fifth largest, about the size of the United States, Mexico, and Central America. Located in the southern polar region it is 700 miles from the nearest land mass, the tip of South America, and over 2000 miles from any of the other continents.[b] As a result of its position over the South Pole and its generally high elevations, about 99.7% of the land is covered by an ice cap. The areas that remain uncovered, about 3,000 square miles, contain unique ecosystems with relatively few organisms. Compared to the relatively wet Arctic, which is a large ocean basin surrounded by heavily vegetated land; the Antarctic, because of its high ice cap and low precipitation, is a true polar desert.[d] In winter, because the continental land mass is almost entirely within the Antarctic Circle, virtual darkness prevails, and then in the summer a continual light period occurs. In addition, in winter the sea is frozen for hundreds of miles surrounding the continent. The combination of snow and ice cover and darkness makes it impossible for plants to grow during the winter. In spring, the sea-ice breaks and open water is present along the coast, where snow-free areas now exist and plants grow. Except for the few inland mountains, upon which some hardy lichens and mosses grow, the vegetation of Antarctica is restricted to the coastal region.[b]

Early Biological Expeditions

The history of the earliest Antarctic expeditions is available in many popular writings, as well as in a few technical historical treatments. The scientific results were usually published in extensive reports frequently appearing over many years.[d]

Frontispiece Credits: Two Antarctic Lichenologists, Dodge and Lamb. Carroll W. Dodge (1895–1988). (From Rudolph *Mycologia* 82: 161. 1990)

Title page of *Lichen Flora of the Antarctic Continent* . . . . (From Rudolph's copy in the Herbarium Library of The Ohio State University)

I. Mackenzie Lamb (1911–1990). (From the Hunt portrait collection; permission to reproduce photograph courtesy of Hunt Institute for Botanical Documentation, Carnegie Mellon University, Pittsburgh; also in Llano, *The Bryologist*, 94: 315–320)

Title page of *Index Nominum Lichenum* . . . . (From the copy in the J. N. Couch Biology Library–Botany Section, University of North Carolina, Chapel Hill)

Llano (1965)[20] has written a brief summary of early biological exploration of Antarctica.

> The circumnavigation of the Southern Ocean by James Cook [(1728–1779)] in 1772–75 presaged the era of south polar navigation. Johann R. Forster [(1729–1798)], naturalist to the expedition, brought back the first [E]mperor penguin and James Cook's report spoke of the abundance of seals on the island he named South Georgia. However, the southern fur seal, *Arctocephalus*, was already the object of a wide hunt by English sealers and, in 1790, United States vessels from Connecticut joined the English in the Falklands and South Georgia. By 1819, the South Shetland rookeries were discovered, bringing together more ships than have ever been seen since in these waters. These seal hunting voyages resulted in many notable discoveries and led, inevitably, to the sighting of Antarctica in 1820. James Eights, a naturalist from New York State, accompanied one such U.S. expedition in 1829-31 and became the first American scientist to visit the Antarctic regions. Fuller observation[s] and reports on the plants of the region were given by James D. Hooker, surgeon, who visited Antarctic waters with the British Erebus and Terror Expedition of 1839–43. In the following 60 years, little of biological interest from that part of the world appeared in the scientific literature, notwithstanding that during this time the southern fur seal was hunted almost to extinction and whaling in the Southern Ocean came to the fore.

Biological Studies in the Twentieth Century[d]

At the beginning of the twentieth century, land-based exploring expeditions began studies of organisms on the Antarctic continent. Many of the national expeditions in the early part of the century preserved collections of biological organisms and descriptions of them were later published by home-based scientists. Rarely were trained biologists included in such expeditions, whose main purpose was geographical exploration. This trend continued throughout most of the first half of the twentieth century. The accomplishments of those biologists who accompanied these expeditions have been published in the expeditionary reports. Today they form a sound and technical basis for extending inquiries into Antarctic biota.

In the past quarter of the twentieth century, the entire approach to the study of Antarctic biology has radically changed. This change is the result of an increase in detailed scientific investigations rather than survey exploration. In this period, international cooperation with scientific objectives became dominant. This cooperation began with the Norwegian-British-Swedish Expedition (1949–1952) and continued through the International Geophysical Year (1957–1958), which had a strong Antarctic component, but was not very biological. As a result of the latter Antarctic program and its success, twelve nations signed a treaty in 1959 that set the continent aside for cooperative scientific research. Many more biologists have gone to Antarctica and studied its ecosystems in the 16-year period between 1959–1975 than during the previous one hundred or more years. Because of its unique ecosystems and its recent availability for prolonged scientific study, and because of government support of scientific investigations at permanent stations, field camps, and on ships, Antarctica has surged ahead as a source of biological interest and discovery.

Books about recent Antarctic research activity are most often written by journalists or popular writers, military logistics personnel, civilian support personnel, and expedition leaders, but only rarely by the scientists themselves. A great need exists to obtain information about the research activities from the scientists who obtained the field collections, made the observations, and analyzed the data. Some information is available in the scientific literature, but in-depth data are needed to record and interpret the personal experiences and motivations of the scientists as they relate to their work in that distant inhospitable continent.

Because Antarctica is a continent isolated and geographically discrete, and because it supports unique terrestrial ecosystems, it is easy to isolate and identify the total effort in Antarctic terrestrial biology. Since 1957, all scientific programs have been supported by various governments, and since 1961 activity has been governed by international treaty with inspection.[29,30] Many of the original twelve signatories of the treaty and some other nations have conducted terrestrial biology research programs. Scientists from Argentina, Chile, Great Britain, Japan, the Soviet Union, and the United States, however, have contributed the majority of the investigators. The United States Antarctic Research Program established three biological laboratories, one each at McMurdo Station and Hallett Station in the New Zealand sector of the continent (Ross Dependency), and one at Palmer Station, Anvers Island, in the Antarctic Peninsula region.

The biological sciences in Antarctica have gone through three stages in the decade of the 1960s.[e] First, some descriptive approaches have become highly relevant for current and future ecosystem studies, as well as for the evaluation of biological resources. Second, ecosystem investigations with multidiscipline coordination have become the most effective means for understanding broad biological marine phenomena. Third and currently, a growing emphasis is placed on experimental research, and such efforts should assure the promise for an exponential development in useful information and knowledge. Understanding how organisms function has been one of the most fruitful attacks on all Antarctic biological problems. Advances in experimental biology are in small measure due to the modern facilities and equipment provided to physiologists, biochemists, microbiologists, and other scientists who work with live organisms in their natural environment at the McMurdo and Palmer stations.

Early Lichenological Studies

The plants of Antarctica are almost exclusively non-vascular cryptogams, except for two angiosperms on the Antarctic Peninsula near South America, below the Antarctic Circle. The flowering plants are *Colobanthus crassifolius* (d'Urv.) Hook., a member of the pink family, and *Deschampsia antarctica* Desv., a grass.[27] Although both species are in the Climatic Zone, their distribution includes many subantarctic islands and South America. In addition, a number of close relatives of these species are in New Zealand, Australia, and South Africa.[a,b] In Antarctica, these flowering plants grow in clumps in the wettest, most protected places.[b]

The lichens are the major vegetational element, and the flora contains about 350 species. In contrast to the flowering plants, they are in the least protected and driest habitats, distributed throughout the continent wherever rocks are exposed, as far south as 86°. Many lichen species are restricted entirely to Antarctica.[b]

As stated above, the collection and study of biological organisms from Antarctica began as early as the 1820s when sealers and whalers visited the Antarctic Peninsula near the southern tip of South America. Initially, Capt. William Napier collected one lichen in 1821, and James Eights, M.D. (1798–1882), of Albany, New York, obtained specimens of three lichens, a moss, and a grass in the South Shetland Islands as early as 1829.[4] Over 120 years ago, the British botanist Joseph D. Hooker (1817–1911) studied plants when he accompanied Sir James Clark Ross (1800–1862) on the famous Antarctic voyages of 1840–1843. In 1841, on the way to the Ross Ice Shelf at latitude 78°S, this expedition explored the Ross Sea, stopping at two small islands off the Victoria Land coast, but no plant collections were made and, indeed, no plants were seen. Hooker[9] wrote:

> The coast was lined with ice, but interspersed with fallen masses of stone, rocks and sand, and it was impossible to advance a yard into the interior; but far as [the] eye could reach and glasses could range, not a particle of vegetation existed.[c]

In the Antarctic Peninsula, Hooker had more botanical success. On Cockburn Island, off the northeastern tip of the Antarctic Peninsula, latitude 64°12'S, longitude 56°50'W. Hooker[30] collected cryptogamic plants only:

> The flora of Cockburn Island contains nineteen species, all belonging to the orders, *Mosses*, *Algae*, and *Lichens*. Twelve are terrestrial; three inhabit either fresh water or very moist ground; and four are confined to the surrounding Ocean. . . . The greatest amount of novelty is found here . . . for example, of the Mosses, two out of five are new. There are seven Algae, and two of them, or less than a third, are new. Of six species of Lichen, four are already described.[c]

Further plant collections were not made until fifty-two years later, when plants on the Antarctic continent proper were seen for the first time. In 1895 Carsten E. Borchgrevink (1864–1934) gathered some lichens at Cape Adare in Victoria Land at latitude 71°17'S,[1] and from then on most expeditions to the Antarctic continent collected plants. The men of these expeditions were not, for the most part, botanists, but it became apparent to the botanists who worked on the collections from the expeditions that a terrestrial flora of Antarctica was present in some of those areas where rocks, gravel, or sand are, in summer, snow-free.[c]

Taxonomic Studies

Taxonomic studies of Antarctic lichens have been conducted by a number of botanists since the beginning of the century, and this progress was summarized by Llano (1959, 1961).[17,18;b] Today, the botanical exploration of Antarctica continues in a systematic way to cover the many areas of the continent as yet unexplored. Identification and correlation of collections are still major problems, and a need exists for detailed taxonomic studies. A number of papers summarize the current state of knowledge of the various plant groups in Antarctica.[3,6,7,8,16,18,19,22,28,32;c] The continuing studies by Carroll W. Dodge (1895–1988)[g] of the University of Vermont, who prepared a flora for the continent, and I. Mackenzie Lamb (1911–1990)[h] of Harvard University, who studied the lichens of the Antarctic Peninsula and adjacent islands, provided descriptions of many lichens.[b]

[Thirty years have passed since the above paragraphs were written. Since then, the work of Dodge and of Lamb, now both deceased, is finished, and the following statements about their work on Antarctic lichens were written by Rudolph (1990)[25] and by Llano (1991),[21] respectively].[f]

Carroll W. Dodge and His Antarctic "Lichen Flora"

Carroll W. Dodge, whose lichenological studies were mainly taxonomic and floristic dealing with exotic floras, became the American authority on Antarctic and tropical lichens at a time when few Americans were studying lichens. All of the early significant collections of Antarctic lichens were studied by him, sometimes together with his graduate students (Rudolph, 1990).[25]

Dodge's interest in lichens began while he was an undergraduate student and continued through his life. Because of his efforts in securing original lichen species descriptions from the literature and examination of types, and using these to create keys, he was willing and able to name lichens from all parts of the world. He based his identifications on morphology and chemical tests. Species concepts were for him more narrow than those of some other lichenologists, but they were always based upon careful observation and detailed descriptions. He described as new to science over 650 species and varities of lichens, of which at least 100 were from the Antarctic continent. His studies of the Antarctic lichen flora were supported in the 1950s and 1960s by grants from the National Science Foundation, and they are listed in Rudolph (1990).[25] These studies resulted in a book-length treatment *Lichen Flora of the Antarctic Continent and Adjacent Islands* (1973). His extensive herbarium and library were donated several years before his death to the Farlow Library and Herbarium of Cryptogamic Botany at Harvard.[25]

I. Mackenzie Lamb and His Studies of Antarctic Lichens

I. Mackenzie Lamb had few peers in lichenology. He was a keen collector, sharp observer of lichen habitats, a careful and thorough albeit somewhat conservative systematist, and a scholar of historical materials. His travels brought him in contact with the leading lichenologists of the countries he visited, and he spoke several languages. His classic works on the Antarctic lichen flora (*Neuropogon,*[10,12] *Buellia, Rinodina, Usnea, Ramalina, Himantormia, Alectoria, Cornicularia*),[14] and the pyrenocarpous lichens[11] are reliable, and his monograph on the largely southern hemisphere genus *Placopsis* will remain a standard for knowledge of the genus for a long time to come (Llano, 1991).[21]

Lamb's work on the difficult genus *Stereocaulon* occupied much of his research time for over 25 years. Publication of the originally planned full scale monograph of the genus *Stereocaulon* was not possible due to economic reasons.[15] Only the keys and a 'conspectus' of the data obtained on chemistry of the specimens seen, distribution and nomenclature were published. All of the illustrations to the entire monograph, which was never published, were lost.[21]

Lamb's careful studies of the Antarctic lichen flora were never completed and published. The papers he did publish are listed by Llano (1991).[21] Lamb's manuscript notes on the Antarctic lichen flora began in the 1950s. They consist of typed and beautifully handwritten notes, complete with descriptions, annotated references to type and other material, detailed drawings, photographs, etc., for numerous species of 28 Antarctic genera contained in three volumes of 470 pages, many double-sided, entitled *Manuscript Notes on Antarctic Lichens,* Vols. I–III, now in the British Antarctic Survey archives.[21]

Lamb's book *Index Nominum Lichenum* (1963),[13] the title page of which is illustrated here, continues the coverage of Alexander Zahlbruckner's *Catalogue Lichenum Universalis* and provides new lichen names published between 1932 and 1960. According to the publisher 415 genera, 8,205 species, and 6,789 infraspecific taxa are cited. In his review of the book, lichenologist William Culberson stated that this compilation is a "landmark book."[2]

Ecological Studies

Since so few botanists were in Antarctica prior to the scientific programs of the 1960s, very little information had been amassed concerning lichen ecology. Siple

(1938)[26] described in broad terms the environment in which the lichens, that he collected on the Second Byrd Expedition, were growing. Lamb wrote on several associations of lichens on the Antarctic Peninsula. In recent years Llano (1959, 1961),[17,18] Rudolph (1963, 1967),[23,24] and Follmann (1964)[5] have made further ecological contributions. It appears from these data that the ecology of Antarctic lichens is as complex and potentially challenging as the ecology of Arctic, temperate, and tropical areas. Lichens grow in various microhabitats and their abundance appears to be related to various environmental factors such as temperature, water, nitrogenous materials from birds, wind, frost action on the substrata, and exposure. The published reports mentioned above are mainly descriptive, rather than experimental.[b]

The challenges and excitement of investigations in biology and lichenology in Antarctica will continue to lure scientists to this continent. Lichens, yet to be discovered and described, and experimental studies, yet to unravel ecological phenomena in this extreme environment, await future studies.

Fig. 29.5. Hallett Station, Antarctica.

(Permission to reproduce photograph courtesy of Clifford Wetmore, taken by him during season 1961–1962; in the collection of RLS)

References Cited

1. Bull, H. J. 1896. *The Cruise of the 'Antarctic' to the South Polar Regions*. Edward Arnold, London and New York. 243 pp.

2. Culberson, William. 1963. [Book Review: I. Mackenzie Lamb. *Index Nominum Lichenum*. The Ronald Press, New York. xi, 809 p.]. *The Bryologist* 66: 90–91.

3. Drouet, Francis. 1961. Chapter II. A brief review of the fresh-water algae of Antarctica, pp. 10–12. *In Science in Antarctica. The Life Sciences in Antarctica. Part I of a Report by the Committee on Polar Research.* Publication 839. National Academy of Sciences—National Research Council, Washington, D.C. xi, 162 pp.

4. Eights, James. 1833. Descriptions of a new crustaceous animal found on the shores of South Shetland Islands, with remarks on their natural history. *Transactions of the Albany Institute* 2: 53–69.

5. Follmann, G. 1964. Das Pflanzenleben der Antarktis. Botanische Ergebnisse der 17 chilenischen Antarktis-Expedition. *Die Umschau* 64: 100–103.

6. Greene, S. W. 1964. Problems and progress in Antarctic bryology, pp. [173]–179. *In Biologie Antarctique. Premier Symposium organisé par le S.C.A.R., Paris 2–8 septembre 1962.* Comptes-rendus. Actualités scientifiques et industrielles 1312. Hermann, Paris. 651 pp. [Comité de Rédaction: Robert Carrick, Martin Holdgate, [et] Jean Prévost].

7. _____. 1964. [Chapter] XIV. Plants of the land, pp. 240–253. *In* Sir Raymond Priestley, Raymond J. Adie, and G. de Q. Robin (eds.), *Antarctic Research. A Review of British Scientific Achievement in Antarctica.* Butterworths, London. xi, 360 pp.

8. _____ and D. M. Greene. 1963. Check list of the sub-Antarctic and Antarctic vascular flora. *Polar Record* 11: 411–418.

9. H[ooker], [Sir] W[illiam] J. 1843. Notes on the botany of H. M. Discovery Ships, Erebus and Terror in the Antarctic voyage with some account of the tussac grass of the Falkland Islands. *London J. Botany* 2: 247–329. (Quote, p. 270). [Also issued by H. Bailliere, London, with the title, *Notes on the Botany of the Antarctic Voyage Conducted by Captain James Clark Ross . . . in Her Majesty's Discovery Ships Erebus and Terror, with Observations on the Tussac Grass of the Falkland Islands.* 1843. 83 pp.].

10. Lamb, I. Mackenzie. 1939. A review of the genus *Neuropogon* (Nees & Flot.) Nyl. with special reference to the Antarctic species. *The Journal of the Linnean Society of London (Botany)* 52: 199–237.

11. _____. 1948. Antarctic pyrenocarp lichens. *Discovery Reports* 25: 1–30.

12. _____. 1948. Further data on the genus *Neuropogon. Lilloa* 14: 139–168.

13. _____. 1963. *Index Nominum Lichenum. Inter Annos 1932 et 1960 Divulgatorum.* The Ronald Press Company, New York. xi, 809 pp.

14. _____. 1964. *Antarctic lichens. I. The genera Usnea, Ramalina, Himantormia, Alectoria, Cornicularia. British Antarctic Survey, Scientific Reports* No. 38. British Antarctic Survey, London. 34 pp. + Appendices & 9 pls.

15. _____. 1977. A conspectus of the lichen genus *Stereocaulon* (Schreb.) Hoffm. *The Journal of the Hattori Botanical Laboratory* 43: 191–355.

16. Llano, G. A. 1956. Botanical research essential to a knowledge of Antarctica, pp. 124–133. *In* A. P. Crary [and others] (eds.), *Antarctica in the International Geophysical Year*. American Geophysical Union of the National Academy of Sciences—National Research Council, Publication No. 462. Washington, D.C.

17. [_____]. 1959. Antarctic plant life. *Transactions, American Geophysical Union* 40: 200–203.

18. _____. 1961. Chapter III. Status of lichenology in Antarctica, pp. 13–19. *In Science in Antarctica. The Life Sciences in Antarctica. Part I of a Report by the Committee on Polar Research.* Publication 839. National Academy of Sciences—National Research Council, Washington, D.C. xi, 162 pp.

19. _____. 1965. Chapter Thirteen. The flora of Antarctica, pp. 331–350. *In* Trevor Hatherton (ed.), *Antarctica. A New Zealand Antarctic Society Survey*. Frederick A. Praeger, Publishers, New York, Washington. xvi, 511 pp.

20. _____. 1965. United States Antarctic Research Program in Biology. *BioScience* 15: 254–258.

21. _____. 1991. I. Mackenzie Lamb, D. Sc. (Elke Mackenzie) (1911–1990). *The Bryologist* 94: 315–320.

22. Murray, J. 1963. Lichens from Cape Hallett area, Antarctica. *Transactions of the Royal Society of New Zealand. Botany* 2: 59–72.

23. Rudolph, Emanuel D. 1963. Vegetation of Hallett Station area, Victoria Land, Antarctica. *Ecology* 44: 585–586.

24. _____. 1967. Lichen ecology and microclimate studies at Cape Hallett, Antarctica, pp. 900–910. *In* S. W. Tromp and W. H. Weihe (eds.), *Biometeorology. Proceedings of the Third International Biometeorological Congress, Held at Pau, S. France, 1–7 September 1963, Organized by "The International Society of Biometeorology,"* Vol. 2, Pt. 2. Symposium Publications Division, Pergamon Press, Oxford.

25. _____. 1990. Carroll William Dodge, 1895–1988. *Mycologia* 82: 160–164.

26. Siple, Paul A. 1938. The Second Byrd Antarctic Expedition–Botany I. Ecology and geographical distribution. *Annals of the Missouri Botanical Garden* 25: 467–514.

27. Skottsberg, C. 1954. Antarctic flowering plants. *Botanisk Tidsskrift* 51: [330]–338.

28. Steere, William C. 1961. Chapter IV. A preliminary review of the bryophytes of Antarctica, pp. 20–33. *In Science in Antarctica. The Life Sciences in Antarctica. Part I of a Report by the Committee on Polar Research.* Publication 839. National Academy of Sciences—National Research Council, Washington, D.C. xi, 162 pp.

29. Taubenfeld, Howard J. 1961. A Treaty for Antarctica. *International Conciliation* 531: 243–322.

30. Turrill, W. B. 1953. *Pioneer Plant Geography. The Phytogeographical Researches of Sir Joseph Dalton Hooker. Lotsya-A Biological Miscellany.* Vol. 4. Martinus Nijhoff, The Hague. x, 267 pp. (Quote, p. 137).

31. United States Department of State. 1959. Twelve nations sign treaty guaranteeing nonmilitarization of Antarctica and freedom of scientific investigation. *Department of State Bulletin* 41: 911–917.

32. Wille, N. 1928. Süsswasseralgen von der Deutschen Südpolar-Expedition auf dem Schiff 'Gauss,' pp. 373–445. *In* Erich von Drygalski (ed.), *Deutsche Südpolar-Expedition 1901–1903 im Auftrage des Reichsministeriums des Innern,* VIII. Band. Walter de Gruyter & Co., Berlin und Leipzig.

Publication Notes

Compiled and edited from Rudolph's unpublished manuscript[a] on the ecology of lichens, two published papers[b,c] on the history of the vegetation, a grant proposal[d] on the history of terrestrial biology, an undated unpublished manuscript[e] on the continent's biology, and excerpts from the books by Dodge (1973)[g] and by Lamb (1966).[h]

[a] [1964]. ["The ecology of lichens during three austral summers in Antarctica, 1961–1964"]. Typewritten, 6 pp. [Manuscript].

[b] 1965. Antarctic lichens and vascular plants: Their significance. *BioScience* 15(4): 285–287. [+ References, Typewritten, 3 pp. Mimeograph].

[c] 1966. Terrestrial vegetation of Antarctica: Past and present studies, pp. 109–124. *In* J. C. F. Tedrow (ed.), *Antarctic Soils and Soil Forming Processes. Antarctic Research Series*, Vol. 8. American Geophysical Union of the National Academy of Sciences—National Research Council. [Washington, D.C.].

[d] 1976. A history of Antarctic terrestrial biology, 1948–1975. Typewritten, 29 pp. [Grant, not funded]. [see pp. 2–5].

[e] [undated]. ["On the Biology of Antarctica"]. Typewritten, 4 pp. [Manuscript].

[f] Editors' note in text.

[g] Carroll W. Dodge. 1973. *Lichen Flora of the Antarctic Continent and Adjacent Islands.* Phoenix Publishing, Canaan, New Hampshire. xxii, 399 pp.

[h] I. Mackenzie Lamb. 1966. *Index Nominum Lichenum. Inter Annos 1932 et 1960 Divulgatorum.* The Ronald Press Co., New York. xi, 809 pp.

Introduction No. 30

A Century of Plant Science (1991)

During 1991, the year of the centennial of The Ohio Academy of Science, Rudolph was invited by editor Lee Meserve of *The Ohio Journal of Science* to write a paper summarizing the major events in the past 100 years in the plant sciences (Fig. 30.1). As one of several invited papers covering the various disciplines represented in the Academy, Rudolph wrote a concise, thoughtful, and well-researched review of the major events in the plant sciences. The paper, "One Hundred Years of Plant Science 1891–1991, with an Ohio Perspective" (1991), emphasized those contributions made by Ohioans to Ohio botany. The portion of his paper[b] that pertains to the history of plant science at the national level is modified and reprinted here. **(Paper No. 30).**

Fig. 30.1. Emanuel D. Rudolph's friends who were past presidents of The Ohio Academy of Science and the term of their service. l. to r. Allen G. Noble (1989–1990), Richard P. Goldthwait (1958–1959) deceased, T. Richard Fisher (1980–1981) deceased, Charles M. Vaughan (1977–1978) deceased, Paul M. Daniel (1985–1986), John L. Crites (1981–1982), Jerry H. Hubschman (1978–1979), Andrew M. White (1987–1988), Robert E. Madaffer (1984–1985), Richard H. Mahard (1982–1983), Ralph J. Bernhagen (1975–1976) deceased.

(Taken 2 May 1992, by Ronald L. Stuckey, at the 101st Annual Meeting of The Ohio Academy of Science, Gardner Student Center, The University of Akron, Akron, Ohio; in the collection of RLS)

Rudolph's Publication Activity

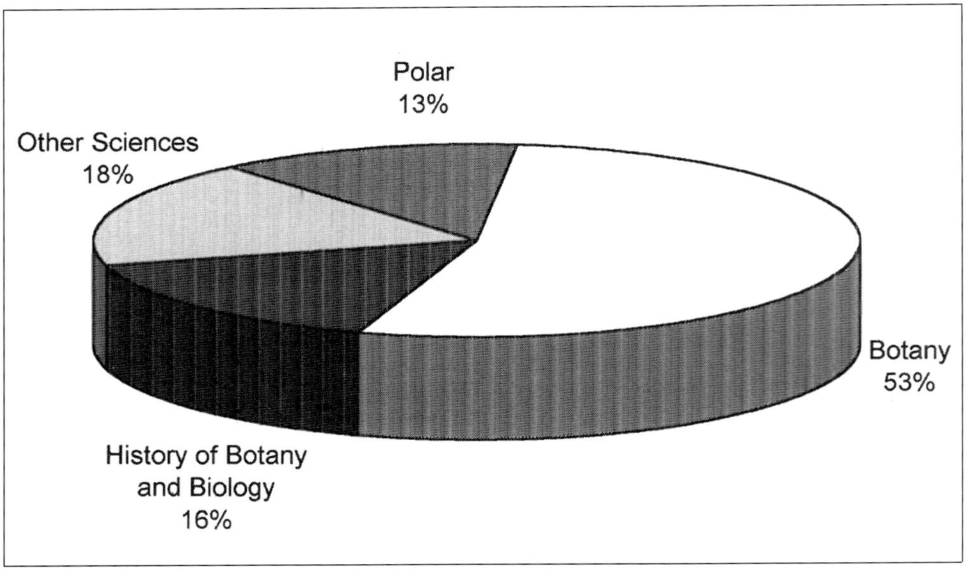

Kinds of books for which Rudolph wrote book reviews, based on a total of 195

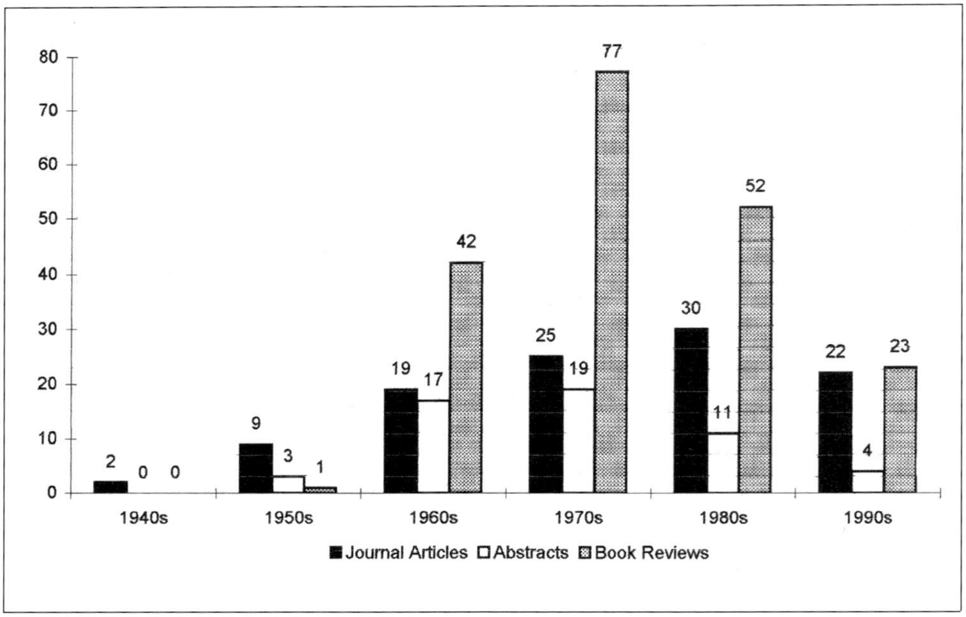

Numbers of papers grouped into three categories: 107 journal articles, 54 abstracts, 195 book reviews

Paper No. 30 (1991)

One Hundred Years of Plant Science (1891–1991)

The major revolution in twentieth century biology, the mid-century molecular biology revolution, has been well-documented.[8,13,18] Like the mid-nineteenth century Darwinian revolution, the growth of molecular biology has changed the way many biologists think. The predominance of molecular biology makes it easy to ignore the other major biological developments during this century that were less spectacular and less dramatic.

A primary concern for botanists a century ago was the nature of plant inheritance. The standard explanation, a complete blending of parental characteristics, while seemingly obvious, was of limited value in analyzing specific cases, or predicting their mechanisms. Several investigators—for example, Hugo de Vries (1848–1935) and William Bateson (1861–1926)—challenged this standard explanation. In 1900 the rediscovery of the particulate, or non-blending, explanation of Gregor Mendel (1822–1884) set in motion the development of a new discipline, modern genetics.[2,12,16] Quickly, plant scientists studied the mechanisms of Mendelian genetics in theoretical and practical ways. Once understood, these studies led to profound improvement of crop plants and to a better understanding of the mechanisms of speciation and evolution. For the first time, American biologists took the center of the international scientific stage. The coupling of earlier knowledge of cell structure with the new genetics, in placing emphasis on the chromosomes, was also a major accomplishment.

Ecology is a part of biology that has drastically changed in the past hundred years.[10,19] Plant ecologists in the late 1890s were concerned with precisely describing and comparing plant communities, and with explaining how individual plants managed to survive in their environments. Early in this century, American botanists began to explain the dynamic nature of plant communities, and later to model and quantify their changes. They became concerned with prehistoric changes, and with energy and substance flows in communities. They investigated and experimented with plant-animal relationships and with the ways that plant and animal populations evolved in particular environmental conditions. In recent years, plant ecologists have been studying environmental damage and human perturbations of plant communities and populations. Ecology is more experimental and predictive than it was a century ago.

Frontispiece Credits: Rudolph's Publication Activity. Two original graphs prepared by William R. Burk and David Romito.

Little-heralded but immensely important changes occurred in cytology in this century. Better optics, and then electron microscopy, drastically changed how biologists understand cells, tissues, and organs.[3,4] The ability to visualize the vast array of organelles and membrane systems in cells has provided opportunities for interpreting how structures can be wedded to functions. Dynamic biochemistry is now a vital part of cell biology. Plant cells have provided special challenges to those using experimental techniques and tissue cultures.

Plant physiology also has made major strides in the past hundred years. How growth and development are regulated by enzymes and hormones has been clarified,[1,21] and molecular biology has aided in the interpretation of development. The importance of trace elements and the mechanisms for water and food transport are better understood. The steps in the photosynthetic pathways are wonderfully clear and the energy transfers are becoming ever clearer. Recently, the use of transgenic mutants has moved plant physiology into a new arena.

Taxonomy and systematics of plants have had a rebirth of vigor and excitement in this century because of the incorporation of new kinds of data and the development of new methods of data analysis.[7,17] Systematics has been transformed by (1) introducing experimental methods for studying variation and speciation; (2) using secondary product chemistry, cytology, and macromolecules for comparing and contrasting organisms; and (3) involving phenetic and cladistic relationships with computers to analyze much larger data sets. New systems for coevolution have been generated. Major floras of large areas of the world have been or are being written with an urgency stimulated by the rapid loss of many natural areas and individual species.

Some branches of plant science, such as morphology, paleobotany, phycology, and mycology, had strong beginnings in the nineteenth century, and twentieth-century investigators built significantly on these foundations.[6,9,11,14,15,20] A better understanding of life cycles and developmental stages has resulted. Ultrastructural studies have made it possible to better relate stages of development in algae, fungi, and fossil plants. Recent interpretations for these groups of organisms have resulted in the removal of the fungi and some algae from the plant kingdom into other kingdoms.

The molecular biology revolution mentioned earlier has indeed transformed how biologists interpret plants. A better understanding of nucleic acids and other macromolecules in relation to the form, function, and regulation of plants has been rapid. Every aspect of plant biology has been influenced, from cytology and genetics to morphology and physiology, and even systematics and ecology.[5] Sophisticated instrumentation has become part of the major center of plant biological research in this century. It truly has been a century of biological revolutions.

References Cited

1. Barton-Wright, E. C. 1933. *Recent Advances in Plant Physiology*. 2nd Ed. J. and A. Churchill, London. x, 341 pp.

2. Bowler, Peter J. 1989. *The Mendelian Revolution, the Emergence of Hereditarian Concepts in Modern Science and Society*. Johns Hopkins University Press, Baltimore. viii, 207 pp.

3. Bradbury, S. 1967. *The Evolution of the Microscope*. Pergamon Press, Oxford. x, 357 pp.

4. _____ and G. L'E. Turner, (eds.). 1967. *Historical Aspects of Microscopy*. W. Heffer and Sons, Cambridge, [England]. 227 pp.

5. Crawford, Daniel J. 1990. *Plant Molecular Systematics, Macromolecular Approaches*. John Wiley and Sons, New York. xii, 388 pp.

6. Ewan, Joseph (ed.). 1969. *A Short History of Botany in the United States*. Hafner Publishing Co., New York. ix, 174 pp.

7. Huxley, Julian (ed.). 1940. *The New Systematics*. Oxford University Press, London. viii, 583 pp.

8. Judson, Horace Freeland. 1979. *The Eighth Day of Creation, Makers of the Revolution in Biology*. Simon and Schuster, New York. 686 pp.

9. Mayr, Ernst. 1982. *The Growth of Biological Thought; Diversity, Evolution, and Inheritance*. Harvard University Press, Cambridge, Massachusetts. ix, 974 pp.

10. McIntosh, Robert P. 1985. *The Background of Ecology, Concept and Theory*. Cambridge University Press, Cambridge, [England]. xiii, 383 pp.

11. Morton, A. G. 1981. *History of Botanical Science. An Account of the Development of Botany from Ancient Times to the Present Day*. Academic Press, London. xii, 474 pp.

12. Provine, William B. 1971. *The Origins of Theoretical Population Genetics*. University of Chicago Press, Chicago. xi, 201 pp.

13. Olby, Robert. 1974. *The Path to the Double Helix*. Macmillan, London. xxiii, 510 pp.

14. Sirks, M. J. and Conway Zirkle. 1964. *The Evolution of Biology*. Ronald Press, New York. v, 376 pp.

15. Steere, William Campbell (ed.). 1958. *Fifty Years of Botany. Golden Jubilee Volume of the Botanical Society of America*. McGraw-Hill Book Co., New York. xiii, 638 pp.

16. Stubbe, Hans. 1972. *History of Genetics, from Prehistoric Times to the Rediscovery of Mendel's Laws*. MIT Press, Cambridge, Massachusetts. xi, 356 pp.

17. Stuessy, Tod F. 1990. *Plant Taxonomy. The Systematic Evaluation of Comparative Data*. Columbia University Press, New York. xvii, 514 pp.

18. Tiley, N. A. 1983. *Discovering DNA. Meditations on Genetics and a History of the Science*. Van Nostrand Reinhold Co., New York. xxiii, 288 pp.

19. Tobey, Ronald C. 1981. *Saving the Prairies. The Life Cycle of the Founding School of American Plant Ecology, 1895–1955*. University of California Press, Berkeley. x, 315 pp.

20. Wagner, W. H., Jr. [and others]. 1974. 25 years of botany, 1947–1972. *Annals of the Missouri Botanical Garden* 61: [1]–261.

21. Weevers, T. 1949. *Fifty Years of Plant Physiology*. Schletema and Holkema, Amsterdam. xi, 308 pp.

Publication Notes

Revised and edited from a portion of Rudolph's manuscript[a] for an invited paper[b] commemorating the centennial anniversary of the founding of The Ohio Academy of Science and published 1991 in *The Ohio Journal of Science*. (Permission to reproduce Rudolph's published paper courtesy of the Academy)

[a] [1991]. "One Hundred Years of Plant Science 1891–1991, with an Ohio Perspective." Handwritten, 23 pp. + 61 pp. of analyses; typewritten, 13 pp. + 3 tables. [Manuscript].

[b] 1991. Centennial Note: One hundred years of plant science 1891–1991, with an Ohio perspective. *The Ohio Journal of Science* 91: 104–107. [See p. 104].

APPENDIX

RELATIONSHIPS

BETWEEN SCIENCE AND RELIGION

Talks in the Wellesley College Chapel (1955–1960)

Given by Emanuel D. Rudolph

Introduction Nos. 31–38

Talks Given by Emanuel D. Rudolph in the Wellesley College Chapel (1955–1960)

One of Emanuel D. Rudolph's generally unknown contributions was a series of talks on the relationships between science and religion that he presented in the Wellesley College Chapel from 1955–1960 (Fig. 31.1). The first seven talks are preserved as short typescripts and the eighth talk as a nine-page handwritten manuscript. These talks were identified by a 3x5 card that Rudolph attached to the file folder containing these papers. On this card in his handwriting style of very recent time, he wrote: "Given in the Wellesley College Chapel between 1955–1960." Treatments of this topic were not expected in Rudolph's archives, and by adding the card, he made clear to the finder that these essays were a part of his writings.

None of the typescripts had a date or title, the latter of which is provided here by the editors. The eight papers are arranged in the order in which they appeared in Rudolph's file, except the fifth one, which because of its subject matter, is placed first in this sequence. Presentation number 7, on "Taking Inventory of Belief" is missing its first two pages. The editors have retained the textual modifications that Rudolph had made to the quoted passages of these talks, and we have not confirmed their citations. These presentations provide an insight into the formative thinking of young Professor Rudolph, and reflect the manner in which he developed and lived his life as a scientist. **(Paper Nos. 31–38)**.

Science and Religion, Represented by Two Buildings at Wellesley College

Houghton Memorial Chapel

Science Building

Paper No. 31 (1955–1960)

Relationships between Science and Religion I. Complementary Components

It is often heard that science and religion are antagonistic, or that they are totally unrelated. It could be pointed out with substantiating historical evidence that scientific findings have many times clashed with religious dogma; however, it is my belief that the scientific attitude does not clash with the spirit of religion, but rather that they are complementary.

Science attempts to understand and also to control natural phenomena. In Psalm 19, the psalmist affirms the belief of our religious heritage that God had given man control over nature. Without this belief, science might well never have developed in our western civilization. It never did develop in the East where this belief is not in their religious tradition.

Since the beginnings of modern science, scientific thinking has complemented the spirit of religion in at least four ways:

1. Religious and scientific thinking both proclaim the orderliness of Nature. [The following two quotations appear to be taken from either Isaac Newton (1642–1727) or Albert Einstein (1879–1955) or both, but not made clear by Rudolph. Editors].

• • •

> The scientist is possessed by the sense of universal causation. The future, to him, is every whit as necessary and determined as the past. There is nothing divine about morality, it is a purely human affair. His religious feeling takes the form of a rapturous amazement at the harmony of natural law, which reveals an intelligence of such superiority that, compared with it, all the systematic thinking and acting of human beings is utterly insignificant reflection.
>
> Enough for me the mystery of the eternity of life, and the inkling of the marvelous structure of reality, together with the single-hearted endeavor to comprehend a portion, be it ever so tiny, of the reason that manifests itself in nature.

Frontispiece Credits: Science and Religion, Represented by Two Buildings at Wellesley College. Houghton Memorial Chapel on the campus of Wellesley College. (Permission to reproduce photograph courtesy of Wellesley College Archives)

Science Building on the campus of Wellesley College, Wellesley, Massachusetts. (From a photograph by Emanuel D. Rudolph; in the Rudolph collection) Note: The Rudolph photograph collection, although maintained separately, is part of the Ronald L. Stuckey photograph collection.

Fig. 31.1. Entrance to the Houghton Memorial Chapel on the campus of Wellesley College.

(Permission to reproduce photograph courtesy of Wellesley College Archives)

2. Religion and science are complementary in their acceptance of a universal truth. Both believe a "super guide" exists. As Claude Bernard (1813–1878), the father of modern physiology, has said:

> . . . I may say that our experimenter puts questions to nature; but that as soon as she speaks, he must hold his peace; he must note her answer, hear her out and in every case accept her decision. It has been said that the experimenter must force nature to unveil herself. Yes, the experimenter doubtless forces nature to unveil herself by attacking her with all manner of questions; he must never answer for her nor listen partially to her answers by taking, from the results of an experiment, only those which support or confirm his hypothesis. . . . An experimenter must not hold to his idea, except as a means of inviting an answer from nature. But he must submit his idea to nature and be ready to abandon, to alter or to supplant it, in accordance with what he learns from observing the phenomena which he has induced.

3. Science and religion are complementary when the willingness of a true scientist or a truly religious person willingly changes his/her thinking after a new scientific discovery is made, or when in his search he discovers a new aspect of God. Francis Bacon (1561–1626) had this to say in 1620:

> Others fear, from past example, lest motion and change in philosophy should terminate in an attack upon religion. . . . There are others who appear anxious lest there should be something discovered in the investigation of nature to overthrow, or at least shake, religion, particularly among the unlearned. These two apprehensions appear to resemble animal instinct, as if men were diffident, in the bottom of their minds and secret meditations, of the strength of religion and the empire of faith over the senses, and therefore feared that some danger awaited them from an inquiry into nature. But anyone who properly considers the subject will find natural philosophy to be, after the work of God, the surest remedy against superstition, and the most approved support of faith.

Alfred North Whitehead (1861–1947) expressed this value of the changing, growing mind in religion when he said:

> Religion is the vision of something which stands beyond, behind, and within the passing flux of immediate things; something which is real, and yet waiting to be realized; something which is remote possibility, and yet the greatest of present facts; something that gives meaning to all that passes, and yet eludes apprehension; something whose possession is the final good, and yet is beyond all reach; something which is the ultimate ideal, and the hopeless quest. . . . The worship of God is not a rule of safety—it is an adventure of the spirit, a flight after the unattainable. The death of religion comes with the repression of the high hope of adventure.

4. The scientist is a person who has certain moral and spiritual obligations. Scientists are more and more beginning to realize their responsibility to their fellow men. A statement from a report of the interim committee on the social aspects of science of the American Association for the Advancement of Science expresses this responsibility:

> This appears to be a critical time for review of the general state of science and its relation to society. We are now in the midst of a new and unprecedented scientific revolution which promises to bring about profound changes in the condition of human life. The forces and processes now coming under human control are beginning to match in size and intensity those of nature itself, and our total environment is now subject to human influence. In this situation it becomes imperative to determine that these new powers shall be used for the maximum human good, for, if the benefits to be derived from them are great, the possibility of harm is correspondingly serious. As scientists we are particularly concerned with determining how we should meet this situation, both as individuals and through our organizations.

Science and religion are complementary in their appreciation of the orderliness of nature; in their acceptance of universal truth; in their willingness to search for new insight into understanding; and in their belief in the responsibility of man as a human being.

Fig. 31.2. Whitin Observatory and Sage Hall, on the campus of Wellesley College.

(From a post card in the Rudolph Collection)

Paper No. 32 (1955–1960)

Relationships between Science and Religion II. Asking Important Questions

Last summer as I was driving along some of the back roads in the Middle West, I became aware of many hollyhocks growing wild along the roadsides. These now somewhat unfashionable garden flowers were originally brought by the early settlers of the central part of our country, much as the still earlier settlers of the East Coast had brought them to New England from Europe to remind them of their former homes. This desire to create familiar surroundings in a strange new place is probably as old as human society.

The now wild hollyhocks in Illinois, Missouri, Indiana, and Ohio tell a tale of man's attempt to preserve **THE** familiarity of his surroundings. In much the same way, our thoughts and ideas are like these hollyhocks brought by us from our past, perhaps from home or school, and kept and cultivated to provide a comfortable, familiar atmosphere for us when faced with the strangeness of a new problem. Usually we prefer to take the answers provided for us by our parents, teachers, ministers, and friends rather than really to think out the answers for ourselves. Instead of thinking about the answers with which we can truly live, we will gladly do all kinds of busy-work. In fact, it is often desirable to be so busy that we have no time for the perplexing and searching thought that would have to go into answering fundamental questions. As Sir Joshua Reynolds (1723–1792) once put it:

> A provision of endless apparatus, a bustle of infinite enquiry . . . employed to evade and shuffle off real labor . . . the real labor of thinking.

It is difficult to search ourselves for answers to fundamental problems when we can easily go on living with the hand-me-down answers of others. Do you remember the disturbing injunction in the Gospel of John, Chapter 3, "Except a man be born again, he cannot see the kingdom of God?" This concept of spiritual rebirth has been stated in another way by the well known psychoanalyst Erich Fromm (1902–1980), who said:

> Man after his physical birth has to go through a continuous process of birth. Emerging from the mother's womb is the first act of birth; from her breast is the second; from her arm the third. From here on the process of birth can stop; a person can develop into a socially adjusted and useful person and yet remain stillborn in a spiritual sense. If he is to develop into what he potentially is as a human being, he must continue to dissolve the primary ties of soil and blood. He must proceed from one act of separation to the next. He must give up certainty and defenses and take the jump into the act of commitment, concern, and love.

I have not meant to imply, nor do I think Dr. Fromm meant to imply, that the solutions of those who have asked the same important questions and provided answers are of no value for others. However, unless provided answers have been thought through and incorporated as a vital and committed part of one's thinking, they are of negative value, and prevent one from really attempting to solve the questions. One must be willing to accept wholly an answer after careful thought and then take a stand with it; be it popular or not. It is very likely that the answer to a spiritual problem will be different for different people, or that it will be different for the same person as he grows. What is important is the constant willingness to think about the questions, and try to find a direction.

I am reminded of a Hindu story. The story says:

> Two friends, while strolling about, happened to pass a place where the word of God was being preached. One of them said, 'Brother, let us go there for a while and hear the good words spoken.' The other replied, 'No friend, what is the use of hearing the word of God? Let us spend the time in yonder public house in amusement and pleasure.' The first one did not consent to this. He went to the place where the word of God was being read and began to hear it. The other went to the public house, but did not find the pleasure that he had anticipated there and was thinking all the while, 'Alas, why have I come here? How happy is my friend hearing all the while the sacred life and deeds of the Lord.' Thus he began to meditate on the Lord even though in a public house. The other man, who was hearing the word of God, also did not find pleasure in it. Sitting there, he began to blame himself, saying, 'Alas! Why did I not accompany my friend to the public house? What great pleasure he must be enjoying there at this moment.' The result was that he who was sitting where the word of God was preached meditated on the pleasure of the public house and acquired the fruit of the sin of going to the public house because of his bad thoughts; while the man who had gone to the public house acquired the merit of hearing the word of God because of his good heart. The spiritual gain of a person depends upon his sentiments and ideas, and proceeds from his heart and not from his visible actions.

We continually have the opportunity of being spiritually born again. All that we need is the willingness to ask the important questions anew and look for direction that has meaning for us in terms of our times.

Paper No. 33 (1955–1960)

Relationships between Science and Religion III. Habitual and Conscious Thinking

Several years ago, I heard one of the foremost interpreters of Zen Buddhism tell a story that, when you first hear it, is rather perplexing. That story has returned to me many times and caused me to think about its meaning.

A young man presented himself at a monastery and said that he wanted to become a monk and learn about Zen. He was accepted and after a year, he appeared before the master and said, "A year ago I came here to learn about Zen and never once have you sat down and explained Zen to me; I therefore want to leave." The master answered him, "You may leave if you wish, but you are wrong because I have instructed you in Zen. Each morning when you get up, I say good morning to you; that is Zen. Then you bring me my breakfast and I thank you; that is Zen. Later you bring me my tea and I thank you; that is Zen." Thus he enumerated many of the everyday tasks the doing of which he said was Zen.

One of the meanings that this story has for me is that what we do, particularly in our everyday acts as they affect others, makes us what we are as individuals and as children of God. For we are especially favored by being able to make the choice in our actions and thoughts.

As educated persons, we perhaps have a special responsibility. I believe that this responsibility is to think! To think about what we do. For as is written in that familiar passage from Micah (Chapter 6, verse 8), "He hath shewed thee, O man, what *is* good, and what doth the Lord require of thee, but to do justly, and to love mercy, and to walk humbly with thy God."

How can we tell what actions to take in order to do justly, unless we think? Justice is not simply something doled out by judges; it is something each one of us can make part of his life—justice for others as well as justice for ourselves. How can we do ourselves justice, not to mention others, if we fail to use our mental powers to our utmost ability?

Discussions are often heard on the need and value of weighing all aspects of a matter before making decisions or choices. This approach is what one should certainly do. Discussed below are two matters that do not seem to require thought or seem to really require choice. That is their danger. We do not even think about them, and that is where we miss the chance to do ourselves and others justice.

The first situation involves habit. On the surface one would say that habit as a way of acting is neither of itself good nor bad, but rather depends upon what type of action it is that is being performed. Just think of how long it would have taken to get dressed, eat breakfast, and walk here this morning if we had to think about each single movement we made in doing these acts. Habit can thus be good because it allows us to do routine actions without time-consuming thought. On the other hand, habit in thinking, where a continually fresh approach is needed for every situation, is not a good approach. By our very nature as human beings, we tend to form habits in thinking. We put information into pre-existing pigeon-holes. How many times have we met someone for the first time and within a few seconds placed him or her into some existing category in our mind, and thus prevented ourselves from giving that person a chance to make his or her own impression and have his or her actions unbiasedly appraised? I believe that unrealized habitual thinking is dangerous because it prevents us from acting justly.

The second pitfall is what might be called action according to some code of behavior. Codes of behavior or codes of honor are all man-made and thus subject to human fallibility. Good and bad codes exist depending upon the principles of the people who have made them. The Wellesley Code of Honor is without doubt a good one; those of certain teen-aged gangs without doubt are bad. What is important, however, is the danger of accepting such a code without thinking carefully about it simply because it is "the thing to do." We should continually be thinking about our actions. It is difficult to be different and not to follow without question the accepted pattern of behavior in one's own social circle, but remember: we are conscious, thinking individuals who have the capacity for choice. Think then about the ways prescribed by the current code of behavior of your group and constantly take stock of them. Are they the just ways that the Lord requires of you?

The way may not be easy, because thinking is difficult and perplexing and at times unpopular, but it is well worth the effort if some wisdom and understanding are gained. For as it says in Proverbs (Chapter 23, verse 23), "Buy the truth, and sell *it* not; also wisdom, and instruction, and understanding."

Beware then the trap of saying, "I can tell she is one of those crazy-mixed-up kids who cannot be counted on to do much;" or, "That is the way we do things around here and if you do not like it, it's just too bad." Let us take just a little time to think about what we do and say. If we take the advice of the master in the Zen monastery and think carefully about our everyday actions, particularly those that seem to require the least thought, we will be doing better what the Lord requires of us. We need to use our minds so that we may do that which is just in the sight of the Lord. We should perform our daily tasks with ever-increasing wisdom and understanding.

Paper No. 34 (1955–1960)

Relationships between Science and Religion IV. In Praise of Doubt

Doubt is what one feels about his understanding of religion and what it means to a person and how it impels a person to act. We are all conditioned in our thinking about religion by our experience. When we were children, religion and its meaning were simplified for us by our parents and teachers to make it understandable. It is natural that as we grow older, these simple answers to the problems that religion places before us are not sufficient. Here is where doubt enters.

We must then ask, What is the nature of religious experience? Although one could argue the point, I believe that it is ultimately an emotional, or call it spiritual, response. However, the problem comes in attempting to explain the meaning to others and then in carrying these convictions into action. This approach calls for one that is rational or logical. Religious experience is the putting of spiritual or mystical insight into human rational form.

It is the basic emotional response, really, that must be taken without proof or simply upon faith. No one can prove God's existence by pure reason to someone who does not want to believe in God. It is thus in working out one's own rational approach to religion that the doubt surfaces. I believe that is where the doubt should surface. We are continually changing as we grow older, and if our interpretation of religion and its meaning does not change with us, then religion will soon become a static, dust-covered ritual about which one thinks little and does even less to implement.

Faith is the emotional basis of religion upon which we must then build our changing, doubting, rational actions. Perhaps because I am a botanist, I am fond of those passages in Matthew, which pin-point faith as the basis of religion. It says in Matthew (Chapter 13, verses 31 and 32):

> The kingdom of heaven is like to a grain of mustard seed, which a man took, and sowed in his field:
>
> Which indeed is the least of all seeds: but when it is grown, it is the greatest among herbs, and becometh a tree, so that the birds of the air come and lodge in the branches thereof.

and later (Chapter 17, verse 20):

> Verily I say unto you, if ye have faith as a grain of mustard seed, ye shall say unto this mountain, Remove hence to yonder place; and it shall remove; and nothing shall be impossible unto you.

I like to think of the minute mustard seed, which to us is a pest because of its weedy habit, but to the Biblical peoples was an article of commerce used as an oil source. Given only one of these seeds, in a growing season thousands more seeds will be formed by the prolific plant, and these new seeds in several seasons could cover the area with mustard plants. So it is with faith; all one needs is a very little and soon it will multiply, because then one's actions will be fashioned by it.

We hear so much about the dangers of the overly emotional approach to religion, and it is a danger. Several weeks ago, I attended a talk given by a man who went to collect and study the hallucinogenic mushrooms eaten by some of the Indians in parts of Mexico. Perhaps you have read about these amazing plants which when eaten produce a beautiful vision. They transport one to exotic places and even make one feel that he is truly in heaven. Is this what we should ask from religion? If so, then when the drug companies start to sell this hallucination producing chemical, all we will have to do is take a pill a day to have our emotional religious experience. This approach is absurd. It is not religion any more than pure reason is. The purpose of religion is not personal happiness alone.

My contention this morning is that first we must accept God on faith. This approach is the first step, and once it has been achieved then the remainder comes in the constant growth, with doubt, of our attempts to carry out God's will. Religion must be sincere, and to be sincere it must be somewhat different for each one of us because no two of us are exactly alike. Then too, to be sincere it must keep pace with our intellectual growth.

The sincerity of religion from an unidentified Japanese writer is pertinent:

> Your true self is spirit and this is sincerity itself. Sincerity when it works naturally is truth. Truth, therefore, is a principle of your spirit. And since your spirit is part of the spirit of God, the principle of your spirit is the principle of Heaven. Thus, the truth of the heart of sincerity is the truth of God Himself, and sincerity which is revealed in the truth is communion with God. "Of all the thoughts and acts which come before me" says God, "if only there be a little sincerity, if only a bit of the Truth of Heaven, then quickly will I accept it, and quickly will I reward thee." For sincerity and God are one and inseparable . . . .

Let us then be sincere in our efforts to carry out the Truth of God by not seeking the easy way out, but rather by doubting our understanding of this truth and by doubting our methods of carrying it out in our own lives.

Paper No. 35 (1955–1960)

Relationships between Science and Religion V. On Being Creative

In a familiar poem by Edwin Arlington Robinson (1869–1935), Miniver Cheevy continually complains about his lot, but takes no action to improve it. Here is a man who refuses to face the problems of the world in which he is living, while dreaming about what he might have done if he had lived in some previous age.

We all have times when it is comforting to complain, and I believe that value exists in this complaining if it is constructive agitation leading to some action. As long as we believe that everything in our world is wrong and, like Miniver Cheevy, do nothing about it, or if we believe that everything is well with the world and with our actions, how can there be any real improvement? In the end, man's creative activity is the only way to improve our world.

Creativity is most certainly a very personal attribute of man, the one most difficult to comprehend. We in this age of individual and national insecurity are anxious to foster scientific creativity. Most thoughtful people feel that true scientific creativity cannot be bought with increased technological training, but must come from a general increase in creative activity.

A number of years ago, I attended a conference concerned with the nature of creativity. At this conference some of our most productive artists, writers, scientists, and humanists tried to explain the nature of the creative process. What really interested me was that whatever the nature of this process, they all agreed it was independent of field or discipline. In other words, creativity in science or art, in the opinion of these people, basically was a sudden inspirational insight. Naturally, those individuals trained in a particular discipline allow for insight into some problem needing interpretation in that field.

Albert Einstein (1879–1955) once said:

> Man tries to make for himself in the fashion that suits him best a simplified and intelligible picture of the world; he then tries to some extent to substitute this cosmos of his for the world of experience, and thus to overcome it. This is what the painter, the poet, the speculative philosopher and the natural scientist does, each in his own fashion. He makes this cosmos and its construction the pivot of his emotional life, in order to find in this way the peace and security which he cannot find in the narrow whirlpool of personal experience.

Thus, Einstein recognizes the dissatisfaction of the creative person with his "narrow whirlpool of personal experience". Without this dissatisfaction, he would not be led to speculate nor to reinterpret his environment.

You may say, all of this rhetoric is fine, but what can we, who are not Miltons, Bachs, Newtons, or Einsteins, do? I believe that each of us can do something as a responsible individual. This action rests upon our recognition of the value of constructive agitation. It means that we have a keen awareness of the limitations of our narrow individual circles and attempt to use to the utmost the creative powers within us, however great or small they may be, within us to improve our understanding of our world. As thinking individuals, we can also help to encourage others to be creative and try to be tolerant of the creative activities of other people even when we disagree with the results. Better yet, we can allow the creative activities of others to be stimuli for our thinking.

Many of you soon will be in positions in communities where you will help to determine, among other efforts, educational policies. You will then need to decide whether to take the view of continual non-constructive complaining, of happy complacency with inconsequential surface physical improvements, or of thoughtful and creative discontent. Let us hope that you will choose the last approach.

All of us need strength and help to use our God-given creative powers. What the Hellenistic thinker Philo said about God's creative power can be an inspiration for making our thoughts live. He wrote:

> It is the property of God to make . . . . He himself never ceases to rest . . . for whereas things produced by human arts when finished stand still and remain as they are, the products of Divine skill, when completed, begin again to move.

Paper No. 36 (1955–1960)

Relationships between Science and Religion VI. Responsibility

Responsibility seems especially timely to me for recently in New York, while at the meetings of the American Association for the Advancement of Science, I attended two important sessions: a symposium, "The Scientist's Role in the Community: New Responsibilities in the Nuclear Age," and an address by Sir Charles P. Snow (1905–1980) on "The Challenge to Science of World Conditions Today." Both of these sessions pointed out the special responsibilities of scientists to society. Sir Charles P. Snow seemed to believe strongly that the scientific attitude and training which so much emphasize the search for the truthful answer to problems, require that the scientist speak out truthfully from his scientific background about matters of government or society in general, as they relate to his special background. He must do this even though his statements may be contrary to the prevailing public opinion. This kind of responsibility is probably relevant to only a few of you, and I mention it only to show you how I decided to focus on the problem of the responsibility of the individual.

Scientists in the early part of our century characteristically avoided concern or responsibility for the applications that were made of scientific discoveries. "Science for Science's sake" was the old catch-phrase. Suddenly, not so long ago, as a result of the development of nuclear weapons and all their ramifications, scientists were confronted with the problem of moral responsibility. A scientist could easily slip out of the responsibility by saying that scientific discoveries in themselves were neither good nor bad. It is only in how they are applied by others that these values can be attached. Some scientists, perhaps because they had pangs of conscience, felt moral responsibility for the ways of using scientific discoveries.

Perhaps closer to our situation is the problem of our responsibilities as students and teachers in a college. Several years ago, I discovered the interesting appendix to C. P. Snow's novel *The Masters* (1951) in which is given a thumb-nail sketch of the development of the college system at Cambridge University, as noted from this passage:

> The students liked their life, but no one else did. Certainly not the townspeople; nor the students' parents; nor the teachers; nor possibly the more bookish and domesticated of the students themselves. So, almost from the origin of the university, there were attempts to get them out of their lonely lodgings into boarding-houses. Boarding-houses were cheaper, they could live four or five to a room and have meals in common—the salt meat, salt fish, beer and bread of a medieval Cambridge winter. It was possible to get a university teacher to live in the same house and keep an eye on them.

> These boarding-houses had nothing to do with teaching; the students just lodged there, and went off in the morning to schools. They were simply a sensible means of keeping those youths from the wilder excesses. Some of them were given money, rules, and became known as colleges, but their purpose remained the same.

American colleges are really not like the Cambridge or Oxford colleges, but rather are more like the universities in other parts of England and Europe. In his book *The Rise of Universities* (1923) Charles Homer Haskins (1870–1937) discusses the origins of universities during the Middle Ages, and he makes this comment:

> It is in institutions that the university tradition is most direct. First, the very name university, is an association of masters and scholars leading the common life of learning. Characteristic of the Middle Ages as such a corporation is, the modern world has found nothing to take its place. Next, the notion of a curriculum of study, definitely laid down as regards time and subjects, tested by examination and leading to a degree, as well as many of the degrees themselves—bachelor, as a stage toward the mastership, master, doctor, in arts, law, medicine, and theology. Then the faculties, four or more, with their deans, and higher officers such as chancellors and rectors, not to mention the college, where-ever the residential college still survives. The essentials of university organization are clear and unmistakable, and they have been handed down in unbroken continuity. They have lasted more than seven hundred years. Universities are at times criticized for their aloofness or their devotion to vocationalism, for being too easy or too severe, and drastic efforts have been made to reform them by abolishing entrance requirements or eliminating all that does not lead directly to bread and butter; but no substitute has been found for the university in its main business, the training of scholars and the maintenance of the tradition of learning and investigation. The glory of the medieval university, says Rashdall, was the consecration of Learning, and the glory and the vision had not yet perished from the earth. The medieval university it has been said was the school of the modern spirit.

Our responsibility, if we have any, as students or teachers, is thus to this tradition of learning and investigation. It is not to the social or living activities of the university. Our responsibility to the university tradition is worth frequent reflection.

Let us consider our responsibility as human beings. If we lack responsibilities as scientists in a nuclear age, or even if we believe that the responsibilities owed to the university tradition are not ours, we cannot deny our human nature and thus cannot evade responsibilities that are part of that nature. We have a limited time on this earth, and if we have any duties or responsibilities, they must relate to ourselves and to others who share our finiteness. Our actions, if they are not purely selfish, many believe relate us to the infinite force of God. Martin Buber (1878–1965) wrote:

> It is senseless to ask how far my action reaches, and where God's grace begins; there is no common border-line; what concerns me alone, before I

> bring something about, is my action, and what concerns me alone, when the action is successfully done, is God's grace. The one is no less real than the other, and neither is a part-cause. God and man do not divide the government of the world between them, man's action is enclosed in God's action, but it is still real action.

As Buber further pointed out:

> Our actions do have a meaning in the larger context of the totality of God's plan.

In his book *Flight to Arras* (1942), Antoine de Saint-Exupéry (1900–1944) made the following observations about responsibility of the individual:

> It was the contemplation of God that created men who were equal, for it was in God that they were equal. This equality possessed an unmistakable significance. For we cannot be equal except we be equal in something. The private and the captain are equal in the nation. Equality is a word devoid of meaning if nothing exists in which it can be expressed. This equality in the rights of God—rights that are inherent in the individual—forbade the putting of obstacles in the way of ascension of the individual; and I understand why God has chosen to adopt the individual as His path. But as this choice also implied the equality of the rights of God "over" the individual, it was clear that individuals were themselves subjected to common duties and to a common respect for law. As the manifestation of God, they were equal in their rights. As the servants of God, they were also equal in their duties. I understand finally why the love of God created men responsible for one another and gave them hope as a virtue. Since it made of each of them the ambassadors of the same God, in the hands of each rested the salvation of all. No man had the right to despair, since each was the messenger of a thing greater than himself. Despair was the reflection of God within oneself. The duty of hope was translatable thus, "And dost thou think thyself important? But thy despair is self conceit!"

In these terms our individual actions as they relate to ourselves and others must have meaning. Some people believe that if they cannot single-handedly save the world they cannot do anything worth while. I think that it is rather in everyday actions, done responsibly, that we can accomplish some good and that we can become prepared for the time when we might be called upon to participate in some greater activity. Buber has written:

> Existence will remain meaningless for you if you yourself do not penetrate into it with active love and if you do not in this way discover its meaning for yourself. Everything is waiting to be hallowed by you; it is waiting to be disclosed in its meaning and to be realized in it by you. For the sake of this your beginning, God created the world. He has drawn it out of himself so that you may bring it closer to him. Meet the world with the fullness of your being and you shall meet him. That he himself accepts from your hands what you have to give to the world is his mercy. If you wish to believe, love!

Let us find meaning in life through responsible actions pursued thoughtfully. We need to be ready to accept responsibilities, and fill our actions with love so that our actions will be done according to the will of God.

Fig. 36.1. Entrance to the College Library, on the campus of Wellesley College.

(From a post card in the Rudolph photograph collection)

Paper No. 37 (1955–1960)

Relationships between Science and Religion VII. Taking Inventory of Beliefs

One of our more profitable pursuits is asking: What are my fundamental beliefs? As you take inventory of your "belief structure," I hope you will find somewhere the conviction that doing a task, be it menial or exalted, is worth doing well. Far too often, it seems we have accepted the belief that expediency or reward are the only ends for which we will labor.

I make a plea for the belief that a job done with one's whole effort is the best and only way to do it. If this conviction is in one's "belief hierarchy," it makes the tasks of life more difficult, but also more challenging and in the long run more rewarding. Doing a job better than was thought possible before starting is a truly gratifying experience. How much comfort or exhilaration can one get from saying, "I did the job as well as Jane Doe and why should I knock myself out?" Or how much stimulus for self-growth can one get from thinking, "This is the way the world is, what can I do about it? It's all the fault of the Industrial Revolution with its subjugation of human individuality." I am sure we could all expand these rationalizations *ad infinitum*, but do they really do us any good?

In January, when most businesses take inventory of their stock, it would be a good idea to take inventory of the fundamental beliefs of our nation and of ourselves. With this approach we will be in a better position to see where the stock needs filling-in. You may be labeled as an idealist, in our age of practicality, for believing in something that does not bring tangible rewards. Remember, goods and position can be taken away, but not ideals and beliefs.

As Jose Ortega y Gasset (1883–1955) wrote:

> Human existence . . . must begin by an ordered inventory of its system of convictions and to this end it must establish before all else which belief is fundamental, decisive, sustaining and breathing life into all the others.

If we were to put our society to the test that Ortega suggested, I wonder what our fundamental beliefs would be? I will not attempt that evaluation because it would take someone much wiser than I. Just imagine how to include in the summation the beliefs of the status-quo seekers and the radicals; the truly religious and the pseudo-religious; the pacifists and the massive retaliationists; the self-righteous and the Beats; and to not leave out all of the others, even you and me.

Fig. 37.1. Emanuel D. Rudolph teaching a class in field botany at Wellesley College, Massachusetts, March 1959.

(In the Rudolph photograph collection)

Paper No. 38 (1955–1960)

Relationships between Science and Religion VIII. The Uncertanity of Science

As a part of growing up, most of us have had the experience of having some of our illusions about life shattered. I remember very well while in high school how disillusioned and hurt I was when I discovered people deriding the pacifism of an elderly teacher whom I respected very much. She was, to be sure, not very realistic about the problems of the world, but she was sincere in her beliefs and lived up to her ideals in her actions. She did not seem to mind being laughed at; in fact, she probably was too busy to notice it.

I remember too, a shipping clerk I came to know when I was an undergraduate student. It did not matter to him who you were, a small child, a college student, or a distinguished visitor; all received the same open welcome, the same instructing assistance, and the same interest and encouragement. I remember how angry and disillusioned I was with the supervisor when he took advantage of the assistant's good nature. The supervisor gave him a job that was more than he could handle physically, but one which was done without complaint. The extra effort took its toll because it led to a fatal heart attack. I learned a great deal from this gentleman, probably the least of which was how to tie a package that would stay tied all the way to India.

All of us have been assaulted more or less by the realities and disillusionments of life and death. Our placid idealized universes are constantly being attacked. While growing up, we learn that we cannot live in the world and yet escape its workings.

Our time has been called the age of uncertainty, and it is not an unfair analysis. In this uncertain world, science seems to those who are not scientists to be one of the most certain areas of human endeavor. A hundred years ago even the scientists believed in the certainty of science. When I was a freshman in college I had the illusion that science was a field in which a number of laws provided certainty. Discoveries lead to truths, and what could be more certain than truth? It was a shock to discover that science was filled with uncertainties. As Jacques Barzun wrote in his book, *Darwin, Marx, Wagner* (1941):

> Science as a Delphic oracle exists only in the popular imagination and the silent assumptions of certain scientists. At any given time there are only searchers who agree or disagree. The march of science is not an orderly army parade, but rather a land rush for the free spaces ahead. This means a degree of anarchy. Besides, Fogyism, Sadism, love of stability, self seeking, personal likes and dislikes, and all other infirmities of mind play a decisive

> a part in science as in any other cultural enterprise. This inevitable condition should neither be deplored nor hushed up. It is, in fact, what makes us admire the power of the methods which by creating conventions and testing them enable men to measure, plot, and control events.

Dr. Karl R. Popper (1902–1994), professor of logic and scientific method at the University of London, has put the relationship of science and certainty in another way. He wrote:

> Science is not a system of certain, or well established statements; nor is it a system which steadily advances toward a state of finality . . . . The old scientific ideal of epistémé—of absolutely certain, demonstratable knowledge—has proved to be an idol. The demand for scientific objectivity makes it inevitable that every scientific statement must remain *tentative forever*. It may indeed be corroborated, but every corroboration is relative to other statements which again, are tentative. Only in our subjective experiences of conviction, in our subjective faith, can we be absolutely certain.
>
> The empirical bases of objective science has thus nothing "absolute" about it. Science does not rest upon rock bottom. The structure of its theories rises, as it were, above a swamp. It is like a building erected on piles. The piles are driven down from above into the swamp, but not down to any natural or "given" base; and when we cease our attempts to drive our piles into deeper layers, it is not because we have reached firm ground. We simply stop when we are satisfied that they are firm enough to carry the structure, at least for the time being.

Thus, as these men have indicated, science in its growth, like each of us in our own development, has experienced the growing-pains of disillusionment. I do not think that this disillusionment is bad. It can, however, be a danger if it leads to abandonment of important ideals that make our lives, or the life of science, worth living. It becomes so easy to substitute pragmatic rules for basic ideals. Rules of behavior that have standards such as: "Might makes right," or "Life is meaningless" so it makes no difference how I behave; or so long as I act as though I believe in religion it makes no difference how I really feel.

You might well ask: "Why have ideals anyhow?" The answer to this question, like the answer to the question "How certain is science?" is a relative one. We must be satisfied that our base is firm enough to carry the structure of our lives through good times and bad, through our current stage of development and through our future stages. We have before us the examples of others who have made their lives meaningful by living by such ideals as doing their best in the job that needs doing, which really is the job at hand. We can also know that God will provide the direction and the strength so long as we do not skeptically ask for final certain answers.

INDEX

C

D

E

F

G

H

M

S

T

U

About the Editors

Ronald L. Stuckey, born 9 January 1938 in Bucyrus, Crawford County, Ohio, is a country farm boy who lived his first 18 years in Lykens Township. He graduated as valedictorian from Lykens High School (1956), earned a B.S. *cum laude* (1960) in Biology from Heidelberg College, Tiffin, Ohio, and an M.A. (1962) and Ph.D. (1965) in Botany from The University of Michigan, Ann Arbor.

From 1965–1991 he served as Professor of Botany in The Ohio State University, Columbus, where he taught courses in Local Flora, Aquatic Flowering Plants, and Plant Nomenclature. At the University's Stone Laboratory in Lake Erie, Stuckey enjoyed 25 summers (1966–1991) of teaching and research in field botany and vascular aquatic plants, and for eight of those summers he served as a part-time administrator for the teaching program. Stuckey is an internationally recognized authority on the identification and geographical distribution of aquatic and wetland plants in North America. He has also written extensively on the botanical history and exploration of eastern North America.

Stuckey is past president of the OSU Chapter of Sigma Xi, the Ohio Academy of Medical History, and The Ohio Academy of Science. The author of more than 100 scientific journal papers and several books, Stuckey has interests in bluegrass music, photography, geography, genealogy, and old books.

William R. Burk, originally of Norwalk, Connecticut, is currently Biology Librarian at the University of North Carolina in Chapel Hill. After earning his graduate degrees, M.S. (Botany) and A.M.L.S. (Library Science), he served as Science Cataloger first at the Marriott Library, University of Utah and then at Davidson Library, the University of California—Santa Barbara. A Fellow of The Ohio Academy of Science and cofounder of the Triangle Area Mushroom Club in North Carolina, he studies and publishes on the history of North American botany and also writes book reviews for the electronic review service, E-STREAMS. He published a bibliography on North American Gasteromycetes (fungi) in 1980 and expects to complete another volume of this work soon.

While in graduate school, Burk became an avid collector of books and mycological posters and charts. In 1997 he donated his over 2,000 volume library of children's natural history books to the Botanical Research Institute of Texas, Fort Worth. Named the Oliver G. Burk Memorial Lbrary in honor of his horticulturist father, the library is used in teaching at the Institute. Burk gave his over 200 mycological posters and charts to the LuEsther T. Mertz Library, New York Botanical Garden, Bronx, New York, in 2000.

Colophon

One thousand copies were printed using soy-based inks via offset lithography.

This book was printed and bound in Texas by *Millet the Printer, Inc.*

The cover of this book is acid free, recycled 80 lb/216 gsm Fortune Matte Cover.

Text pages of the book are acid free, recycled 70 lb/104 gsm Fortune Matte Text.

All illustrations for the book were scanned by *Millet the Printer, Inc.*

Cover design by StuartBacon, Fort Worth, Texas.

Computer page layout by Ricki C. Herdendorf,
EcoSphere Associates, 1507 Cleveland Road East, Huron, Ohio 44839-9503, USA.

The text is set in 10 and 11 pt. *Times* with 12 and 13.2 pt. leading.

Titles are set in 11 to 18 pt. *New Century Schoolbook* with 13.2 to 21.6 pt. leading.

This book was electronically produced on the Power Macintosh 8500, using Adobe® Pagemaker®, version 6.5.

Professors of Botany at The Ohio State University planning a new approach to the teaching of general botany; seated near the amphitheater on the oval of the campus at OSU, Columbus. l. to r. David Lee (post-doctoral student), Elton F. Paddock (1913–1992), Douglas G. Fratianne, Emanuel D. Rudolph (1927–1992), Tod F. Stuessy, John A. Schmitt (1925–1994), Gary B. Collins (Taken 1971; in the collection of RLS)

Emanuel D. Rudolph and Elva Lawton (1896–1993), Seattle, Washington (Taken September 1962, by Ann Rudolph; in the Rudolph photograph collection)

Emanuel D. Rudolph with Hildur Krog and Eilif Dahl (1916–1993) collecting lichens on Crescent Beach, Olympic Peninsula, Washington (Taken September 1969, following the International Botanical Congress in Seattle; in the Rudolph photograph collection)